Edward Dobson

The Students's Guide to the Practice of Measuring and Valuing Artificers' Works

Sixth Edition

Edward Dobson

The Students's Guide to the Practice of Measuring and Valuing Artificers' Works
Sixth Edition

ISBN/EAN: 9783337812102

Printed in Europe, USA, Canada, Australia, Japan

Cover: Foto ©Lupo / pixelio.de

More available books at **www.hansebooks.com**

THE
STUDENT'S GUIDE

TO THE

PRACTICE OF MEASURING AND VALUING

ARTIFICERS' WORKS.

THE

STUDENT'S GUIDE

TO THE

PRACTICE OF MEASURING AND VALUING ARTIFICERS' WORKS.

CONTAINING

DIRECTIONS FOR TAKING DIMENSIONS, ABSTRACTING THE SAME,
AND BRINGING THE QUANTITIES INTO BILL; WITH TABLES
OF CONSTANTS, FOR VALUATION OF LABOUR, AND FOR
THE CALCULATION OF AREAS AND SOLIDITIES.

ORIGINALLY EDITED BY

EDWARD DOBSON, ARCHITECT.

REVISED; WITH CONSIDERABLE ADDITIONS ON MENSURATION AND
CONSTRUCTION, AND A NEW CHAPTER ON DILAPIDATIONS,
REPAIRS, AND CONTRACTS.

BY E. WYNDHAM TARN, M.A., ARCHITECT,

AUTHOR OF "THE SCIENCE OF BUILDING," "PRACTICAL GEOMETRY FOR ARCHITECTS,"
"A TREATISE ON ROOFS," ETC.

SIXTH EDITION.

INCLUDING

A COMPLETE FORM OF A BILL OF QUANTITIES.

With Eight Plates and Sixty-Three Woodcuts.

LONDON:
CROSBY LOCKWOOD AND SON,
7, STATIONERS' HALL COURT, LUDGATE HILL.
1889.

PREFACE.

THE following work is based upon one published many years since, which was originally written expressly for the rising student by an eminent architect and surveyor of upwards of fifty years' experience; and the manuscript having been left at his death in an imperfect state, was carefully arranged for publication, with much additional matter, by Mr. E. Dobson. For a long time it has been the only standard guide to the methods employed by surveyors in their measurement of builders' works, and the various rules which are laid down have been generally accepted as the safest that can be adopted with a view to obtaining an accurate estimate of their cost.

In the present edition the work has been entirely revised, and a large amount of information added with respect to the technicalities and modes of construction employed in the several trades; for no one can be a skilful measurer of work done by artificers unless he makes himself thoroughly and practically conversant with the processes which they adopt in executing their works.

With this object in view the editor has endeavoured not only to make the student acquainted with most of the technicalities used in building operations, but also to explain, as far as the scope of the work would allow, the manner in which those operations are carried on. The student will also find rules given for ascertaining the strength of substances used in building, as well as the weights of most of those which come under the title of *building materials*.

The Introduction, explaining the object and plan of the work, was written by the original author, and the remarks therein contained are left without alteration as they appeared in the first edition.

The first chapter contains an explanation of the methods employed by surveyors for reducing superficial and cubical measurements by help of the branch of arithmetic called duodecimals. Tables will be found in the Appendix at the end of the work, by help of which much of the labour of calculation is saved. It is, however, recommended to the student to practise the art of 'squaring dimensions,' without any aid from the tables, and only to refer to them for the purpose of testing the accuracy of his calculations. In this chapter are also given rules for finding the area, circumference, and solidity of various geometrical figures commonly met with in buildings; these will be found useful in measuring many branches of artificers' work.

The second chapter treats entirely upon the digging and well-sinking required in building operations, explaining how the quantity of earth excavated can be ascertained, and the value of the labour thereon estimated.

In the third chapter are described the various modes of forming foundations previous to the operations of the bricklayer, a full description of whose work is next given, as also of the materials used by him in his work, and the mode in which that work is executed : the several terms used by bricklayers are also explained, and the ordinary system of measuring and valuing employed in this branch of trade is set forth. The modes of manufacturing concrete and the several uses to which it is applied, are described in this chapter, together with the rules for its measurement. The processes of laying and measuring the tiling and slating to the roofs of buildings are also given, with the sizes, weights, &c., of the several varieties of tile and slate employed.

The work of the carpenter and joiner is treated upon in the fourth chapter, the several processes used by those artificers being described, and the terms which they employ explained. Under the head of 'Carpentry' will be found a description of the modes of framing floors, roofs, shoring, &c., and rules are given for ascertaining the strength of timbers and their proper scantlings according to the manner in which they are strained. Under the head of 'Joinery' will be found a notice of the several terms used by joiners in fitting pieces of wood together, and a description of the methods they employ for that purpose. The measurement and valuation of joiners' work is more intricate than that of any other trade: it is here fully gone into and, it is hoped, made clear to the student; but great care is necessary on the part of the measurer in this branch of work, especially when he is taking off the 'quantities' from a set of plans for a building not yet erected. The work must be thoroughly analysed in order to get at the true value of each part.

In the fifth chapter the student will find a description of the methods of working in stone, to which the term 'Masonry' is applied, with a short account of the various kinds of stone usually employed in England, their qualities as to durability, hardness, weight, and strength. The principles by which the surveyor is to be guided in measuring masons' work are fully laid down. Rules for finding the transverse strength of stone are also given.

The sixth chapter is devoted to an explanation of the several works which come under the name of 'Plastering,' together with the methods by which the cost of such work is usually estimated.

The employment of iron in the erection of buildings of every description has increased very greatly during the last few years, rendering it necessary that the student of architecture should be thoroughly conversant with its several qualities. In the seventh chapter will be found an account of the various forms in which that valuable material is now employed in building operations, especially in the framing of roofs, together with rules for ascertaining its strength under different arrangements of load, and its weight according to shape and size. Several branches of engineers' work are also considered, including methods of warming and ventilating, coiling shutters, lightning conductors, lifts, &c. Tables and instructions for the valuation of iron-work are also given. Ironmongery and bell-hanging are treated upon in the same chapter.

In consequence of the extensive introduction of slating as a covering for the roofs of houses, lead has been nearly discarded for that purpose except for those which have a very low pitch, or are nearly flat; and even then other materials, such as asphalte and cement, are often used in preference. In a few ancient buildings which have escaped the ravages of time, we find roofs covered with lead of very great thickness; but generally the lead has been stripped off and melted down on account of its value, especially in times of war and revolution, when the great demand for that metal has enhanced its price; and even in peaceful times it is no uncommon thing to find lead stolen off the roofs of buildings. What lead-work is now done on roofs is explained in chapter eight, under the title of 'Plumbing,' which term, moreover, includes everything connected with laying-on water throughout a building, water-closets, &c. The employment of zinc for roofing purposes is increasing considerably, its peculiarities and the proper manner of treating it being now understood; and for an inexpensive and light kind of roofing there is no material which can compare with it. As the mode of using zinc is very similar to that of using lead for roofing purposes, its description is introduced in the chapter on plumbing. A short account of gas-fitters' work is also appended to the same chapter, and a notice of electric lighting.

The finishing or ornamenting of the interior of a house is placed in the hands of the 'Decorator,' whose work begins after all the other workmen have left. The work of the decorator, which is treated of in chapter nine, includes painting, colouring, papering, fresco, scagliola-work, mosaic-work, looking-glasses, and also the application of variously coloured marbles, as in panelling, paving, or chimney-pieces. This branch of work is usually executed in first-class houses by persons who have nothing to do with the erection of the building; and before being commenced the building should be allowed to stand some months after it has left the builder's hands, in order that it may get thoroughly dry and the work settled. The various kinds of glass used for building purposes are also described in this chapter, under the heading of glaziers'-work.

The tenth chapter is devoted to several subjects which

constantly come under the notice of the building surveyor, and with which it is essential that he should make himself familiar. The subject of 'Dilapidations' is one upon which several treatises have been written, but of which only an outline of the general principles which must be adopted in assessing their value can here be given. In the notice of 'Repairs' a specification is given of the works generally considered necessary to be done by a tenant under the covenants of a lease. The subject of 'Building Contracts' is touched upon, a form of contract being given with the general conditions agreed upon between the Institute of Architects and the London Builders' Society.

The measurement of the obstruction of 'Ancient Lights' is a subject that every architect who builds in London or other large towns has constantly to consider; the notice here given on this subject is chiefly taken from a paper read by the present Editor before the Institute of Architects, and published in the 'Transactions' for 1870.

The prices of labour vary considerably in different parts of the country, the wages of the men being generally higher in London and other very large towns, where board and lodging are dear, than in the smaller towns and villages, in which they are comparatively cheap. It is therefore necessary, in valuing artificers' work, to take into consideration the locality in which it is executed, and not to attempt to fix one arbitrary price for the whole country. The cost of building also greatly depends on the goodness or badness of the roads leading to the work; and if goods can be brought near the site, either by canal or railway, much expense is saved; as also, if the heavier materials, as stone, brick, lime, sand, or timber, can be readily obtained on the spot or in the immediate locality, since the *leading* or carting of these greatly affects the cost of the building. Inasmuch as the use of water enters largely into the several works, a good supply should be obtained before commencing any building operations, the cost of carrying water any great distance amounting to a considerable item in the expenses; for which purpose, when a building is erected in the country, the sinking of a well is the first thing to be attended to, if there is no other supply of water close at hand.

It may be taken as a general rule, in erecting buildings in various parts of the country, that the materials, whether stones or bricks, slates or tiles, which are found most convenient to hand should be made use of as far as possible, and that the local workmen should be employed thereon, as they are generally better acquainted with the peculiarities of climate and locality, and the working-up of the local materials, than men brought from a distance can be. Until very recent times every county in England may be said to have had its own especial style of building, arising partly from the nature of the materials found in the locality, and partly from peculiarities in the climate. In the present day, owing to the rapidity and ease with which various materials are conveyed from place to place, these peculiarities are to a great extent disappearing in the better class of buildings, and a greater similarity is found to pervade them than formerly was the case; and as the workmen employed on such buildings can now be moved about much more easily than in former times, their provincial prejudices and peculiarities have become considerably modified. On this account the methods of building and of measuring artificers' works have become greatly assimilated all over the kingdom, so as to admit of one general standard being everywhere adopted, with slight modifications as circumstances require. It is therefore believed that the present work will be found of equal value to students, whether training in the offices of provincial surveyors and architects or in those of London practitioners.

E. W. T.

LONDON,
December, 1883.

PREFACE TO THE SIXTH EDITION.

IN the present Edition a few corrections have been made where found necessary. Some additional information on the subject of measuring will be found in the text, and also in the appendix, to which a complete form of a Bill of Quantities as taken off from plans has been added.

E. W. T.

September, 1888.

CONTENTS.

INTRODUCTION.

	PAGE
Preliminary Observations	1
On Measuring	5
On Abbreviation	6
On Rotation	6
On Abstracting, and bringing into Bill	7
On Valuation	7
On Constants of Labour	8

CHAPTER I.

MENSURATION, ETC.

	PAGE
Squaring dimensions	10
On the use of decimal fractions	12
To find the area of a triangle	13
Ditto, ditto, square and oblong	14
Ditto, ditto, parallelogram and trapezium	14
Ditto, ditto, an irregular figure	15
Ditto, ditto, and length of a side of any regular polygon	15
To find the circumference and area of a circle	17
Ditto, area of square inscribed in a circle	17
Ditto, ditto, circular ring	18
Ditto, area of a segment and sector of a circle	19
Ditto, centre of an arc of a circle	20
Ditto, area of an ellipse	20
Ditto, length and area of any curve	21
Ditto, solidity of a right-solid	21
Ditto, ditto, prism, cylinder, or cone	21
Ditto, ditto, frustum of a cone	21
Ditto, ditto, hollow cylinder	22
Ditto, surface and solidity of a sphere	23
Table of English measures	23
Explanation of table of the area and circumference of circles	25
Table of ditto, ditto	26

CHAPTER II.
EXCAVATING, WELL-SINKING.

	PAGE
EXCAVATING	27
Measurement of digging	27
Ditto, wheeling and carting	28
Ditto, planking and strutting	29
Weights of various kinds of earth	29
Claying of vaults	29
Measuring digging in sideling-ground	30
WELL-SINKING	31
Dry-steining	31
Boring	32
Pumping water out of wells during the process of sinking	32
Example of measurement	33
Abstract	34
Rotation for bringing the quantities of digging into bill	35
Constants of labour	35

CHAPTER III.
FOUNDATIONS, BRICKWORK, TILING, SLATING.

FOUNDATIONS	36
Concrete	36
Piling	38
Hollow iron cylinders	39
BRICKWORK	39
Varieties of bricks	39
Beds of brickwork, courses, headers, stretchers	40
Brick bond, English bond, Flemish bond	41
Thickness of brick-walls, as fixed by Act of Parliament for different heights and lengths	43
Flues	46
Party walls, cross walls, flank walls	46
Battering	47
Herring-bone work	47
Footings of walls	47
Damp-proof course	47
Hollow walls	47
Sleeper and fender walls	48
Trimmer-arches to fireplaces	48
Inverts	48
Pargeting	48
Gauged-work	49
Skew-back, chase, bird's mouth	49
Reveals of windows	49
Wall-plates, bond timber	49
Iron-hooping used as bond	50

CONTENTS. xv

PAGE

BRICKWORK—*continued*.
    Pointing, striking joints 50
    Splays 50
    Bricknogging 51
    Underpinning 51
    Brick paving 51
    Grouting to walls 51
    Concrete, made with sulphate of lime 51
    Artificial stone 51
    Retaining walls 52
    Table of thickness of ditto 53
    Asphalte 53
    Terra cotta 54

MEASUREMENT of brick-work 55
    Ditto   brick-drains 57
    Ditto   pipe-drains 57
    Ditto   chimneys, ovens, and coppers 57
    Ditto   circular work 57
    Ditto   facings 58
    Ditto   gauged arches, bricknogging, paving, &c . . 58
    Ditto   brick on edge coping 58
    Ditto   repairs to old work 58

TILING, technical terms and explanations 58
    Pan-tiling 59
    Plain-tiling 59
    Tile-creasing, hip, ridge, and valley tiles 59
    Cement-filleting 60
    Measurement of tiling 60
        Ditto   tile paving 60
        Ditto   facing with glazed tiles 60
    Example, showing the method of keeping the measuring of brickwork 61
    Directions for abstracting the measurement of brickwork . . 64
    Form of abstract 66
    Rotation to be attended to in bringing the quantities of concrete, brickwork, and tiling into bill 67
    Valuation of bricklayers' work 68
    Table of the size and weight of various articles 68
    Quantities of materials required for a rod of brickwork . . 68
    Table, showing the number of bricks or tiles required for a yard of paving 69
    Table of quantity of materials required to execute a square of tiling 70
    Calculation of labour: Table of constants, for concrete, brickwork, paving, tiling 70
    Example 1. To find the value of a cubic yard of concrete . . 71
        Ditto 2. Ditto, ditto, rod of brickwork . . 71
        Ditto 3. Ditto, ditto, foot of malm facing . . 72
        Ditto 4. Ditto, ditto, yard of paving . . . 72
        Ditto 5. Ditto, ditto, square of plain tiling . . 72
    Table, showing the value of reduced brickwork per rod, according to the price of bricks, &c. 73

xvi   *CONTENTS.*

	PAGE
SLATING	73
Description of slates, gauge, lap	73
Directions for measuring slating	74
Slate slabs, shelves, and cisterns	75
Valuation of slaters' work : table of materials and labour	75
Example.—To find the value of a square of Duchess slating, copper nailed	75
Ditto, ditto, measurement of slating	76
Ditto, ditto, abstract ditto	76
Ditto, bill of quantities, ditto	77

CHAPTER IV.

CARPENTRY AND JOINERY.

Definitions	78
CARPENTRY : Naked flooring	78
Herring-bone strutting	79
Wall-plates, sleepers	79
Scantlings of joists for different spans	79
Trimmers, trimmer-joists	79
Mortice and tenon	79
Double-floors, binding-joists, bridging-joists, ceiling-joists	80
Scantlings of binding-joists for different spans	80
Framed-floors, girders	80
Scantlings of girders for different spans	81
Lintels, bressummers	81
Flitch-girders	81
Quartered-partitions	82
Roofs, lean-to	82
V-roofs	83
Hip-roofs, valleys	84
Collar and tie-beam roofs	84
King and queen post roofs	85
Purlins, rafters	86
Battening and boarding for slating	86
Gutters	86
Ashlering	87
Curb or Mansarde-roof	87
Domical and cylindrical roofs	88
Gothic collar roof	89
Ditto hammer-beam roof	89
Barge-boards	90
Halving and scarfing timbers	91
Centerings for arches	91
Fences with feather-edged paling	92
Timber, as cut out of the tree, sapwood, knots, &c.	92
Strength of timber beams and pillars	93
Resistance to stretching	96
Strains on the several timbers forming a truss	96

CONTENTS.

xvii

PAGE

CARPENTRY—*continued*.
    Scantlings of several timbers forming a truss 97
    Shoring 97
    Needling 98
    Weather-boarding 98
    Sound-boarding 99
    Scaffolding 99
    Wood bricks 99
    On measuring carpenters' work 99
    Measurement as timber and labour 100
        Ditto   of cube, fir framed 100
        Ditto   labour and nails to roofs, floors, &c. . 100
        Ditto   cube fir, no labour 101
        Ditto   roofs 101
        Ditto   floors 102
        Ditto   centerings 102
        Ditto   bracketing to cornices, cradling . . . 103
        Ditto   gutters, watertrunks, doorcases . . 103

JOINERY : planks, deals, and battens 104
    Grooving, rebating, mortising, tongueing, mitres . . . 104
    Dovetailing, clamping, housing 106
    Staircases, handrails, balusters 106
    Skirtings, dados 108
    Architraves, boxings, angle-staffs, matched-boarding . . 108
    Sashes and casements 109
    Dormers, skylights, shop windows, &c. 111
    Floors, floor-boards 111
    Folding-floors, straight-joint floors, tongued-floors, dowelled-
        floors 112
    Parquetry-floors 113
    Doors, shutters, and framing 113
    Ledged doors 113
    Panel-doors and framing 113
    Mode of finishing the panelling 114
    Door linings 115
    Shutters 115
    Folding shutters 115
    Lifting   ditto 115
    Shop   ditto 115
    Window blinds 116
    Water-closet fittings 117
    Abbreviations 118
    Measuring joiners' work 119
        Ditto   flooring 120
        Ditto   mouldings 120
        Ditto   doors, linings, &c. 121
        Ditto   skirtings, plinths 122
        Ditto   pilasters 122
        Ditto   dado 122
        Ditto   sashes, frames, shutters, &c. 123
        Ditto   staircases 125
    Rotation in which carpenters' and joiners' work is measured . 126

b

JOINERY—*continued*.
- Abstracting carpenters' and joiners' work . . . 127
- Form of abstract of carpenters' work . . . 128
  - Ditto   joiners' work 129
  - Ditto   ironmongery 130
- Rotation to be attended to in bringing into list . . 130
- Valuation of carpenters' and joiner's work . . 132
- Memoranda 132
- Number of boards to a square of flooring . . 132
- Calculation on the value of timber . . . 132
- Table of constants for labour and nails to roofs . . 133
  - Ditto,   ditto,   naked floors . . 133
  - Ditto,   ditto,   quarter partitions . . 133
  - Ditto,   ditto.   on fir timber . . 134
- Calculation of the value of deals 134
- Table of constants of labour on deals . . . 135
  - Ditto,   ditto,   and nails to battening . . 135
  - Ditto,   ditto,   ditto, weather boarding . . 135
  - Ditto.   ditto,   ditto, rough boarding . . 136
  - Ditto,   ditto,   ditto, deal floors . . 136
  - Ditto,   ditto,   ditto, batten floors . . 136
  - Ditto,   ditto,   ditto, framed grounds . . 136
  - Ditto,   ditto,   ditto, skirtings . . 137
  - Ditto,   ditto,   ditto, gutters and bearers . . 137
  - Ditto,   ditto,   ditto, door linings . . 137
  - Ditto,   ditto,   ditto, ledged doors . . 137
  - Ditto,   ditto,   ditto, framed partitions . . 137
  - Ditto,   ditto,   ditto, deal mouldings . . 137
  - Ditto,   ditto,   ditto, doors hung complete . . 138
  - Ditto,   ditto,   ditto, window linings . . 138
  - Ditto,   ditto,   ditto, window backs, elbows, and soffits . . 138
  - Ditto,   ditto,   ditto, boxings to windows . . 138
  - Ditto,   ditto,   ditto, inside window shutters . . 139
  - Ditto,   ditto,   ditto, sashes and frames hung complete . . 139
  - Ditto,   ditto,   ditto, staircases . . 139
  - Ditto,   ditto,   ditto, outside strings to stairs . . 140
  - Ditto,   ditto,   ditto, wall-strings . . 140
  - Ditto,   ditto,   ditto, dados . . 140
  - Ditto,   ditto,   ditto, columns and pilasters . . 140
- SAWYERS' WORK 141

CHAPTER V.

MASONRY.

- TECHNICAL TERMS AND EXPLANATIONS . . . 142
  - Description of stones used in building . . 142
  - Granites: their qualities 142
  - Ditto   modes of working 144
  - Sandstones; grits, flags: their qualities . . 144

CONTENTS.

	PAGE
TECHNICAL TERMS AND EXPLANATIONS—*continued*.	
Mansfield stone : its quality, weight and strength	145
Red sandstones	145
Limestones ; rags, oolites : their quality, weight and strength	145
Magnesian-limestones : their quality, weight, strength	146
Preservation of stone from disintegration	146
Foundations of stone walls	147
Rubble-walling	147
Flint-walling : batter	148
Ashlar-facing	148
Quoin-stones	148
Sawing : measurement	149
Moulded-work : measurement	149
Hewn-stone : measurement	149
Joggled-joints : throating, splay, chamfer	149
Arches	149
Gothic arches and vaulting	150
Measurement of arch-stones	151
Flagging : landings	151
Cutting and pinning	151
Coping	151
Ditto, measurement of	152
Sills	152
Sinks : how measured	152
Cramps ; dowels ; lewis	152
Banker ; beam-filling	152
ON MEASURING STONE-MASONS' WORK	153
Measurement of stone steps	154
Ditto slabs	156
Ditto labour on stone	156
Abbreviations used in measuring	157
Measurement of staircases	157
Ditto landings	158
Ditto square steps to entrance doors, &c.	158
Ditto coping	159
Ditto string-courses	160
Ditto square plinths	160
Ditto window sills	161
Ditto curbs	161
Ditto columns	161
Ditto architraves over columns	163
Ditto blockings and cornices	163
Ditto niches	164
Ditto stone facings	165
Weight of stone	166
Valuation of labour	166
Table of constants for the different descriptions of masons' work	166
Labour on statuary or vein marble	166
Mode of measuring walls in stone-countries	167
Rotation to be attended to in bringing into bill	168
Transverse strength of stone	168

CHAPTER VI.
PLASTERING.

	PAGE
TECHNICAL TERMS AND EXPLANATIONS	170
Nature and description of plasterers' work	170
Mode of forming cornices	170
Marble and Parian cements	171
Ornamental work to ceilings	171
Ventilation by openings in ceilings	171
Outside stucco	172
Mode of forming cement cornices	172
Portland cement	172
Selenitic cement	172
Cement angles to rooms	173
Claircolle, whitening, lime-whiting, distemper	173
Blue-lias stucco	173
Mode of executing internal plastering	174
Rough-cast	174
Pugging to floors	175
ABBREVIATIONS used in measuring	175
Rotation in bringing into bill	175
Directions for measuring plasterers' work	175
Form of abstract	177
Rotation to be attended to in bringing into bill	178
Valuation of plasterers' work	178
Calculation of materials and labour	179

CHAPTER VII.
SMITHS' WORK, ENGINEERS' WORK, IRONMONGER, BELL-HANGER.

	PAGE
SMITHS' WORK	180
Cast-iron : its use and strength	180
Cast-iron beams : form and strength	180
Ditto columns : their strength	181
Story-posts	183
Measurement and valuation of cast-iron work	183
Table of the weights of cast-iron cylinders	184
Rain-water-pipes : how measured	184
Eaves-gutters	185
Cast-iron railings, gates, gratings, &c. : how measured	185
Wrought-iron	186
Wrought-iron columns : resistance to crushing and bending	186
Rolled joists ; fire-proof floors	187
Riveted-plate-beams	188
Wrought-iron roofs	188
Corrugated iron	190
Iron tanks and cisterns	191

CONTENTS.

SMITH'S WORK—*continued.*

	PAGE
Weight of wrought-iron in various forms	191
Galvanising	191
Chimney-bars, railings	191
Saddle-bars, casements, rivets	192
Table of weight per lineal foot of rolled angle-iron	192
Ditto ditto ditto tee-iron	192
Ditto ditto ditto I-joists	192
Ditto ditto ditto flat bar-iron	193
Ditto ditto ditto round bar-iron	193
Ditto per foot superficial of sheet-iron	193
Fittings to stables, cowhouses, and piggeries	193
Spiral staircases	196
Measurement of balconies, railings, hoop-iron, ties, straps, &c.	196

WARMING AND VENTILATING 197
   Open stoves or grates 197
   Ventilating valves let into flues 197
   Ventilating by admission of cold air at top of a room . 198
   Close or pedestal stoves 198
   Warming by hot air apparatus 198
   Hot-water pipes laid under floors 198
   Drawing-off the vitiated air from a room . . 199
   Warming and ventilating prison cells . . . 199
   Ventilators for rooms 199
   Ventilation by the action of sun-burners . . 200
   Stoves and ranges 201

REVOLVING-SHUTTERS 202
   Table of space required for revolving-shutters . . 202
   Self-coiling shutters 203
   Table of space required for self-coiling shutters . 203
   Measurement of revolving-shutters . . . 203

LIGHTNING-CONDUCTORS 203

LIFTS 204

IRONMONGERY 205
   Nails, screws, bolts, and hinges 205
   Locks 207
   Furniture 207
   Latches 208
   Shutter-bars, lifts, &c. 208
   Hat-pins, meat-hooks, brackets 208
   Cabin-hooks, sash and shutters fastenings, sash-lines . 209

BELL-HANGING 209
   Church and turret bells : how hung . . . 209
   House bells : mode of hanging 210
   Electric bells 210
   Pneumatic bells 211

CHAPTER VIII.

PLUMBING, ZINC-WORKING, GAS-FITTING.

	PAGE
PLUMBING	212
Milled lead : used in roofs and flats	212
Ditto ditto hips, valleys, and ridges	213
Ditto ditto flashings	214
Ditto ditto gutters	214
Ditto ditto cisterns	214
Table of thickness of milled lead according to weight	215
Lead to trap-doors and skylights	215
Ditto to dormer windows	215
Measurement of lead-work	215
Lead pipes for water	216
Table of the weight of lead pipes, according to their diameter	216
Iron pipes and tin-lined pipes for water	216
Socket pipes, funnel pipes, soil pipes	216
Traps: syphon traps, bell traps, D and P traps	217
Cocks	217
Pumps	218
Water-closets	218
Waste-preventers	219
Earth-closets	219
Hot-water supply to baths	220
Copper for roofs	220
ZINC-WORKING	221
Weight and qualities of zinc	221
Zinc used for covering roofs, flats, and gutters	221
Corrugated zinc for roofs	221
Zinc gutters and rainwater pipes	222
Galvanising iron, perforated zinc	222
GAS-FITTING	222
Materials used in gas-fitting	222
Gas-meters	223
Quantity of gas supplied by pipes of various sizes	223
Gas-stoves	223
Electric lights	223

CHAPTER IX.

PAINTING, GLAZING, PAPER-HANGING, DECORATING.

	PAGE
PAINTING	225
Preparation of wood and other substances for painting	225
Knotting and priming	225
Flatting	226
Painting on plastered walls	226
White-lead, zinc-white	226
Mastic cement	226

CONTENTS.

PAINTING—*continued*.
Clearcole; distemper	226
Marbling and graining	227
Staining, varnishing, and polishing	227
Cleaning painted work	227
Proper time for executing outside work	227
Abbreviations used in measuring	227
Rotation in bringing into bill	228
Directions for measuring	229
Form of abstract	230
Rotation to be attended to in bringing into bill	231
Valuation of painters' work	231

GLAZIERS' WORK 232
Crown-glass: different qualities	232
Sheet-glass: qualities, weights, and thicknesses	232
Patent-plate: qualities, weights, and thicknesses	233
British-plate: qualities, valuation	233
Bending glass	233
Rough plate-glass, fluted and plain	233
Rolled cathedral glass	233
Glazing, or fitting glass into sashes, &c.	233
Fixing glass in stone-work	234
Lead lights	234
Measurement of glaziers' work	234
Calculation of the produce of a crate of glass	234
Value of labour and putty, per foot superficial	235
Calculation of the value of glazing per foot from the prime cost per crate	235

PAPER-HANGING, DECORATING 235
Preparation of walls for papering	235
Measurement of paper-hanging	236
Ditto pumicing and preparing walls	236
Covering damp walls with tin-foil and tar-paper	236
Mouldings	236
Distempering	237
Fresco-painting	237
Scagliola	237
Mosaic decoration	237
Enamelled-slate	238
Marble	238
Polished granite	239
Silvered-glass; gilding	239

SUMMARY 240

CHAPTER X.

DILAPIDATIONS, REPAIRS, CONTRACTS, ANCIENT LIGHTS.

	PAGE
DILAPIDATIONS	241
Under covenants of lease	241
Assessment of value	242
Ecclesiastical	242
REPAIRS	243
Under covenants of lease	244
Specification of Work to be done under different trades	244
Valuation	248
CONTRACTS	249
Form of building contract	249
General conditions for building contract	250
ANCIENT LIGHTS	255
Measurement of obstructions	255
Tables of relative values of light	257
Example of application of the Tables	259

APPENDIX I.

HOPPUS'S TABLES

TABLE I.	Square of unequal-sided timber	261
TABLE II.	Solid or cubical measure of timber	265
TABLE III.	Superficial Measure	284

TABLE IV.	Weight of various Building Materials	309

APPENDIX II.

Taking off Quantities from Plans	311
INDEX	327

THE PRACTICE

OF

MEASURING AND VALUING ARTIFICERS' WORK.

INTRODUCTORY CHAPTER.

PRELIMINARY OBSERVATIONS.

THE AUTHOR,* having retired from the profession, has been enabled to devote considerable time to the preparation of the present work, which is intended for the information of the young student in a department which, in some respects, is not the most pleasant part of the architect's duty; more particularly when it is one to which he does not feel himself perfectly competent, which is the case if he has not had the opportunity, or has neglected to avail himself of the means, of obtaining the requisite information. It is therefore strongly recommended to the student, that, after he has acquired sufficient knowledge of construction for making out working drawings correctly, he should attend to the rules by which, in due time, he may become qualified to measure and value the work when performed. The disinclination often felt by young gentlemen of education for the study of these rules, and of the mechanical part of

* The MS. of this work was originally prepared by an eminent surveyor of fifty years' experience, after whose decease it was arranged for publication by Mr. E. Dobson.—E. W. T.

the profession, make it the more necessary to impress on their minds the absolute necessity of studying these essential qualifications,—which can only be done with any probability of success, by commencing at the lowest, and rising gradually to the higher departments. If the student neglects the operative part, he must never expect to be capable of making working drawings without incurring the ridicule of the mechanic; and when he commences business on his own account, if he also neglects the measuring department, he will be obliged to employ persons to make out his specifications, and to measure and value his works when completed. The expense incurred by thus employing others to do what he is incapable of, is a minor consideration; for it is imperative on the young architect to reflect that he will be the responsible agent between the gentleman and the builder, and that if, during the erection of an edifice, he allows the work to be insecurely performed, or suffers his employer to be imposed on, not only is his character at stake, but he is also amenable to the laws of his country (and very properly); so that following the profession of an architect, not being duly qualified, may be attended with the most serious consequences; for whether an architect allows his employer to suffer from inattention on his own part, or from the ignorance or dishonesty of the persons employed by him, it is precisely the same in effect, he being professionally employed, and receiving his commission on the cost of the building, which is paid him for designing, directing, and superintending its construction, and seeing that the whole is performed in a proper and workmanlike manner, examining and passing the accounts, and making every arrangement for their final settlement. Consequently, in case of failure in any respect, he is answerable, from whatever cause it may arise, except the improper interference of his employer. Independently of this serious responsibility, if he does not qualify himself in the operative part, it is impossible that he can ever follow his profession with any comfort or satisfaction. Even in passing over or through his own buildings, he is obliged to be most careful of giving any directions, fearful lest he should commit himself before the common mechanic, who very soon discovers if the architect has practical knowledge, and consequently in what manner the work may or must be done, and acts accordingly.

It may be stated that architects of extensive practice cannot attend to all these things themselves. True; but be it remembered, that young men do not very soon get into such practice, particularly if they are not well qualified; and when they do, it is the more essential that they should perfectly understand the practical part of their profession,—that they may select proper assistants, and having chosen them, that they should know from their own experience if they perform their duty with ability and integrity.

This treatise was commenced originally for the purpose of giving the pupils studying under the author, who had an extensive country practice, a correct idea of measuring, abstracting, bringing into bill, and valuing the different artificers' works, agreeably to the methods considered by London surveyors as the most correct and expeditious. The great talent and extensive practice of metropolitan surveyors must be allowed as sufficient authority for concluding that the rules laid down by them are superior to any others that can be adopted. Independent of which, it being the practice for the architect, or his clerk or surveyor, to meet the surveyor appointed by the tradesman to take the dimensions, abstract their contents, make out the quantities into bill, and value the work together, it is absolutely necessary that a regular system should be adopted and strictly adhered to in every part of the business, or much confusion would arise, as is generally the case whenever London surveyors have to meet country practitioners; and it is consequently of the utmost importance to establish the same system throughout the kingdom. The great improvements made in travelling, and the velocity with which we are now conveyed, will soon place every part of this country within a few hours' journey from the metropolis; and the natural consequence of these increased facilities of communication must be, that our habits and methods of doing business will proportionally assimilate.

It is not intended, in this part of the work, to explain the methods of manufacturing any materials, as bricks, tiles, &c., or the methods of performing the respective works, except so far as to enable the young student to describe the work which he is about to measure, and to ascertain if it be executed in a proper and workmanlike manner. But a perfect knowledge of this department can only be obtained by great attention, perseverance, and practice. The method is shown of valuing all the leading articles in each trade, by first ascertaining the fair price to be allowed for the materials, according to the prime cost thereof, and by adopting what the author considers the ne plus ultra; viz., a decimal, by which, if correctly ascertained, the amount of labour thereon at all periods may be immediately found, by multiplying that decimal by the rate of wages allowed: this is the only method by which perpetual prices can be formed. Materials and labour are continually, but not proportionally, fluctuating; consequently the value of work can only be determined by first ascertaining the cost of the materials expended, and making the requisite allowances for profit and waste, and then the amount of labour in executing it.

As the tradesmen's bills must be passed and signed by the architect, the prime cost of materials may in most instances be obtained without much difficulty, and in all cases may be demanded before he allows the prices charged. The quantities required per rod, perch, square, or yard, according to the description of work, the architect ought, agreeably to certain rules, to be capable of determining. But many difficulties arise, and the greatest attention is requisite to ascertain correctly the fair average of time to be allowed between the common and best workmen, and also between what men can, and what they will, do. The decimal must therefore be calculated agreeably to our respective judgments, and from the best information we can obtain; the correctness of which depends on the attention we have paid to the subject, and the opportunities we have had of arriving at our conclusions. Those which are now submitted to the public will be found as correct as they can be made in the compilation of a work like the present. It is anticipated that the professional man may, in his advice to the student, be induced to place this subject properly before him, and establish rules by which every description of work may be valued, according to the prime cost of materials and the rate of wages, at any time and place when and where the work has been performed.

ON MEASURING.

In order to illustrate the principle of measuring the different artificers' works, drawings of reference are given, as the only means of conveying to the architectural student, who has never attended to the admeasurement of work, the correct method of proceeding. The description of book generally used for measuring is shown, with lines ruled according to the old practice: few modern surveyors, however, think of ruling the columns for the dimensions, any more than they would rule lines to write by, it not being more requisite to those who are in the constant practice of measuring work; but it is always customary to insert the date and the name of the person met, and also for whom, and where the work is done, in the manner hereafter described.

In entering dimensions in the measuring-book, observe that the number of times is always stated on the left of the dimensions; and in measuring brick-work, the number of bricks in thickness on the right side; leaving another space or column for the amount the dimensions square to. Also be particular in entering the wastes in the book; that is, the manner in which the length and width of each dimension is made out; which is frequently done by collecting several together; and likewise the particular situation of the work; so that the student may be able to account for or make out how every dimension was taken, should any misunderstanding arise at a distant period, and he be called upon to give the necessary explanation respecting the way in which he has taken the work: he will then be as ready and quick as it is necessary to be correct.

ABBREVIATION.

Every method that can be adopted to expedite the taking of dimensions with accuracy is most desirable. It is recommended to the young student to attend to the following practice; viz., using a kind of shorthand or abbreviation in describing the different works, which greatly facilitates the operation, and gives time for more attentively observing the measuring rods, to know from ocular demonstration that the dimensions are taken and called correctly; which all who have had much practice in measuring find to be very essential in correcting inaccuracies, from whatever cause they may occur. Although it may appear that this method of adopting initials is not sufficiently explanatory, they will, with a very little practice, be read and understood with as much ease and certainty as if the words were written at full length. In this, as in the other departments, details are given to each respective trade.

ROTATION.

No profession can be successfully pursued without adopting a regular system; and in no department is this more essential than in measuring the multifarious works in a building, which can only be accomplished with any degree of accuracy by invariably taking the respective works in regular succession, by which it is scarcely possible to omit any part of the work, which would constantly occur if some positive and undeviating rule were not attended to. In the following pages, the regular rotation to be adopted in measuring each particular description of work is given under the heads of the respective trades.

ON ABSTRACTING,
AND BRINGING THE QUANTITIES INTO BILL.

The form of the abstract is drawn out for each trade, and also the rotation that should be observed in placing the particular kinds of work, which, if constantly attended to, will greatly facilitate the operation, as it is always known in what part of the abstract any description of work will be found: this more particularly alludes to the abstract for carpenters' and joiners' work, where there are so many different heads as to make it absolutely necessary to pay the greatest attention to their order and regularity. This and the peculiarities to be attended to in each trade, are more particularly described at the commencement of their respective abstracts. The student is to observe that, before he begins to take out the quantities, he prepares the abstract, by considering what articles he will have, and writes the heads of them in their proper columns, according to the rotation to be observed in bringing them into bill. On this subject examples are given in each trade; but the general rule to be attended to in such trades, where some of the work is valued by the rod, perch, yard, or square, is to place these first, and next the work valued by the cube foot, commencing with the quantities on which there is the least labour, and so in regular rotation to those that have the most. Next proceed with the articles that are valued by the superficial foot, commencing with the lowest, and, as before stated, to those of most value; having entered all those by the foot superficial, then take those by the foot run in a similar manner, and next those that are numbered, as is more particularly described after their respective abstracts.

VALUATION.

In entering on this department, it is imperative to impress on the mind of the young student the absolute necessity of being circumspect and correct. If he intends to maintain his independence and be respected, he must make a point of conscientiously doing his duty with strict integrity; to accomplish which it is not only essential that he be honest in his intentions, but that he should be qualified for the business he undertakes. Whether an act of injustice arises from ignorance or intention, it is precisely the same in effect; it therefore behoves him on every account to be qualified for acting on his own judgment. But he cannot consider himself competent to measure and value artificers' work unless he understands the nature of that work, the manner in which it is executed, the time required to perform the same, and can ascertain the prime cost of the materials used thereon at the period when the work was done. It is only possible to state the time and materials that should be expended in the several works taken on an average, but which will vary according to the description and execution thereof, both as regards the materials used and the ability of the workmen employed. It is the duty of the architect to take all these circumstances into consideration before he affixes a value on the work; consequently, in this department, the greatest care, attention, and judgment are requisite, to do justice to all parties. To give the student the necessary impetus for acquiring these essential qualifications, was the author's principal motive in offering this work to the aspirant.

CONSTANTS OF LABOUR.

These constants represent the time requisite to perform a given quantity of work, of the kind specified, in days and decimal parts of a day*; the factor to be applied, being the rate of wages per diem* for one or more men, according to the nature of the work.

These decimals are calculated, in all the trades, for the price per day* allowed the master in his day bills, consequently with his profit thereon, being the only rate that can be ascertained, the master of course paying each

---

* In the present edition all these constants have been altered to decimals of *an hour*, wages being now generally calculated by the *hour* instead of by the *day*, which latter is rather an indefinite period of time; so that the factor to be applied is the *rate of wages per hour*.

man per week according to his abilities and industry; therefore the full value of the labour, including the master's profit, will be found by multiplying the decimal by the rate of wages, as shown in their respective tables. Likewise, in all cases it must be understood that the prices stated in the table for labour and nails include fixing; and when added to the price of deals, calculated as shown in p. 134, will give the value of the work fixed complete, including labour, nails, and materials, according to the prime cost of materials and rate of wages allowed.

CHAPTER I.

MENSURATION, ETC.

SQUARING DIMENSIONS are the arithmetical calculations required to be made in reducing the measurements of artificers' work, and are mostly performed by means of *duodecimals*, or fractions in which the denominator (understood but not expressed) is always a power of 12. Thus, the area of a surface whose lineal dimensions are 5 ft. 7 in. by 2 ft. 5 in. is found by multiplying these dimensions together and expressing the result in square feet, 12ths of square feet, and 144ths of square feet, as shown below.

```
              5  7
              2  5
          ───────────
5'7"×2'=11    2
5'7"×5"=  2   3  11
          ───────────
```

By addition, 5'7"×2'5"=13 5 11=13 sq. ft., 5-12ths of a sq. ft., and 11-144ths of a sq. ft.

The object of Table III. (see Appendix) is to save the greater part of the above calculation, which can be performed by addition only in the following manner:— Thus we have 5 ft. × 24 in. = 10 ft.; 7 in. × 24 in. = 1 ft. 2 in.; and also we have 5 ft. × 5 in. = 2 ft. 1 in.; and 7 in. × 5 in. = 0 ft. 2 in. 11 pa. Therefore, by addition, we obtain the same result as before; thus

```
     10
      1   2
      2   1
          2  11
    ─────────────
     13   5  11
    ─────────────
```

Tables are also formed by which such results as the above can be obtained by simple reference without any calculation at all (see Hawkings' 'Tables of Superficial Measurement'), the length and breadth being expressed in inches and the superficial area in square feet, twelfths of square feet, and one hundred and forty-fourths of square feet. Thus we find opposite length 67 in. and breadth 29 in., the area 13.5.11, as obtained above.

Some materials, as timber, stone, &c., being valued according to their *cubical contents*, we have first to multiply together any two of the lineal dimensions as before described, obtaining a result expressed in square feet, 12ths, and 144ths. This result has then to be multiplied by the third dimension, and the cubical contents are found in cubic feet, 12ths of cubic feet, and 144ths of cubic feet. As an example, we find the cubical contents of a piece of timber 11 ft. 9 in. by 1 ft. 2½ in. by 1 ft. 4 in. :—

```
                           11  9
                            1  2  6
                 ─────────────────
           11'9"× 1' = 11  9
           11'9"× 2" =  1 11  6
           11'9"× ½" =     5 11
                      ─────────────
By addition, 11'9" × 1'2½" = 14  2  5
                             1  4
                      ─────────────
           14'2"5 × 1" = 14  2  5
           14'2"5 × 4" =  4  8 10
                      ─────────────
By addition, 14'2"5 × 1'4" = 18 11  3 = 18 cub. ft., 11-12ths of a cub.
                          ft., 3-144ths of a cub. ft.
```

The same result may be obtained approximately by means of Tables I. and II. (see Appendix). Thus, in Table I. we find that a piece of timber 14½ in. × 16 in. is equivalent to a piece 15¼ in. square; and by Table II. we find 11 ft. length of 15¼ in. square contains 17.9.2 cub. ft. ; and at the bottom of the same page we find that three-quarters of a foot of 15¼ square contains 1.2.6 cub. ft.; and adding these two together we get a total of 18.11.8 cub. ft. This is rather more than the result obtained by calculation, which difference arises from the square of 15¼ in. being only an *approximation* to the product of 14½ × 16. A more accurate result may be obtained by using Table III. only substituting *inches* long for *feet* long in the vertical column, and inches, or rather 12ths of square feet, in the tabulated result. Thus 16 × 14½ gives 19.4, or 1.7.4, which has only to be multiplied by the length 11 ft. 9 in. in order to obtain the cubical contents required.

Decimal Fractions are those which have 10, or a power of 10, for their denominator (understood but not expressed), and are commonly employed for purposes of calculation, as alluded to at page 8 for finding the value of artificers' work according to the price of labour. In decimals the fraction is indicated by a point before the figure: thus, ·4 signifies 4-10ths; ·24 signifies 2-10ths and 4-100ths; ·358 signifies 3-10ths, 5-100ths, and 8-1000ths, and so on; if there are no 10ths but only 100ths and 1000ths, a cypher is put after the decimal point, thus ·058; if there are no 100ths, then two cyphers, thus ·008, and so on. Multiplication of decimals is performed in the same way as for any other figures: thus to multiply ·358 by 7, we proceed as in common multiplication and obtain the number 2506; then put as many figures to the right of the decimal point as in the number multiplied, giving as the result 2·506, that is 2-units, 5-10ths, 0-100ths, 6-1000ths.

We will apply the decimals to the calculation of the labour in making and hanging a 2 in. four-panel square door, 6 ft. 6 in. by 3 ft. or 19½ sq. ft. in area. We find at page 138 the constant for a two-panel door is ·68 for every square foot, to which is to be added for the two additional panels the fraction ·17, making the number to be ·85; which means that the time allowable for making each superficial foot of the door is 8-10ths of an hour and 5-100ths of an hour; if we multiply ·85 by 19½ we obtain the number 16·575, by which we find that the making of the door takes 16 hours, 5-10ths of an hour, 7-100ths of an hour, and 5-1000ths of an hour. Now, suppose a joiner's wages to be 10 pence per hour, then the value of the labour on the door is found by multiplying 16·575 by 10, which gives 165·75 pence, or 165 pence, 7-10ths of a penny, and 5-100ths of a penny, which is equivalent to 13s. 9¾d.

In measuring artificers' work it is often necessary to find the area, perimeter, or solidity of certain geometrical figures; it is therefore proposed to give a few of the rules by which such measurements can be most readily made.

To find the area of a TRIANGLE.

Let A B C be the triangle (fig. 1). Drop the perpendicular A D upon the base B C. Multiply the length B C in feet by half the height A D in feet, and the product is the area of the triangle in square feet. If the lengths are in yards, the area will be in square yards. This rule is expressed as follows:

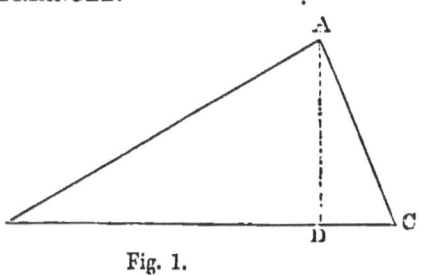

Fig. 1.

Area of triangle = ½ A D × B C;

which is half the area of an Oblong or Rectangle having the same base and altitude.

To find the area of a triangle when the lengths of the three sides are given:

Let s be the half sum of the three sides: deduct from s the length of each side in succession; multiply together all the three quantities thus formed, and their product by the value of s; extract the square root of this last product, and the area of the triangle is found. This rule is expressed thus: let a, b, c be the lengths of the three sides, then $s = \dfrac{a+b+c}{2}$; form the quantities $s-a$, $s-b$, $s-c$; then we have

Area of triangle = $\sqrt{\{s(s-a)(s-b)(s-c)\}}$

Example.—To find the area of a triangle whose sides are 12 ft., 15 ft., and 19 ft. Here, $s = \frac{1}{2}(12 + 15 + 19) = 23$; $s - a = 23 - 12 = 11$; $s - b = 23 - 15 = 8$; $s - c = 23 - 19 = 4$;

Area of triangle = $\sqrt{23 \times 11 \times 8 \times 4} = 4\sqrt{506}$.
= 90 square feet.

The area of a SQUARE is found by *squaring* the length of a side; that is, multiplying the length of a side by itself. If the side is measured in feet, the area will be in square feet. The length of the *diagonal*, or line joining two opposite corners of a square, is found by multiplying the length of a side by the square root of 2, which is 1·414 if expressed in decimals, or 1.5.0 in duodecimals.

The area of an OBLONG, or four-sided figure having opposite sides equal and all its angles right angles, is found by multiplying the length by the breadth. If the lengths are measured in feet, the area will be in square feet. The length of the diagonal of an oblong is found by adding together the squares of two adjacent sides, and extracting the square root of the sum: that is, if l is the length, b the breadth, then diagonal of oblong = $\sqrt{b^2 + l^2}$. This figure is also called a RECTANGLE.

The area of any PARALLELOGRAM (fig. 2) A B C D, or four-sided figure having opposite sides equal and parallel, is found by dropping the perpendicular A E on the base D C; and the area is the product of the length D C into the height A E.

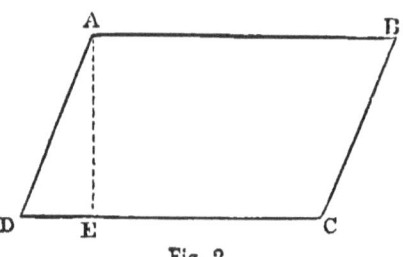
Fig. 2.

The area of a four-sided figure, or TRAPEZIUM, having only one pair of sides parallel, namely, A B and D C (fig. 3), is found by adding together the lengths A B and D C of the parallel sides, and multiplying their half-sum by the vertical distance A E between them. That is,

$$\text{Area of trapezium} = \frac{AB + DC}{2} \times AE.$$

Example.—Let the top and bottom of a trapezium be 5 ft. and 7 ft., the vertical distance between them being 12 ft. Then the area is $\frac{5+7}{2} \times 12$, or 72 sq. ft.

To find the area of any irregular figure or POLYGON bounded by straight lines, as A B C D E (fig. 4), divide it into triangles by the dotted lines A C, C E, then drop perpendiculars from the opposite vertices upon these lines, and find the areas of the triangles by the rule above given. The area of the whole figure is evidently the sum of the areas of the several triangles into which it is divided.

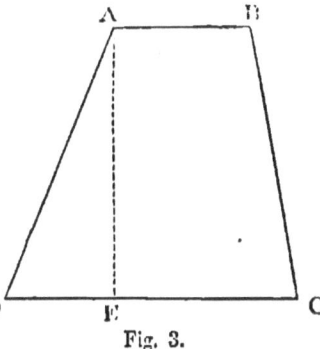

Fig. 3.

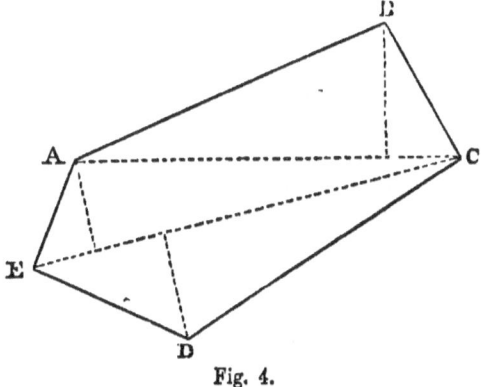

Fig. 4.

To find the area of any regular POLYGON, or many-sided figure having all its sides and angles equal, as A B C D E F (fig. 5); let o be the centre of the figure, draw the diagonals A O D, B O E, C O F, so as to divide the figure into as many equal triangles as there are sides; then the area of the polygon is the sum of the areas of these triangles. The point o, where the diagonals all intersect, being equidistant from all the sides of the figure, is the centre of the inscribed circle, or circle which touches all the sides; and if M O N is drawn perpendicular to the opposite sides A B and D E, it is the diameter, and M O or N O the radius of the inscribed circle. The area of one of the triangles, as A O B, is A B multiplied by half M O, or the length of a side multiplied by half the radius of the inscribed circle; therefore the area of the figure is found by multiplying the length of a side by the radius of the inscribed circle, and the product by *half* the number of sides in the polygon.

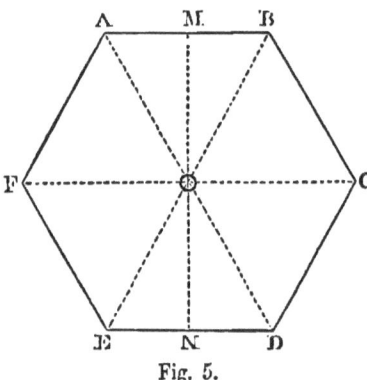

Fig. 5.

In the HEXAGON or figure of *six* equal sides, the length of a side is found by multiplying the diameter of the inscribed circle, M N, by the number ·577. And the area is the square of the radius or half-diameter, o N, multiplied by 3·462.

Example.—Let the diameter of the inscribed circle be 10 ft.; then the length of a side is ·577 × 10, or 5·77 ft., and the perimeter is 6 × 5·77 or 34·62 ft. The area is 5^2 × 3·462, or 86·55 sq. ft.

In the OCTAGON, having *eight* equal sides, the length of a side is found by multiplying the diameter of the inscribed circle by the number ·414. And the area is the product of the square of the half-diameter into the number 3·314.

Example.—Let the diameter be 10 ft.; then the length of a side is ·414 × 10, or 4·14 ft., and the perimeter is 8 × 4·14, or 33·12 ft. The area is 5^2 × 3·314, or 82·85 sq. ft.

In the DECAGON, or *ten*-sided figure, the length of a side is the diameter multiplied by ·3249. The area is the square of the half-diameter multiplied by the number 3·249.

Example.—Let the diameter be 10 ft.; then the length of a side is 3·249 ft., and the perimeter is 32·49 ft. The area is $5^2 \times 3·249$, or 81·22 sq. ft.

In the DODECAGON, or polygon of *twelve* equal sides, the length of a side is the diameter multiplied by ·268. The area is the square of the half-diameter multiplied by 3·215.

Example.—Let the diameter be 10 ft.; then the length of a side is 2·68 ft., and the perimeter is 32·16 ft. The area is $5^2 \times 3·215$, or 80·38 sq. ft.

To find the circumference of a CIRCLE, whose diameter is given, multiply the diameter by 22, and divide the product by 7. Or, if greater accuracy is required, multiply the diameter by 355, and divide the product by 113. The same result will be also obtained in decimals, if the diameter is multiplied by the number 3·1416, for which the Greek letter π is generally used.

Example.—The diameter of a circle is 10 ft.; then the circumference is $\frac{22 \times 10}{7}$, or 31·4 ft. More accurately, the circumference is $\frac{355 \times 10}{113}$, or 31·416 ft., which is the same as $10 \times 3·1416$.

To find the area of a circle whose diameter is known, multiply the *square* of the half-diameter by 22, and divide the product by 7. If greater accuracy is desired, multiply the square of the half-diameter by 355, and divide the product by 113; or, multiply the square of the half-diameter by 3·1416.

Example.—The diameter of a circle is 10 ft.; then the area is $\frac{22 \times 5^2}{7}$, or 78·6 sq. ft. More accurately, the area is $\frac{355 \times 5^2}{113}$, or 78·54 sq. ft.

The area of the square inscribed in a circle, that is, with all its angles touching the circumference, is equal to twice the square of the radius.

The area of a circle can also be found by multiplying the square of the diameter by ·7854.

The area of a CIRCULAR RING is found by deducting the area of the inner circle from that of the outer one.

The Table page 26 has been formed to facilitate the calculations of areas and circumferences of circles, and to find the squares of numbers. If the *square* of any number in the first column is required, it is found in the second column. The third and fourth columns give the areas of circles corresponding to the diameters given in the first column, expressed in decimals and duodecimals. The fifth and sixth columns give the corresponding circumferences of circles to the diameters in the first column, both in decimals and duodecimals. The area and circumference of circles of greater diameter than the Table extends to, may be readily found by remembering that areas of circles increase as the *square* of the diameter, and circumferences increase *as* the diameter; that is, if circle B has its diameter double that of circle A, the circumference of B is double that of A, and the area of B is four times that of A; and so on.

Example.—To find by the Table the area and circumference of a circle whose diameter is 15 ft. Referring to the number 7½ (the half of 15) in the first column, we find the area 44·1786, and the circumference 23·5619; multiply the former number by 4, and we have the area of the circle of 15 ft. diameter, namely, 176·7144 sq.ft. Multiply 23·5619 by 2, and we have the circumference of a circle 15 ft. diameter, namely, 47·1238 ft.

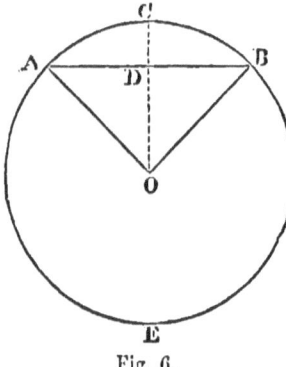

Fig. 6.

Let A C B E (fig. 6) be a circle having its centre at O. Any line, as A B, drawn across it is called a *chord*, and the part A C B is called an *arc*. If the straight lines A O and B O are drawn, the figure A C B O is called a *sector*, and the figure A C B between the chord and arc is called a *segment*. If we take D the middle of the chord A B, and draw O D C, D C is the height or *versed-sine* of the arc. The angle A O B is called the angle *subtended* by the arc A C B, and is always proportional thereto; that is, if the arc A B is twice the arc A C, then the angle A O B is double the angle A O C. The area of the *sector* A C B O is to the area of the whole circle in the proportion of the angle A O B to four right angles, or 360°; so that if the angle A O B is known, then the area of the sector is found by multiplying the whole area of the circle by the value of the angle A O B, and dividing the product by 360. For example: let the angle A O B be 50°, and the radius of the circle 5 ft.; then the area of the circle is $5^2 \times 3\cdot1416$, or 78·54 sq. ft., which multiplied by 50 and divided by 360, gives, 10·9 as the number of square feet in the sector. The length of the arc A C B is to that of the whole circumference in the same proportion as the angle A O B to 360; and in this example the circumference being 31·416 ft., the length of the arc A C B is 31·416 multiplied by 50 and divided by 360, or 4·363 ft.

The area of the *segment* A C B is found by deducting the area of the triangle A O B from the area of the sector A C B O, as found above. Now the area of the triangle A O B is one-half the product of A B into O D. In the above example, where the angle A O B is 50°, and the radius A O is 5 ft., the length of the chord A B is twice the radius O B multiplied by the *sine* of the angle D O B, or 4·226 ft., and the length of O D is the radius O B multiplied by the *cosine* of the angle D O B, or 4·532 ft., hence the area of the triangle A O B is 9·576 sq. ft. The area of the *sector* A C B O having been found above to be 10·9 sq. ft., the area of the *segment* A C B is 10·9 less 9·576, or 1·324 sq. ft.

A simple rule is given by Peter Nicholson, by which a tolerably accurate calculation of the area of a segment of a circle can be made without requiring to know the length of the radius of the circle, the *chord* A B and the height, or *versed-sine*, C D only being given. Multiply the length of the chord A B by the height C D, and to two-thirds of the product add the quotient arising from dividing the cube of the height C D by twice the length of the chord A B. This rule is expressed by the following formula:—

$$\text{Area of segment} = \frac{2}{3} AB \times CD + \frac{CD^3}{2AB}.$$

In the above example A B = 4·226, C D = ·468; hence the area of the segment by this rule is found to be 1·33 sq. ft., which is rather more than that given by the former process; the area, as calculated by Nicholson's rule, being always rather more than the true area, but sufficiently correct for ordinary calculations.

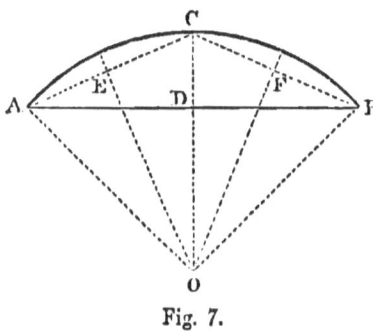

Fig. 7.

When the chord A B and height C D of a segment are given, the centre of the circle and length of the radius can easily be found geometrically in the following manner. Join A C and B C, and take E and F the middle points of the chords A C and B C respectively; draw E O and F O perpendicular to A C and B C; then the point O in which they meet is the centre of the circle. Produce C D to O, then C O is the radius of the circle.

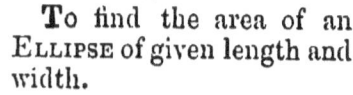

Fig. 8.

To find the area of an ELLIPSE of given length and width.

Let A B (fig. 8) be the length or *major-axis*, C D the width or *minor-axis*, and perpendicular to and bisecting A B; then O is called the *centre* of the ellipse.

The area of the whole ellipse is the product of the half-length A O into the half-width C O, multiplied by 22 and divided by 7. If greater accuracy is required, then multiply the above product by 355, and divide by 113; or multiply the product by the number 3·1416. Since the *axes* divide the ellipse into four equal parts, the area of one of those parts, as A O C, is one-fourth the area of the whole ellipse.

Example.—Let the length A B be 10 ft., and the width C D be 6 ft.; then the area of the whole ellipse is $\dfrac{3 \times 5 \times 22}{7}$, or 47·14 sq. ft.; or, more exactly, 47·124 sq. ft.

The length of a curved line can never be *exactly* measured, but we may approximate to it as nearly as we please by taking a number of points on the curve and joining them by a succession of *chords*, or straight lines subtending portions of the curve. If we measure the length of these chords, we get an approximation to the length of the curve; and the smaller the chords and greater their number the more nearly will the sum of their lengths approach to that of the curve. A similar process enables us to find the area of any figure bounded on one or more sides by a curved line, by drawing parallel lines from the end of all the chords, and measuring the area contained between them and the chords, which will be nearer and nearer to the required area the greater the number of the chords.

A solid having six oblongs or parallelograms for its faces is called a PARALLELOPIPED when each pair of opposite faces are parallel; and if all the angles are right-angles it is called a RIGHT-SOLID. The *solidity, volume,* or cubical content of a *right-solid* is found by multiplying together its three dimensions of length, breadth, and depth.

The solidity of a CYLINDER or PRISM having the top and bottom horizontal and the sides vertical, is found by multiplying the area of the base by the height. The solidity of a CONE is one-third the solidity of a cylinder having the same base and same vertical height.

If the upper part of a *cone* is cut off by a plane parallel to its base, the lower part is called a FRUSTUM of the cone. To find the solidity of a frustum of a cone or pyramid whose axis is vertical and its base a circle or any regular polygon: multiply the area of the base by its half-diameter; also multiply the area of the top by half its diameter, and subtract the latter product from the former; multiply the difference thus found by one-third of the vertical height, and divide by the difference between the two half-diameters. This rule is expressed in the following formula:—

Solidity of frustum of a cone $= \dfrac{1}{3} \dfrac{h}{R-r} (A R - a r)$;

where A is the area of the base, a that of the top; R the radius or half-diameter of the base, r that of the top: h the vertical height.

Example.—Let the frustum of a cone have for its base a circle whose radius is 5 ft., and for its top a circle whose radius is $2\frac{1}{2}$ ft.; the vertical height being 12 ft. In this case $A = 78.54$, $a = 19.63$, $R - r = 2\frac{1}{2}$, $\frac{1}{3}h = 4$; hence the solidity is $\frac{4}{5}$ (78.54 × 5 − 19.63 × 2.5), or 549.76 cub. ft.

To find the solidity of a *hollow* cylinder, first calculate its volume as if it were solid throughout; then calculate the volume of a cylinder having for its diameter the internal diameter of the cylinder; deduct the latter quantity from the former, and the solidity of the cylinder is found.

The calculation of the solidity of hollow or solid cylinders can be easily found by means of the Table p. 26, in which the numbers in the third and fourth columns represent the solid content of cylinders 1 foot in height, of the diameters (also in feet) stated in the first column; and for any other length we have only to multiply that number by the given length to obtain the solid content required.

Example.—To find the solid content of a hollow cylinder whose external diameter is $7\frac{1}{4}$ in., and the internal diameter $5\frac{3}{4}$ in., the height being 7 ft. 6 in.

Referring to the Table, we find against $7\frac{1}{4}$ the number in the third column is 41·2825; and against $5\frac{3}{4}$ the number is 25·9672; subtracting the latter number from the former gives 15·3153, which multiplied by 90 (the length in inches) gives 1378·377 cubic inches as the solidity of the cylinder and this divided by 1728 gives the solidity in cubic feet.

The area of the vertical surface of a cylinder or prism is found by multiplying the circumference of the base by the vertical height. For example, the external circumference of the above cylinder is found in Table p. 26 to be 22·7765, which multiplied by 90 gives 204·9885 square inches for the surface, and this divided by 144, gives the area in square feet.

The sloping or curved surface of a cone is found by multiplying the circumference of the base by half the length measured up the slant: the curved surface of the frustum of a cone is found by adding together the perimeters or circumferences at top and bottom, and multiplying their sum by half the distance between them measured up the slant.

The surface of a SPHERE is equal to four times that of a circle of equal diameter, and is found by multiplying the square of the diameter by the number 3·1416.

The solidity, or volume of a sphere is found by multiplying the cube of the diameter, or the length of the diameter multiplied by itself twice over, by the number ·5236.

VALUES OF ENGLISH MEASURES AND QUANTITIES,
RELATING TO BUILDING AND LAND.

A LINEAL INCH, or inch *run*, is one-twelfth of a lineal foot, and one thirty-sixth of a lineal yard.

A PALM, is 3 lineal inches.

A HAND, is 4 lineal inches.

A SPAN, is 9 lineal inches, or a quarter of a yard, or half a cubit.

A LINEAL FOOT, or foot *run*, is 12 lineal inches, or 3 hands.

A SQUARE INCH, is a square having each side measuring one lineal inch.

A SQUARE FOOT, or foot *superficial*, is a square having each side measuring one lineal foot, or 12 lineal inches, and contains 144 square inches.

A CUBICAL INCH, is a cube having each face one square inch.

A CUBICAL FOOT, is a cube having each face one square foot, and contains 1728 cubical inches.

A CUBIT, is 4 hands and a half, or 1 foot and a half lineal.

A LINEAL YARD, or yard *run*, is 3 lineal feet, or 36 lineal inches, or 2 cubits.

A SQUARE YARD, or yard *superficial*, is a square having

each side measuring one lineal yard, and contains 9 square feet.

A CUBICAL YARD, is a cube having each face one square yard, and contains 27 cubical feet.

An ELL is 1 lineal yard and a quarter, or 45 lineal inches.

A GEOMETRICAL PACE, is 5 lineal feet.

A FATHOM, is 6 lineal feet, or 2 lineal yards.

A SQUARE, is 100 square or superficial feet.

A STATUTE POLE, or *perch*, or *rod*, is 16 lineal feet and a half or $5\frac{1}{2}$ lineal yards.

A CHAIN, is 4 statute poles, or perches, or 22 lineal yards, or 100 links.

A FEN, or *woodland pole*, or *perch*, is 18 lineal feet.

A FOREST POLE, or *perch*, is 21 lineal feet, or 7 lineal yards.

A FURLONG, is 40 statute poles, or perches, or 10 chains, or 220 lineal yards.

A MILE, is 8 furlongs, or 80 chains, or 1760 lineal yards.

A SQUARE STATUTE POLE, or *perch*, is $30\frac{1}{4}$ sq. yds. or $272\frac{1}{4}$ sq. ft.

A SQUARE WOODLAND POLE, or *perch*, is 324 sq. ft.

A ROOD, is 40 square statute poles, or perches, or 1210 sq. yds.

An ACRE, is 4 roods, or 160 perches, or 4840 sq. yds.

A LOAD of rough timber, is 40 cubical ft.

A LOAD of squared timber, is 50 cubical ft.

A LOAD of 1-inch plank, is 600 sq. ft.

A LOAD of $1\frac{1}{2}$-inch plank, is 400 sq. ft.

A LOAD of 2-inch plank, is 300 sq. ft.

A LOAD of $2\frac{1}{2}$-inch plank, is 240 sq. ft.

A LOAD of 3-inch plank, is 200 sq. ft.

A LOAD of $3\frac{1}{2}$-inch plank, is 170 sq. ft.

A LOAD of 4-inch plank, is 150 sq. ft.

A LOAD of statute bricks, is 500.

A LOAD of plain tiles, is 1000.

A LOAD of lime, is 32 bushels.

A LOAD of sand, is 36 bushels.

A HUNDRED of lime, is 35 bushels.

A BUSHEL, is 2218 cubical inches.

A HUNDRED of deals, is 120.

A Hundred of nails, is 120.
A Thousand of nails, is 1200.
A Thousand of slates, is 1200.
A Ton of iron, is 2240 pounds weight.
A Fodder of lead, is $19\frac{1}{2}$ hundred, or 2184 pounds.
A Hundred of lead, is 112 pounds weight.
A Table of glass, is 5 ft., and 45 tables is a case; but of Newcastle and Normandy glass, 25 tables make a case.
A Bundle of 4 feet oak-heart laths, is 120, and $37\frac{1}{2}$ bundles are a load.
A Bundle of 5 ft. oak-heart laths, is 100, and 30 bundles are a load.

N.B.—Fir or Deal Laths are of divers lengths, as 3, 4, 5, and 6 ft.; but all of them are reduced to the standard length of 5 ft.; and so every 150 ft. run of bundles (each bundle containing 100 laths) is a load, being equal to 30 bundles of 5 ft. laths.

TABLE OF AREAS AND CIRCUMFERENCES OF CIRCLES.

By this Table the area and circumference of a circle of given diameter can be found, AT SIGHT, and also the area of the square described upon the diameter.

The first column contains the diameter in inches, feet, or yards, from 1 to 10, increasing by quarters. The second column gives the square of the numbers in the first column, expressed in decimals. The third and fourth columns give the area of the circle whose diameter is found in the first column, expressed in decimals and duodecimals. The fifth and sixth columns give the circumference of the circle whose diameter is found in the first column, expressed in decimals and duodecimals.

If the diameter is given in feet, the column for the areas of circles gives the cubical content of a cylinder 1 ft. high, having the said diameter; so that, to find the cubical content of a cylinder of any other height, the number in that column must be multiplied by the given height; if the diameter is in feet, then the height must be expressed in feet also, and the content will be obtained in cubic feet.

Example.—To find the quantity of earth excavated

from a circular cesspool 4¾ ft. in diameter, and 7 ft. deep. The area opposite to 4¾ is 17. 8. 8., which, multiplied by 7, gives 124. 0. 8. cubic ft. of earth excavated. To find the superficial area of the brickwork required to surround the above cesspool; opposite 4¾ the circumference is 14. 11. 1., which, being multiplied by 7, gives 104. 5. 7. superficial ft. of brickwork.

Diamr. of Circle.	Square of Diamr.	Area of Circle.		Circumference of Circle.	
		Decimals.	Duodecimals.	Decimals.	Duodecimals.
1	1·000	·7854	0 9 5	3·1416	3 1 8
1¼	1·563	1·2272	1 2 9	3·9270	3 11 1
1½	2·250	1·7671	1 9 2	4·7124	4 8 7
1¾	3·063	2·4053	2 4 10	5·4978	5 6 0
2	4·000	3·1416	3 1 8	6·2832	6 3 4
2¼	5·062	3·9761	3 11 9	7·0686	7 0 10
2½	6·250	4·9087	4 10 8	7·8540	7 10 3
2¾	7·562	5·9396	5 11 3	8·6394	8 7 8
3	9·000	7·0686	7 0 10	9·4248	9 5 1
3¼	10·563	8·2958	8 3 7	10·2102	10 2 6
3½	12·250	9·6211	9 7 5	10·9956	10 11 11
3¾	14·062	11·0447	11 0 6	11·7810	11 9 4
4	16·000	12·5664	12 6 10	12·5664	12 6 10
4¼	18·063	14·1863	14 2 3	13·3518	13 4 3
4½	20·250	15·9043	15 10 10	14·1372	14 1 8
4¾	22·562	17·7205	17 8 8	14·9226	14 11 1
5	25·000	18·6350	19 7 7	15·7080	15 8 6
5¼	27·563	21·6475	21 7 9	16·4934	16 5 11
5½	30·250	23·7583	23 9 1	17·2789	17 3 4
5¾	33·063	25·9672	25 11 7	18·0642	18 0 9
6	36·000	28·2743	28 3 3	18·8496	18 10 2
6¼	39·063	30·6796	30 8 2	19·6350	19 7 7
6½	42·250	33·1831	33 2 2	20·4204	20 5 0
6¾	45·562	35·7847	35 9 5	21·2058	21 2 5
7	49·000	38·4845	38 5 10	21·9911	21 11 10
7¼	52·563	41·2825	41 3 5	22·7765	22 9 4
7½	56·250	44·1786	44 2 2	23·5619	23 6 8
7¾	60·061	47·1730	47 2 1	24·3473	24 4 2
8	64·000	50·2655	50 3 4	25·1327	25 1 8
8¼	68·061	53·4562	53 5 6	25·9181	25 11 0
8½	72·250	56·7450	56 8 11	26·7035	26 8 6
8¾	76·563	60·1320	60 1 7	27·4889	27 5 10
9	81·000	63·6173	63 7 6	28·2743	28 3 4
9¼	85·562	67·2006	67 2 5	29·0597	29 0 9
9½	90·250	70·8822	70 10 7	29·8451	29 10 2
9¾	95·060	74·6619	74 7 11	30·6305	30 7 7
10	100·000	78·5398	78 6 4	31·4159	31 5 0

CHAPTER II.

EXCAVATING AND WELL-SINKING.

EXCAVATING.

BEFORE a building can be commenced, it is necessary to dig out the ground to form *trenches* in which the foundations can be laid on as solid a basis as possible; also to excavate the ground for the basement story or cellars, and to dig trenches in which to lay the drains.

Where the excavation has to be carried down to a considerable depth, ground of various degrees of hardness will be met with, and separate items must be made of the different qualities of material excavated; as the quantity of earth that one man can excavate will vary according to the hardness of the soil. The labour of digging and *throwing-out* will also increase with the depth, and a separate price is charged for every additional *throw* of 5 or 6 ft. of depth.

DIGGING is measured by taking first the basement story or cellars, which is stated as digging and throwing-out, or wheeling away; next the excavations to the trenches for footings of walls, and for the drains, cesspools, and well-sinking.

It is customary, in taking the digging to footings of walls to allow about six inches on each side, over and above the width of the footings, for room to work them; but if they are deep, and the ground bad and loose, allow nine inches on each side on account of its falling in. In sunk stories, only allow to the extent of the footings, except in very loose ground. In measuring trenches for concrete foundations, they may be taken the *nett* width of the concrete. Trenches for drain-pipes or brick drains must be at least 12 inches wider than the drains themselves.

In taking the dimensions, the length, depth, and width

must be measured as before described, and reduced to the cubical yard of 27 cubical feet; namely, 3 ft. by 3 ft. by 3 ft. This quantity of 27 cubical feet is called a single load, and contains 21 striked bushels. Two cubic yards equal one double load.

In estimating excavators' work, it is advisable to keep the wheeling, carting, filling-in, and ramming to foundations and drains, separate from the actual digging and throwing-out; also the digging to the basement and cellars, drains, foundations, cesspools, wells, &c., under separate items.

The amount of digging which a man can perform in a day depends so much on the nature of the soil on which he has to operate, that it is almost impossible to fix a constant for this description of labour. The following data may, however, serve as a slight guide:—

In loose ground a man will throw up about ten or twelve cubic yards per day; but in hard or gravelly soils, where *hacking* with the *pick* is necessary, from five to six cubic yards, according to the hardness of the ground, will be a fair day's work.

When the site has to be levelled before the building is commenced, or the top soil has to be removed, the excavation is taken by the yard superficial if the depth does not exceed 1 foot. Under the Metropolitan Building Acts, it is required that "no house, building, or other erection, shall be erected upon any site or portion of any site which shall have been filled up or covered with any material impregnated or mixed with any fæcal, animal, or vegetable matter, or which shall have been filled up or covered with dust, or slop, or other refuse, or in or upon which any such matter or refuse shall have been deposited, unless and until such matter or refuse shall have been properly removed, by excavation or otherwise, from such site. Any holes caused by such excavation must, if not used for a basement or cellar, be filled in with hard brick or dry rubbish."

FILLING INTO BARROWS and WHEELING is estimated by the *run* of 20 yards: an additional charge being made for every 20 yards beyond the first run. A gang of three men, two for filling and one for wheeling, will remove about 30 yards per day to the distance of one *run;* and the labour of removing earth may be calculated according to distance, allowing three men to the first run, and an additional man for every 20 yards of extra distance.

Therefore, to find the price of wheeling any number of cubic yards to any given distance, we have the following rule :—Divide the distance in yards by 20, which gives the number of wheelers; add the two cutters to the quotient, which gives the whole number employed; and the sum, multiplied by the rate of wages per diem, is the price of 30 cubic yards; so that, as 30 cubic yards is to the whole number of yards, so is the price of 30 yards to the entire cost.

FILLING INTO CARTS and CARTING AWAY is estimated by the *mile* of distance from the work; an additional charge being made for every additional mile beyond the first.

PLANKING AND STRUTTING are often necessary in excavating trenches of great depth in loose earth, to keep the ground from falling in, the cost of which must be added to the price for the digging; or it may be measured separately at per foot run, describing the average depth.

A cubic yard of earth in its original position, *before* excavation, will occupy from $1\frac{1}{4}$ to $1\frac{1}{2}$ cubic yards of space *after* being excavated; but it will subside into nearly its original bulk when formed into embankments.

The following table gives the average bulk of one ton weight of different kinds of earth as excavated :—

21 cubic feet of sand weigh one ton.
$20\frac{1}{3}$,, gravel ,,
$18\frac{1}{2}$,, clay ,,
$15\frac{1}{3}$,, chalk ,,
18 ,, night-soil ,,

Night-soil is removed in carts containing 45 cubic feet or $2\frac{1}{2}$ tons.

CLAYING OR PUDDLING OF VAULTS, or clay *tempered* and laid over vaults about 6 in. in thickness, and puddled, is measured by the yard superficial of 9 square ft.; describing the thickness.

In measuring digging in SIDELING GROUND, where the areas of the two ends of the excavation are unequal, the cubic content must be found by the following rule :—

Multiply the sum of the extreme areas, plus four times the middle area, by one-sixth of the length, and the product will be the answer required.

Example.—To find the cubic content of the excavation A B C D E F for the sunk stories of a house to be built on the side of a hill :—

B G A D E represents the natural surface; and C H A D F the levelled surface, obtained by excavating to the depth of 10 ft. at B C, 6 ft. at E F, and nothing at D and A. The whole mass being regarded as a frustum of an irregular pyramid, the triangles A B C and D E F are called the 'extreme areas;' and the triangle N, halfway between and parallel with them, is the 'middle area.'

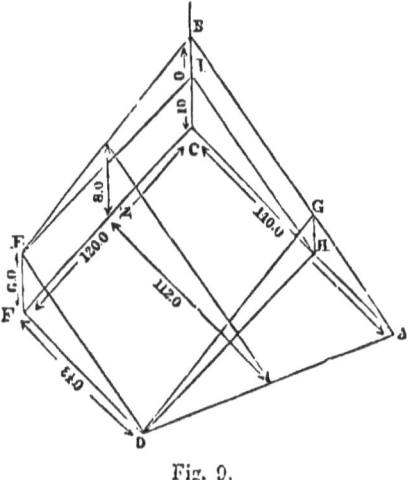

Fig. 9.

$$\frac{C A \times B C}{2} = \frac{140 \times 10}{2} = 700 \text{ square feet in A B C.}$$

$$\frac{F D \times E F}{2} = \frac{84 \times 6}{2} = 252 \text{ square feet in D E F.}$$

$$\phantom{\frac{F D \times E F}{2} = } 952 \text{ sum of extreme areas.}$$

$$4 \times \frac{112 \times 8}{2} = 1792 \text{ four times middle area N.}$$

$$\phantom{4 \times \frac{112 \times 8}{2} = } 2744$$

$$\frac{F C}{6} = \frac{120}{6} = 20 \text{ sixth of the length F C.}$$

27) 54880 (2032 yards 16 feet.
54
——
88
81
——
70
54
——
16

As an illustration of the correctness of the rule, let us

take the same example on a different principle of measurement. The solid A B C D E F may be divided into the two prisms G H I B E D and E F C I H D and the pyramid D A H G. Taking each of these separately, we have—

$$\text{Prism G H I B E D} = \frac{84 \times 4 \times 120}{2} = 20160$$

$$\text{Prism E F C I H D} = \frac{120 \times 6 \times 84}{2} = 30240$$

$$\text{Pyramid D A H G} = \frac{56 \times 4}{2} \times \frac{120}{3} = 4480$$

$$\text{Total} \ldots \overline{54880} \text{ cubic ft.}$$

or 2032 yds. 16 ft., as before.

WELL-SINKING.

WELLS and CESSPOOLS must have their digging kept separate from all other excavations; and being circular, the quantity of earth excavated in each foot of depth can be found by multiplying the square of the half-diameter, in feet, by 22, and dividing the product by 7. The column in Table (p. 26) headed 'Area of Circle,' will also give the quantity of earth in 1 ft. of depth for any diameter from 1 ft. up to 10 ft.; and the cubical content for any depth is found by multiplying the quantity contained in 1 ft. of depth by the number of feet in the total depth of the well.

The excavation down to 30 ft. depth may be put under one item; but beyond that depth the cost increases, and an additional charge is made for every extra 30 ft. beyond the first. The nature of the soil through which the well is sunk must also be taken into consideration.

DRY STEINING, is brickwork laid dry round wells, to keep the earth from falling in. This is generally charged with the digging of the well, at so much per foot of depth.

For wells sunk through loose soils, the brickwork is carried down along with it, a template or *curb* of wood being formed round the well on which the bricks are laid; and as the ground is dug away from underneath, the template is pushed downwards by the weight of the brickwork above, the top of which is always kept level with the surface of the ground.

A better plan however for deep wells is to build rings of brickwork in cement at intervals of from 5 to 12 feet, in 3 courses laid flat. This method obviates the necessity of employing a wood curb, which is always liable to decay. Iron curbs are sometimes employed for wells sunk in sandy soils. Cylinders of cast or wrought iron are also used instead of brickwork, these are sunk down by their own weight, as the soil is excavated from the inside of the cylinder.

When a well has been sunk to a considerable depth without obtaining an adequate supply of water, it is usual, in order to save the expense of further sinking, to perform the operation of BORING into the soil until the water rises into the well in sufficient quantities. A hole is bored a few inches in diameter, into which an iron tube is inserted to prevent the filling up of the hole, when the soil is clay or sand; but when it is hard rock there is no necessity for the tube. For further information on this subject the reader is referred to a "Rudimentary Treatise on Wells and Well-sinking," by J. G. Swindell & G. R. Burnell.

BORING for water depends upon the nature of the soil or rock to be passed through; and the cost per foot of depth increases with every 10 ft.; the price per foot for the second 10 ft. being double the price for the first 10 ft.; for the third 10 ft. it is three times the price for the first 10 ft.; and so on, increasing in arithmetical proportion.

PUMPING water out of wells during the process of sinking is charged separately, by the gallon; there being $6\frac{1}{4}$ gallons in every cubic foot of water.

In order to explain more clearly the method of measuring excavators' work, we will take out the digging for the house shown on Plate I., supposing the digging of the basement story to be taken to the top of the footings of the wall, or 8 ft. 6 in. deep, and the building to be a square on plan 26 ft. each way, the footings projecting 9 ins. all round and to consist of 5 courses, or 15 inches deep, with a concrete foundation 12 ins. thick and 12 ins. wider than the footings.

METHOD OF MEASURING.

28 6 28 6 1 6	1218	5	Digging and throwing out loose stuff to basement story, and wheeling 20 yards.
28 6 28 6 3 6	2842	11	Do., do., gravel, &c., under 5 ft. in depth, and carting 1 mile.
28 6 28 6 3 6	2842	11	Do., do., do., above 5 ft. and under 10 ft. in depth, and carting 1 mile.
95 0 4 9 2 3	1015	4	Do., do., to trenches for foundations and ramming bottom, including planking and strutting, eight-ninths carted away 1 mile.
50 0 1 6 3 0	225	0	Do., do., to drains, and one-third carted away 1 mile.
95 0 1 3 9	89	1	Do., do., filling in only to foundations and ramming.
50 0 1 6 2 0	150	0	Do., do., to drains.
19 8 30 0	590	0	Digging to form well 5 ft. external diameter, not over 30 ft. in depth.
19 8 15 0	295	0	Do., do., over 30 ft. deep.
10 0 6 9	67	6	Claying of vaults.
26 0 5 0	130	0	Removing top soil to a depth not exceeding 12 in., and levelling for area.
30 0	30	0	Well sinking, including steining in ½ bk. 5 ft. diameter, under 30 ft. in depth.
15 0	15	0	Do., do., do., in 1 bk., 5 ft. diameter, above 30 ft. in depth.
10 0	10	0	Boring in clay, under 10 ft. deep.
10 0	10	0	Do., do., over 10 ft deep.

D

ABSTRACT.

Cube.		
Digging and throwing out loose stuff and wheeling 20 yds. 27)1218·5 45·1·4	Carting 1 mile. 2842·11 2842·11 902· 5 75· 0 27)6663·3 246·9·4	Filling in only and ramming. 89·1 150·0 27)239·1 8·10·3
Do., do., in gravel, &c., under 5 ft. deep. 27)2842·11 105·3·1	Digging to trenches, ramming, planking, and strutting. 1015·4 225 27)1240·4 45·11·4	Digging only to form well under 30 ft. deep. 27)590 21·23·0
Do., do., above 5 ft. deep and under 10 ft. 27)2842·11 105·3·1		Do., do., above 30 ft. deep. 27)295 10·25·0

Super.	Run.
Removing top soil not more than 12 in. deep, and levelling. 9)130·0 14·4	Well sinking, steining in ½ bk. 5 ft. diameter, under 30 ft. deep. 30·0
	Do., do., in 1 bk. above 30 ft. deep. 15·0
Claying of vaults. 9)67·6 7·4·6	Boring in clay, &c., depth under 10 ft. 10·0
	Do., do., do., over 10 ft. deep. 10·0

ROTATION.

To be attended to in bringing the quantities into Bill.

EXCAVATOR, WELL-SINKER.

Yds.	ft.	in.	
45	1	4	Cube of digging and throwing out loose stuff to basement story and cellars .
105	3	1	Ditto, ditto, ditto, gravel, clay, or stiff stuff, to ditto, ditto, under 5 ft. in depth . . . . .
105	3	1	Ditto, ditto, ditto, above 5 ft. and under 10 ft. in depth . . . .
45	11	4	Ditto, ditto, ditto, to trenches for foundations and drains, and ramming the bottom, including planking and strutting . . . . .
8	10	3	Ditto, filling in only and ramming to foundations and drains . . .
41	21	11	Ditto, of wheeling 20 yds. distance .
232	24	6	Ditto, carting away to a distance of 1 mile . . . . . .
21	23	0	Ditto, digging only and throwing out to form well under 30 ft. deep . .
10	25	0	Ditto, ditto, ditto, ditto, above 30 ft. deep
7	4	6	Superf. claying of vaults 6 in. in thickness . . . . . .
14	4	0	Ditto, removing top soil not more than 12 in. deep, and levelling . . .
	30	0	Run of well-sinking, including steining in half-a-brick, 5 ft. in diameter, under 30 ft. in depth . .
	15	0	Ditto, ditto, ditto, steining in one-brick, 5 ft. in diameter, above 30 ft. in depth
	10	0	Ditto, of boring in clay, gravel (or other soil), including tools and tackle, with 3½ in. auger; depth under 10 ft. .
	10	0	Ditto, ditto, ditto, ditto, depth above 10 ft. . . . . .

VALUATION OF EXCAVATOR'S WORK.

CONSTANTS OF LABOUR.   Constant to be multiplied by the rate of wages of a Navvy per hour.

Digging in loose soils, 1 throw, per cubic yard . .	·88
Do. clay do. do. . . .	1·00
Do. hard gravel do. do.	1·50
Filling barrow and wheeling, 1 run	·45
Filling into carts	·75
Do. and ramming to walls and drains . . .	·40

CHAPTER III.
FOUNDATIONS, BRICKWORK, TILING, SLATING.

FOUNDATIONS.

THIS term is usually applied to the material which is laid in the earth previous to the brickwork or masonry being commenced. Before any of the walls of a building are erected, it is very essential that the nature of the subsoil should be carefully ascertained, as upon that must depend the kind of foundation to be laid. If we find at a short distance below the surface a hard rock, or a compact bed of gravel several feet in thickness, we have only to make a level bed and can then begin to lay the footings of the walls. When the soil is of clay, sand, or any loose material it will be necessary to form an artificial foundation in order to prevent the weight of the walls from pressing them down into the soil and causing unequal settlements in the building. The usual method is to dig trenches rather wider than the footings of the intended walls, and to fill these with a material called concrete, which becomes hardened and forms into a solid mass. The width of the concrete must depend on the nature of the subsoil; if that is moderately firm it will be sufficient to have the concrete a few inches wider than the footings, but where the subsoil is very loose the concrete should be made wider, and sometimes it is found necessary to spread it over the whole surface of the ground on which the building is to be erected, so as to prevent unequal settlements.

CONCRETE is a kind of artificial stone or rock made by

mixing unslacked lime or cement with gravel or broken-stone and sand, a sufficient quantity of water being added to slack the lime. It is thrown while moist into the trenches for the foundations, and in a few days hardens into a solid mass. By means of concrete, buildings may safely be erected on the softest and most yielding soils, as the weight is uniformly distributed over a large area; and if any sinking takes place, the whole building settles uniformly throughout.

Lime-concrete is made of ground stone lime, and sharp gravel, with a proper quantity of sharp sand, mixed in the proportion of five or six parts of gravel and sand to one of lime, according to the nature of the lime and the proportion of sand mixed with the gravel. Its quality is much improved by the addition of smiths' ashes, or any material containing iron; and for this reason ferruginous gravel is to be preferred whenever it can be obtained. Concrete made with lime expands slightly in slaking; but this expansion is too trifling to be taken into account in framing an estimate. A cubic yard of concrete, containing 27 cub. ft. when mixed, requires 34 cub. ft. of gravel, sand, and lime: therefore, at the proportion of six of gravel to one of lime, a cub. yard of concrete will require 1·1 cub. yard of gravel and sand, and three bushels of lime.

The common method of making concrete is to mix the unslacked lime with the sand and gravel, and to add sufficient water afterwards to slack the lime. It is however better to slack the lime first and make it into a mortar with the sand before mixing with the gravel. In either case it is essential that the whole mass should be turned over several times, so as to thoroughly incorporate the materials before throwing them into the trenches; and this should not be done from a height as is often the case, as the particles are liable to be separated in falling, but should be wheeled in on the level and afterwards well beaten with a rammer. See Burnell on "Limes, Mortars, and Cements."

Cement-concrete is made with *Portland* cement instead of lime, in the proportion of one part cement to seven or eight parts of gravel and sand; care must be taken that the sand and gravel are perfectly free from earthy particles, and the coarser the sand the better. Only a small quantity must be mixed at once, as it sets rapidly. Cement-concrete contracts slightly in setting.

The following are the provisions of the Metropolitan Building Acts with regard to the use of concrete :—

"The site of every house or building shall be covered with a layer of good concrete, at least 6 in. thick, and smoothed on the upper surface, unless the site thereof be gravel, sand, or natural virgin soil.

"The foundation of the walls of every house or building shall be formed of a bed of good concrete, not less than 9 in. thick, and projecting at least 4 ins. on each side of the lowest course of footings of such walls. If the site be upon a natural bed of gravel, concrete will not be required.

"The concrete must be composed of clean gravel, broken hard brick, properly burnt ballast, or other hard material to be approved by the District Surveyor, well mixed with fresh burnt lime or cement in the proportion of 1 of lime to 6, and 1 of cement to 8 of the other material."

It is usual to measure concrete, when used in considerable quantities and not less than 12 in. in thickness, by the cubic yard; but where the thickness is less than 12 in., it may be taken at the superficial yard, the thickness being stated.

PILING is a mode of forming a foundation by means of long pieces of timber called PILES, which are driven vertically into the ground.

Where the site for a building is found to consist of a considerable depth of sand or loose soil, the use of piles is necessitated in order to secure a firm basis on which to build.

Piles are generally used of whole timber about 12 inches square, and are furnished at the bottom with pointed iron shoes, while the top is protected by a circular iron ring to prevent the wood from splitting under the blows received in driving. The length will depend on the nature of the soil and weight of the building to be carried; but if possible, they should be driven down to a hard stratum of earth. The piles are placed about 3 feet from centre to centre, and are driven by means of a heavy weight let down from a height upon their heads. When the required depth is reached, the heads may be cut off level with the ground, and planking fixed across them to keep them in their places. Concrete or masonry can then be laid on the top of the piling.

IRON SCREW PILES are used in soft soils; these consist of wrought iron hollow cylinders having a broad bladed screw at the bottom, and are driven by turning them round at the top.

HOLLOW CYLINDERS of iron are sometimes used to form foundations, the earth being excavated from the inside of the cylinder, which then sinks by its own gravity, as described under the head of well-sinking. When the cylinder is sunk to the required depth, it is filled with concrete, brickwork, or masonry.

For further information on this subject, the reader is referred to Dobson's Treatise on Foundations.

In measuring timber piles, the scantling and length is taken and valued as cube fir; the length to which they are driven being taken separately; also the weight of iron used in forming the rings and shoes, and the labour in pointing, fixing the iron work, cutting off the heads, &c.

BRICKWORK.

The bricklayer's work consists in the building of walls with bricks made of burnt clay, bedded and flushed-up with mortar or cement. The bricks used in this country are usually from $8\frac{1}{2}$ in. to 9 in. long, 4 in. to $4\frac{1}{2}$ in. wide, and about $2\frac{1}{2}$ in. thick; although bricks of other dimensions are occasionally made for special purposes. The varieties of common bricks are *malms* or *marls*, *stocks*, and *place-bricks*. The *best malms* are selected as cutting bricks, for gauged arches, quoins, &c., and are rubbed to the required dimensions and gauge. The *seconds malms* are used for facing the fronts of buildings. *Stocks* are hard, rough looking bricks, in which ashes have been mixed with the clay or loam used in their manufacture, capable of sustaining a great amount of pressure, and consequently are used for the principal walls of a building. *Place-bricks* are those which have not been thoroughly burnt, and are consequently soft and unfit to sustain heavy pressure or exposure to the weather; they are used for internal partitions, sleeper walls, &c., where great strength is not required. There are also *burrs* or *clinker-bricks*, which have become vitrified by being too violently acted on by the fire, so that several bricks have run together into one mass. These are quite unfit for bricklayers' work, but are valuable for mixing with concrete, and also for road making, as they possess great hardness. All these bricks are first dried in the sun upon rows called *hacks* and then burnt in *clamps*, the bricks being built up into a stack in the open air before being burnt, and a fire kindled in the centre. The superior kinds of brick are burnt in enclosed *kilns*, as *red facing* bricks, which owe their colour to the nature of the clay employed upon them; also *moulded* bricks, for forming string-courses, cornices, and other ornamental features in a building.

FIRE-BRICKS are those which are capable of resisting any amount of heat applied to them, and are consequently used for the linings of furnaces and furnace-chimneys, the joints and beds being made with *fire-clay* instead of mortar or cement.

PAVING-BRICKS or *Paviors* are made 1 inch thinner than common bricks, and are of a very hard texture finely tempered. There is also a smaller paving-brick of very hard character, which is of a much smaller size than ordinary bricks, and called a *clinker*.

BEDS of brickwork are those joints which run in horizontal planes; and a *course* is the whole layer of bricks comprised between two adjacent beds. In the erection of walls, when the bricks are laid longitudinally, or with their longest side parallel to the length of the wall, they are called *stretchers*; when laid transversely, or with their longest side perpendicular to the face of the wall, they are called *headers*. Brick *bond* signifies the binding together of two adjacent bricks in the same course by one brick in the next course above pressing on both, or as it is termed, *breaking joint*.

BONDS are systematic modes of arranging bricks so as to insure breaking joint throughout the work. OLD ENGLISH BOND is the term given to brickwork in which *stretchers* only are laid in one course, and *headers* in the next above; and in like manner, headers and stretchers in each alternate course; in which case it is requisite to place quarter-bricks to break the joints at the external angles of the wall; when these are introduced they are termed *closers*. The method of laying two consecutive

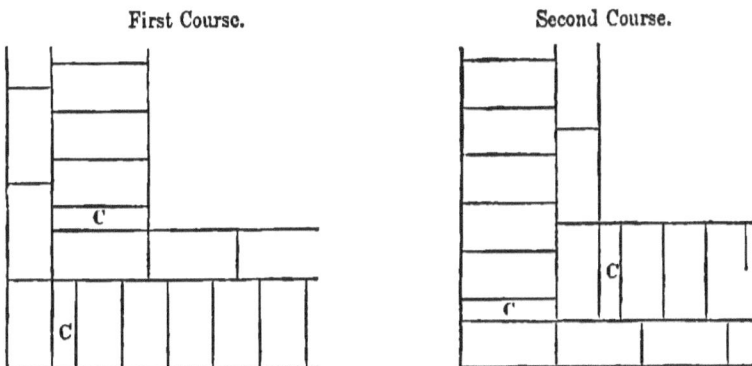

Fig. 10.—ENGLISH BOND.

courses of a 14 in. wall in English bond is shown on figure 10, and the *closers* are marked with the letter c.

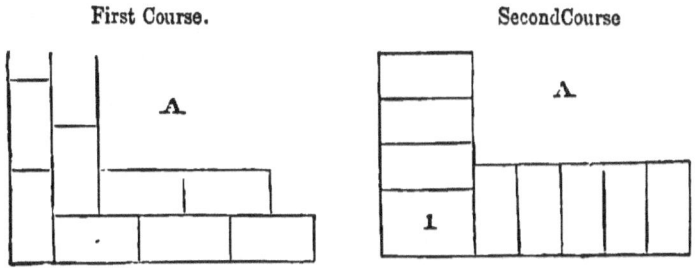

Fig. 11.—ENGLISH BOND.—NINE-INCH WALL.

The use of *closers* might, however, be advantageously superseded by having bricks made one half wider than the ordinary bricks, as shown on figures 11, 12, and 13, and marked with the numbers 1, 2, and 3.

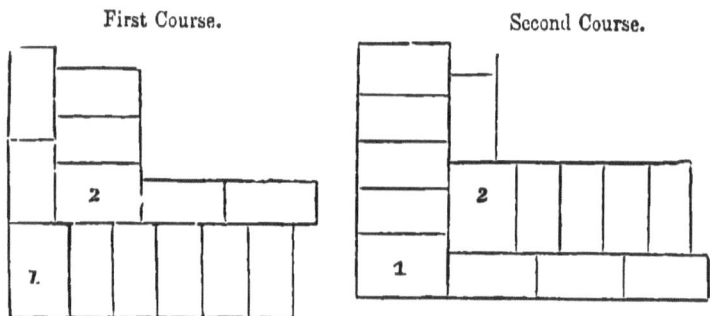

Fig. 12.—English Bond.—Fourteen-inch Wall.

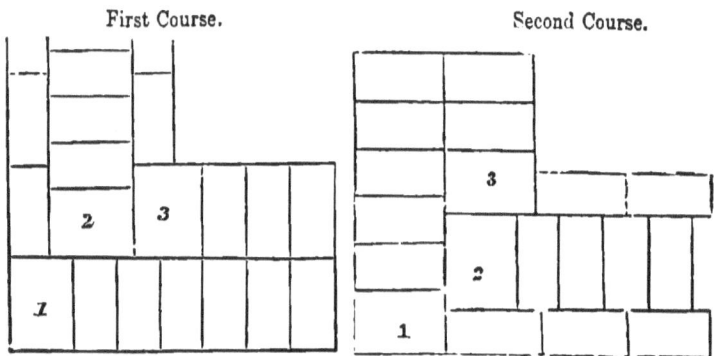

Fig. 13.—English Bond.—Eighteen-inch Wall

Flemish Bond is the term applied to brickwork in which *headers* and *stretchers* are placed alternately in *each* course, as shown for a 14 in. wall in figure 14, the *closers* required to break the joints being marked with the letter c.

This disposition of the bricks is thought by some to be not so strong as the English bond, and it requires more labour to execute. The *closers* might be dispensed with in the manner described for *English* bond (figs. 11, 12,

13,) by using wider bricks, made especially for the purpose.

From the size which bricks are usually made, it follows that the thickness of brick walls must always be a multiple of 4½ in.; thus we have a 9 in. wall; a 13½ in.

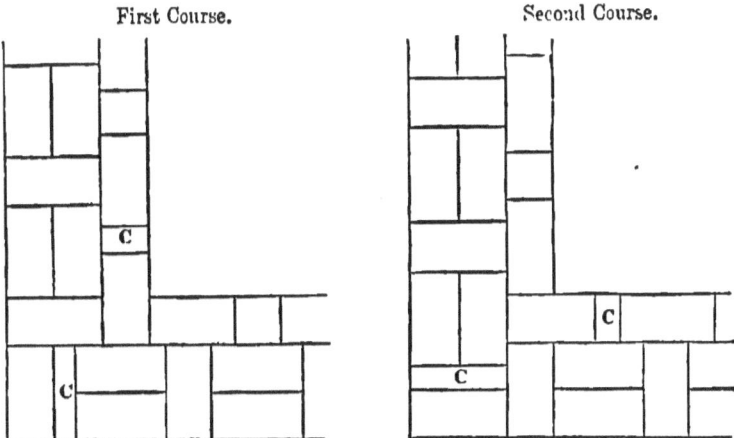

Fig. 14.—FLEMISH BOND.

wall (commonly called a 14 in. wall); an 18 in. wall; a 22½ in. wall; a 27 in. wall; and so on. They are also named according to the number of bricks (in length) which compose their thickness. Thus a 9 in. wall is called a *one-brick* wall; a 14 in. wall a *brick-and-a-half* wall; an 18 in. wall a *two-brick* wall; a 22½ in. wall a *two-and-a-half* brick wall; and so forth.

The following are the thicknesses of external and party brick walls, required by Act of Parliament for buildings erected in London and its suburbs, and may be generally taken as the *minimum* thickness which can with safety be employed in any building in which brick is employed:—

In DWELLING HOUSES, where any external or party wall is less than 25 ft. high and under 30 ft. long, it must be at least 8½ in., or 1 brick in thickness; and where the length of such a wall is greater than 30 ft., the part below the topmost story must be 13 in., or 1½ brick in thickness, and the upper part 8½ in.

When the height is under 30 ft. and the length less than 35 ft., the wall below the two upper stories must be 13 in. thick, and the rest 8½ in. thick; if the length exceeds 35 ft., the whole wall below the topmost story must be 13 in., and the remainder 8½ in.

When the height is under 40 ft., and the length less than 35 ft., the thickness of the wall is regulated as in the last case; but when the length exceeds 35 ft., the wall of the lowest story must be 17½ in., or 2 bricks thick, the rest of the wall below the top story 13 in., and the remainder 8½ in.

When the height is under 50 ft., and the length not more than 30 ft., the thickness of the wall below the top story is 13 in., and the rest 8½ in.; if the wall is not more than 45 ft. long, the thickness of the lowest story is 17½ in., the rest below the top story 13 in., and the remainder 8½ in. For greater length of wall, the lowest story must have walls 21½ in., or 2½ bricks thick; the next story 17½ in., and the remainder 13 in.

When the height is under 60 ft., and the length not more than 30 ft., the lowest story must have walls 17½ in. thick, the rest being 13 in.; if the length is under 50 ft., the two lowest stories must have 17½ in. walls, the rest being 13 in.; if the length exceeds 50 ft., the lowest story must have 21½ in. walls, the two next stories 17½ in., and the rest 13 in.

In walls up to 70 ft. high, having a length not exceeding 40 ft., the thickness must be 17½ in. in the two lower stories, the rest being 13 in.; if the length is under 55 ft., the lowest story must have 21½ in. walls; the two next stories 17½ in., and the rest 13 in. For a greater length of wall, the lowest story must have 26 in., or 3-brick walls, the two next 21½ in., the next story 17½ in., and the rest 13 in.

Walls under 80 ft. in height, with a length of 40 ft., must have 21½ in. thickness in the lowest story, 17½ in. in the two next stories, and 13 in. for the remainder; if under 60 ft. long, the two lower stories must have 21½ in. walls, the two next stories 17½ in., and the remainder 13 in. For greater length, the lowest story must have

26 in. walls, the two next 21½ in., the two next 17½ in., and the remainder 13 in.

In walls up to 90 ft. in height, the length being under 45 ft., the thickness of the two lower stories must be 21½ in., of the two next 17½ in., and of the remainder 13 in. Where the length is under 70 ft., the thickness in the lowest story must be 26 in., in the two next 21½ in., in the two next 17½ in., and the remainder 13 in. If of greater length, the thickness of the lowest story must be 30 in., or 3½-bricks, of the two next stories 26 in., of the next story 21½ in., of the two next stories 17½ in., and of the remainder 13 in.

In walls up to 100 ft. high, having a length not exceeding 45 ft., the two lowest stories must have 21½ in. thickness, the three next 17½ in., and the remainder 13 in. If the length is not more than 80 ft., the two lowest stories must have 26 in. walls, the two next 21½ in., the two next 17½ in., and the rest 13 in. If of greater length than 80 ft., the walls in the lowest story must be 30 in., in the two next 26 in., in the two next 21½ in., in the two next 17½ in., and the remainder 13 in.

In WAREHOUSES, FACTORIES, &c., the thickness of the external and party walls, which do not exceed 25 ft. in height, must be 13 in. at the base. If the height is under 30 ft., and the length not more than 45 ft., the walls must be 13 in. at the base; but if more than 45 ft. long, 17½ in. at the base. In walls not more than 30 ft. high, the thickness of the top story may be 8½ in.; but in all other walls of greater height than 30 ft., the thickness at the top and for 16 ft. down must be 13 in.

When the height is under 40 ft., and the length under 30 ft., the thickness at base must be 13 in.; if the length is under 60 ft., the thickness at base is 17½ in.; and for a greater length, 21½ in. With the height under 50 ft., and length under 40 ft., the base must be 17½ in.; if the length is under 70 ft., the base must be 21½ in.; and for greater length, 26 in. When the wall is under 60 ft. high, and its length not more than 35 ft., the thickness at base must be 17½ in.; if under 50 ft. long, 21½ in.; if of greater length, 26 in.

Where the height of a wall is under 70 ft., and its length not more than 30 ft., the base must be 17½ in. thick: if the length is not more than 45 ft., the base must be 21½ in.; and for greater lengths, 26 in.

If the height is under 80 ft., the length not exceeding 45 ft., the base must be 21½ in. thick. Where the length is under 60 ft. the base must be 26 in.; and for greater lengths, 30 in.

For walls up to 90 ft. high, whose length is not more than 60 ft., the base must be 26 in. thick; if the length is under 70 ft., the base must be 30 in.; and for any greater length, 34 in., or 4 bricks.

Where the height of the wall is not more than 100 ft., the length being under 55 ft., the base must be 26 in. thick. If the length is under 70 ft., the base must be 30 in.; and for any greater lengths, 34 in.

FLUES for chimneys are generally built 14 in. by 9 in. on plan, having a solid *withe* of at least 4½ in. of brickwork between them; and no woodwork should be placed in a wall nearer than 12 in. from the inside of a chimney flue.

Flues must never be made with sharp angles in them, as the soot is liable to accumulate at such points; the angle which one part of a flue makes with another should never be less than 130°, and properly rounded off. If sharper angles are introduced, iron soot doors must be placed at them, so that the soot can be readily removed.

PARTY-WALLS are those which are carried up between two adjoining houses, so as to form a complete separation between them, and to prevent the spread of fire from one house to another. The Building Act requires that party-walls shall be of the same thickness as the external walls, and shall be carried 15 in. high above the roofs, measured at right angles to the slope. No timbers are to come within 4½ in. of the centre of a party-wall; but bond-timbers and wooden plates are not permitted to be laid therein at all.

CROSS-WALL is one that is built as a separation of one part of any building from another part thereof.

FLANK-WALLS of a detached house are those which join the back and front walls together.

BATTERING is sloping the face of a wall inwards from the base towards the top. In brick walls it is measured as extra per foot superficial.

In very thick walls a variety of bond called *herring-bone* is occasionally used. This is formed by a course of stretchers at each face of the wall, filled in with bricks laid diagonally. The wall is then continued again for another course, and the filling-in bricks are laid diagonally, the reverse way of the previous one.

FOOTINGS are all such foundation courses as are wider than the body of the wall above; and being all centrally over each other, and diminishing from the base upwards by half-a-brick, or $4\frac{1}{2}$ in. at a time, they can only leave steps, or *offsets*, of a quarter-brick on each side. See fig. K, Plate I.

The object of the footings is to distribute the weight over as large a surface of the ground as possible, so as to avoid unequal settlements in the building. The projection of the bottom of the footings at each side must be at least one-half the thickness of the wall.

DAMP-PROOF-COURSE—In order to prevent the damp from rising from the soil through the brick walls, it is usual to put a layer of some impervious material above the footings, over the whole thickness of the wall. A coating of asphalte, gas-tar and ashes, or a double course of slates bedded in cement, are the materials most commonly used for this purpose. Perforated glazed bricks have also been manufactured especially to prevent the damp from rising, and at the same time to introduce a current of air under the lowest floor of the building.

HOLLOW-WALLS are walls built with an outer and inner casing, a cavity of 2 or 3 in. being left in the middle so as to prevent weather from driving through. The two parts are tied together either with occasional headers, or with pieces of slate, or with iron straps made for the purpose. This method of building is common in exposed situations.

SLEEPER-WALLS are those which are built at intervals

across the basement, to carry the timbers of the lowest floor.

FENDER-WALLS are those which are built round the fire-places of rooms on the lowest floor, to carry the stone slabs or hearths.

TRIMMER-ARCHES are those which are built to carry the hearths of the fire-places in the upper stories, as shown on fig. 15. The brick trimmer-arch A is thrown across

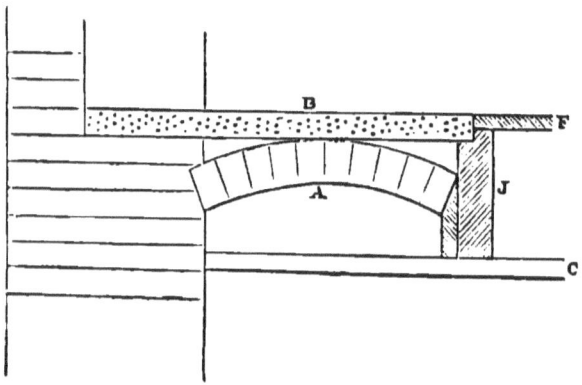

Fig. 15.

from the wall to the trimmer-joist J, on which a skew-fillet is nailed to receive the arch; B is the hearth resting upon the arch, F the floor, C the ceiling of room below. The object of this arch is to prevent hot ashes from getting between the joints of the stone hearth, and setting fire to the floor or ceiling. Trimmer arches are measured by the foot superficial. An iron tie-rod is sometimes introduced to prevent the arch from thrusting out the wooden trimmer. Instead of an arch, a layer of Portland cement concrete, 3 or 4 in. thick, will serve the purpose of carrying the hearth.

INVERTS are arches inverted or turned upside-down, and built in the wall under the openings on the lowest story, in order to distribute the pressure of the piers equally over the whole length of the foundations. They are measured by the foot superficial on the curve.

PARGETING is plastering the inside of chimney-flues

with mortar in which cow-dung is mixed, so as to make a smooth and even surface throughout, the admixture of cow-dung increasing the power of resistance to the action of heat. The flues are afterwards *cored* by a chimney-sweep, to clear out any projections or obstructions. *Parget* is not charged for but *coring* is priced at each.

GAUGED-WORK is when bricks are cut and rubbed upon a piece of stone to a particular *gauge* or size, as for arches over windows or other openings (fig. 16), and set with fine or *putty* mortar.

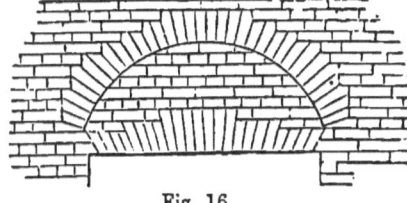

Fig. 16.

SKEW-BACKS are the inclined surfaces necessary to receive the ends of any arch that makes angles with its supports, as in a straight or a segmental arch. See fig. G, Plate I.; also fig. 16, above. Cutting to skew-back 4-in. wide is taken by the foot run, if wider, by the foot super.

CHASE is a vertical channel cut or left in a wall to receive pipes, &c., which it is desired to bury therein. When it is intended to build one wall against another, a chase is sometimes left in the wall first erected, into which the other is afterwards built, forming a plumb-joint the whole height thereof. By the Building Act it is provided that no chase in a *party-wall* shall exceed 14 in. in width, or 4½ in. in depth, nor be nearer than 7 ft. to any other chase in the same side of the wall. Chases are taken by the foot run.

BIRD'S-MOUTH is any re-entering angle that requires the bricks to be notched. See fig. O, Plate I.

REVEALS of windows are the sides or surfaces receding from the outer face of a wall, back to the window-frame, and are required by the Building Act to be at least 4½ in. deep.

WALL-PLATES are the horizontal timbers laid by the carpenter in the wall during its erection to receive the ends of the floor-joists, and to distribute their weight uniformly over its whole length.

BOND-TIMBERS are pieces laid by the carpenter horizontally in the wall during its erection, the thickness of one course of bricks, to tie the work together. Bond-timber is now generally discarded, except for internal brick partitions, on account of its being liable to shrink and decay, and thereby weakening the wall it is intended to strengthen.

IRON-HOOPING is generally used in place of bond-timber for external and party-walls, and bedded in cement.

DRY-STEINING is brickwork built dry round wells and cesspools, to keep the ground from falling in, as described under Well-sinking (p. 31).

POINTING is the filling-up with mortar or cement of the joints of the brickwork on the face, after the wall is built, the mortar in the joints being previously raked out before the pointing is laid in. It is measured by the foot superficial, including in the price quoted, the labour and mortar, and also the scaffolding, if that has to be erected.

Flat-joint pointing is when the mortar in the joints of a brick wall is raked out and filled in again with blue mortar, and the courses are marked with the edge of the trowel.

Tuck-pointing, formerly called *tuck and pat*, is when, in addition to the above flat-joint pointing, plaster or fine mortar is inserted in the joints, with a regular projection, and neatly pared to a parallel width.

Striking the joints is finishing off the mortar joints with the trowel as the work proceeds; it is adopted for inside work when the walls are not intended to be plastered, or are only to be lime-whitened.

OUTSIDE SPLAYS to openings in brick walls have the bricks cut on the slope and are rubbed to show fair. See fig. Q, Plate I.

INSIDE SPLAYS to openings in walls, are only rough cut on a slope to batten or plaster against. See fig. P, Plate I.

BRICK-NOGGING is the term applied to a partition constructed with a row of upright posts or quarters, disposed at 3 ft. apart, and the intervals filled up with brickwork. A partition of this kind is usually 4½ in. thick, or the breadth of a brick.

CHIMNEY-BREAST is that portion of the wall above the chimney opening which projects out into the room, when the wall is too thin to contain the flues.

UNDERPINNING is the cutting away and rebuilding the foundations of an old wall which is in an insecure condition, or which it is required to take down to a lower level. It is a process which requires the greatest care, and only a small portion of the wall must be removed at a time, which is at once filled in with the new work bedded in cement.

BRICK-PAVING is done with stocks, malms, paviors, or clinkers; these are either laid flat or on edge in sand, mortar, or cement, the ground being previously prepared and levelled, or concrete laid.

GROUTING is liquid lime and sand poured into brickwork or over paving, to fill up the interstices in the joints and bind the whole together. The cost of it is included in that of the work that is grouted.

CONCRETE suitable for making fire-proof floors is composed of *sulphate of lime* or *gypsum* mixed with broken bricks, calcined cinders, and other porous material. When used for floors or ceilings the soffit is slightly arched, a rise of 1 in 12 being the *least* that is given to the curved soffit; the spandrils may be filled up so as to form a level surface or floor on the top. When the arch is of the minimum rise a span of 6 to 8 ft. may be covered without any intermediate supports, the concrete being about 4 in. thick at the middle; for greater spans, wrought-iron joists or beams are employed to divide the span, and the concrete arches thrown across them. By giving a greater rise to the arch much greater spans can be covered without intervening supports. The cost of this material is valued by the superficial yard, exclusive of iron work, centerings, or scaffolding.

AN ARTIFICIAL or CONCRETE-STONE has also been

formed in the following manner: finely sifted dry sand is mixed with a small proportion of pulverised stone or carbonate of lime, then is added a solution of a material called the *silicate of soda*, which is obtained by boiling flints in a solution of caustic soda; one gallon of this silicate is mixed in a mill with every bushel of the first-named mixture of sand and powdered stone, and the mass is put into moulds of the required form, well rammed, and allowed to harden; the blocks, on being turned out of the moulds, are saturated with a solution of the *chloride of calcium* obtained by dissolving lime in muriatic acid, by which means the silicate of soda is changed into the silicate of lime, and a deposit of common salt left on the surface, which is washed off with water. The crushing strength of this concrete varies according to the materials used in its composition.

RETAINING-WALLS are those which are built for the purpose of supporting embankments of earth, so that they have to resist the pressure which the earth exerts in endeavouring to assume its *natural slope*. The amount of this pressure will depend on the nature of the soil to be sustained, loose soil having a tendency to form a slope of lower inclination than firm earth or rock. The natural slope of loose earth, as sand or gravel, is about 30° with the horizon, and that of stiff clay about 45°. In the former case, the thickness of the wall at its base must be about one-third of the height, and in the latter about one-fourth of the height; the face of the wall may be made to batter, so as to have the thickness at top about two-thirds that of the base. Retaining-walls should possess perfect cohesion throughout, and the best material for their construction is either stone in heavy blocks, or concrete of lime or cement and coarse gravel.

The following table shows the *angle of repose* or natural slope which different kinds of earth assume when left to themselves, the weight per cubic foot of earth, and the thickness to be given to the base of a retaining-wall when the earth is level with the top, and also when it is sloped up from the top at the angle of repose. The wall is supposed to be built of large blocks of stone or of concrete weighing 140 lbs. to the cubic foot, and the thickness of the wall at top not less than five-eighths of its thickness at the base. The thickness is found for any given height h, by multiplying it by one of the constants in the 4th and 5th column. (See "The Science of Building," by E. W. Tarn.)

Kind of earth	Weight per cubic foot.	Angle of repose.	Thickness at base; earth level at top.	Thickness at base; earth sloped at top.
Compact earth	126 lbs.	55°	$\cdot 20 h$	$\cdot 36 h$
Dry do.	120 ,,	45°	$\cdot 26 h$	$\cdot 44 h$
Shingle	112 ,,	40°	$\cdot 28 h$	$\cdot 46 h$
Dry sand	100 ,,	40°	$\cdot 26 h$	$\cdot 43 h$
Dry clay	120 ,,	40°	$\cdot 29 h$	$\cdot 47 h$
Wet do.	130 ,,	20°	$\cdot 46 h$	$\cdot 61 h$
Gravel	110 ,,	30°	$\cdot 34 h$	$\cdot 51 h$

ASPHALTE is a material largely employed for building purposes, and more especially for flooring, paving, covering of flats, &c. The principal ingredient in its composition is a bituminous limestone, which is reduced to powder and mixed with sharp grit; it is then heated in cauldrons with mineral tar and reduced into a mastic, so that it can be run into moulds in the form of blocks. When required to be used, it is again melted with a small proportion of mineral tar, and poured upon the place previously prepared to receive it; or it is laid on in a powdered state, and rendered solid by the application of hot irons. When used for floors, paving, or flat roofs, it is necessary first to lay from 3 to 6 inches of fine concrete, on which the asphalte is poured and floated over, and just before it sets a fine powder or sand may be sifted over it. In covering flat roofs the joists are fixed 12 inches from centre to centre, and plain tiles laid so as to span the distances between them; over these a thin coating of fine concrete is floated before the asphalte is applied. The weight of one square foot of coarse asphalte $\frac{1}{2}$-inch thick is 6 lbs. $2\frac{1}{2}$ oz.; and that of the finer quality, 6 lbs. 8¾ oz.; the weight of concrete 1-inch thick is 11½ lbs. per square foot. Asphalting is measured and valued by the foot superficial, the concrete being taken separately; channels are measured by the foot run, according to width.

A cheap kind of asphalte or bitumen, which is serviceable for many purposes, can be made with gas-tar and ashes laid on cold, the surface being beaten with a rammer and sprinkled with small pieces of stone or spar. This material is measured by the superficial yard.

Boiling tar poured over a flat roof (previously covered with cement concrete) and sprinkled with ground lime and sand, will form a water-tight covering, and will also serve the purpose of stopping cracks in a roof previously cemented over.

TERRA-COTTA is a building material closely allied to brick, clay being its principal component, and fire being employed to burn it into a condition fit for use. Clays of a peculiar kind, such as are found in Cornwall, Devon, Dorset, Northamptonshire, &c., are mixed with sand, ground glass, china-stone, felspar, and flint, &c., well pulverised. The red, buff, and white colours which belong to terra-cotta, are derived from the clays used in its manufacture. If other colours are required, they are obtained by mixing mineral pigments with the clay. When the clay is properly tempered, it can be moulded into any desired form, left to dry, and then burnt in a kiln. When a number of pieces of the same form, as in a moulded string-course, cornice, or other architectural feature, are required, it is usual to first make a model, and take a plaster mould therefrom, from which the clay is cast, dried, and burnt. When made in blocks of considerable size, it is cast hollow, so as to insure equal hardness and contraction throughout. The material contracts considerably both in drying and in burning, which must be allowed for in preparing the original models. When terra-cotta is well made it is unaffected by exposure to weather, or to the acid gases found in a smoky atmosphere; it also preserves its colour better than stone or common brick, from being less absorbent.

It weighs when solid about 122 lbs. per cubic foot, but being generally used hollow, it is much lighter than this. When used in long moulded courses, there is a difficulty in getting the pieces exactly to fit, the joints and surface of adjoining blocks often requiring to be rubbed down with sharp sand and water, in order to make them true. This defect, however, can in a great degree be avoided by care in the making, drying and burning.

Some kinds of terra-cotta are made from the material called fire-clay, obtained in the neighbourhood of coal pits; this is ground to a fine powder, and mixed with a small quantity of old terra-cotta, also reduced to powder. Other sorts are made from a pure clay, which requires no admixture of any other mineral substance; this will stand any amount of heat that can be applied to it, and produces articles of the hardest description.

MEASUREMENT OF BRICKWORK.

The standard measure for brickwork in London is the rod of 16 ft. 6 in. sq., which dimension being multiplied into itself produces 272 ft. 3 in., but the odd 3 in. are never taken into account. It is therefore always considered as 272 superficial ft., at $1\frac{1}{2}$ brick, or $13\frac{1}{2}$ in. thick, or 306 ft. cube, viz., 272 ft. by 1 ft. $1\frac{1}{2}$ in. All the other thicknesses are *reduced* to this standard, as shown hereafter in the manner of taking the dimensions and abstracting the work.

Brick walls of great thickness are generally measured by the cubic yard.

In measuring brickwork always begin with the foundations, then proceed with measuring each story separately (or as high as the wall continues of the same thickness), as solid work, according to their respective thicknesses; then add for all projections, as breasts of chimneys, &c., deducting the openings, but not the flues, as the extra trouble and the pargeting is deemed equivalent to the deficiency of materials; but deduct the openings of doors, windows, &c.

If the house or building be rectangular, measure two

walls the whole length of the external face, and the other two internally, so as to get the true cubical contents.

But in measuring for labour only, the external face of the work is girt, and multiplied into the height and thickness, to pay for the extra labour of plumbing the angles, and working the returns fair.

In measuring walls that are faced with superior bricks, the walls are first measured as common work, and then the superficial quantity of facing is taken, as hereafter shown, and is valued by considering the facing as two-thirds of a brick thick, and deducting the common brick-work from the price thereof, the same thickness, viz., two-thirds of a brick; by which the value per foot superficial is ascertained.

In measuring circles, or semicircles they are marked accordingly in the measuring-book. Thus:

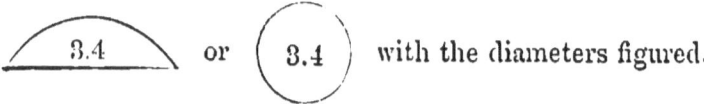

with the diameters figured.

To measure angle chimneys, draw lines on the floor, parallel to the two sides of the room, cutting the parts intersected by the chimney, as shown in the plans, Plate 1; take either side by the height of the story, and half the other (the work forming a triangle) for the thickness, either as the number of bricks, or as cube work, which, by the directions before given, and the example shown in the first chimney taken, proves it to be exactly the same: consequently, if the projection should not amount to any certain number of half-bricks, it would be best to take it as a cube dimension. In all cases it is supposed that the walls, as shown by the dotted lines, are measured before the projecting chimneys are taken, which is the usual custom.

In taking the dimensions of vaults, measure the abutments, or side walls, to the springing of the arch, then bend your rods round the soffit of the arch; and add once and a half the thickness thereof, by which you obtain the average girt of the arch; then take the length clear of the walls; but if the arch is turned over one or both walls, add the thickness thereof to the length of the arch. But in taking the height of the walls, measure to the crown of the arch, without making any deduction for the declivity of the arches, on account of the additional trouble and waste of bricks, in cutting and fitting them to the curved soffit of the arch. Likewise, in deducting openings with circular heads, the dimensions should only be taken to the springing of the arches, on account of the trouble and waste of bricks in fitting them to the arches.

Brick drains of large size to be taken and reduced as common brickwork if built with mortar; ordinary house drains by the foot run, describing the form and size.

Pipe drains are measured by the foot run, and all bends, junctions, traps, &c., are numbered.

Shafts of chimneys are measured as solid work.

Ovens and *coppers* are measured as solid cube brickwork, deducting the ash-holes only.* Tiles and Welsh lumps, and fire bricks, are to be allowed as extras.

In these, or any other brickwork that it is considered best or most convenient to measure by the cubic foot, multiply the solidity by 8, the number of $1\frac{1}{2}$ in. in a foot, and divide it by 9, the number of $1\frac{1}{2}$ in. in $13\frac{1}{2}$ in., which will reduce it to the standard of $1\frac{1}{2}$ brick, or $13\frac{1}{2}$ in. in thickness.

In measuring brickwork no allowance is to be made in quantity for small or difficult works. Timbers inserted in the walls are not to be deducted. When plates are bedded in the walls, two inches to be allowed for ditto where no brickwork is over them. All sills and stone strings are measured in. Brickwork in *cement* is to be kept separate from that executed in mortar.

Circular work may be measured by the girth on the outside, and taken as half as much again as straight work, unless the radius is very small, in which it is best measured nett and kept separate, the length of the radius being stated.

Cuttings of all kinds to be measured superficial; as

---

* This method is in common use amongst surveyors; but it would be far more consistent to measure the actual quantity of brickwork, allowing for the extra labour in price.

outside splays, cut and rubbed to show fair, or inside ditto rough cut for battens, &c. See Plate I., figs. P and Q.

Bird's mouths at per foot run, being notched to fit. See Plate I., fig. O.

Facings of all descriptions to be measured extra by the foot superficial; in which case the reveals are also measured, except where intended to be stuccoed. All openings are to be deducted.

Gauged arches to doors, windows, &c., are also measured by the foot superficial, taking face and soffit.

Groins are measured as common work, only taking the run of cut groins at per foot.

Brick-nogging, by the yard square of 9 ft., including the timbers; all openings to be deducted.

Brick paving, ditto, ditto.

Fascias, *beads* and *quirks*, *dentil* or *plain cornices*, &c., measured and valued by the foot run.

Brick-on-edge coping, if set in cement, to be measured at per foot run, as *extra* to brick-on-edge in cement.

Chimney pots bedded and flanched with cement are numbered.

Setting of stoves and ranges are numbered.

Bedding door and window frames in lime and hair mortar are numbered.

Pulling down old walls, cleaning and stacking the bricks, providing scaffolding, &c., is measured by the *rod* as described for new walls.

Raking out joints of old brickwork, washing down and repointing, including scaffolding, is measured by the foot superficial.

TILING.

TECHNICAL TERMS AND EXPLANATIONS.

EAVES are the lower horizontal edges of the tiling. They are called *dripping* eaves when they overhang the walls.

HIPS are the inclined ridges, like those of a pyramid, formed by the meeting of two tiled surfaces rising from two walls that form a salient angle.

TILING.

VALLEYS are the reverse of hips, being the inclined furrows, like those of a hollow inverted pyramid, formed by the meeting of two roof-faces over a re-entering angle of the plan. Wherever there is a re-entering angle, a valley is unavoidable, while a hip can occur only over a salient angle.

GABLE is the triangle formed by the roof-planes rising from two parallel walls, continued until they meet at right-angles a third or *gabled* wall.

PAN-TILING is a mode of covering a roof with tiles of a rectangular outline or plan, but having a surface both concave and convex, thus ⌣⌢; so that, as they lie side by side, one laps over the other, thereby forming a series of ridges and valleys, alternately running from the top to the bottom of the roof. Pan-tiles are hung on fir laths, which are nailed to the rafters, by means of a ledge formed in their making at their upper ends; and are usually 14 in. long, and 10 in. broad, weighing about 5 lbs. each. They are laid with a *gauge* of 11 or $11\frac{1}{2}$ in.; that is, they show that amount between the bottom of one tile and the bottom of the next above or below, each tile lapping 2 or 3 in. over the one below it; 164 tiles cover one square of 100 superficial feet, when laid to 11-inch gauge.

PLAIN or PLANE TILES are of a rectangular form, with a flat or plane surface; and are usually about $10\frac{1}{2}$ in. long, 6 or $6\frac{1}{2}$ in. broad, and five-eighths of an inch thick. Their weight is from 2 to $2\frac{1}{2}$ lbs. each. Some are made to hang on to the laths with a ledge, as in pan-tiles, while others are made with two holes drilled for nails or pegs. They are laid to a *gauge* of 3, $3\frac{1}{2}$, and 4 inches, with or without a bedding of mortar, 600 tiles covering one square of 100 ft. superficial, laid to a 4-inch gauge.

TILE-CREASING, or WEATHERING, is two rows of plain tiles placed horizontally under the coping of a wall, and projecting about 3 in. therefrom, to discharge the rainwater. It is measured by the foot run.

HIP and RIDGE TILES are of various form and design, and are laid over the edges of the tiled surfaces of the roof, where they meet at a hip or ridge.

VALLEY-TILES are made of a semi-circular section, and are laid under the edges of the tiled surfaces, where they meet at a valley. Hip, ridge, and valley tiles are measured by the foot run.

ORNAMENTAL TILES are plain tiles made into various fancy patterns, and vary in size and design.

CEMENT FILLETING is a luting of cement laid on the tiling where it meets a wall carried up above it. It is measured by the foot run.

Measurement.—Plain and pan-tiling to be measured by the square of 100 ft.

In measuring plain tiling—
    Allow for the eaves 4 in. extra.
    Ditto for dripping do. 6 in. extra.
    Ditto for all cuttings, hips, &c., 3 in. extra.
    Ditto for valleys, 12 in. extra.

In measuring pan tiling—
    Allow for the barge per foot run.
    Ditto for heading to barge per foot run.
    Allow for cutting to hips and splays per foot run.
    Ditto for hips and ridges per foot run.

Number the hip hooks, which should be painted three times in oil.

Ditto T nails, ditto.

Deduct for chimneys and skylights, and deduct and add for dormers.

If the roofs are hipped, take the length at the bottom of the sides, and not measure the end; the two side triangles being equal to the hipped end one.

Tile-paving whether plain or ornamental is measured by the superficial yard, including levelling ground, forming concrete bed with rendering of cement and grouting in cement.

Facing walls with glazed tiles bedded in cement is measured by the foot super. If used in narrow widths, as to form a skirting, it is taken by the foot run.

In order to illustrate the principle of measuring bricklayers' and tilers' work, and bringing it into bill, in Plate I. is given a plan, elevation, and section of the front wall of a house, with the windows to a larger scale, and also plans of different chimneys. The rules before stated are likewise explained, by showing the manner of taking the dimensions in the measuring-book, and the method of preparing the abstract, and entering them therein, together with other imaginary quantities, to make the particular manner of abstracting the work perfectly clear and explicit. B. W. is the abbreviation for "brickwork," and D^{dt} for "deduct."

See the general rules under the head Measuring, viz.:—

BRICKLAYERS' WORK done for A. B., Esq., at his house, Kensington,
By C. D

Measured January 1st, 1870, with Mr. E. F.

ft. in. bks.	ft. in.		ft. n.	
27 6 5	13 9	Brick footing, 2 bottom courses . . .	26 0	front of house.
0 6			0 9	projec. of footings.
		27 6	0 9	do. other end.
		Figs. C and K.	27 6	
26 9 4	20 1	Do. average thickness of the courses above do. .	26 0	
0 9		80 4	0 9	½ B. at each end.
			26 9	
26 0 3	234 0	B. W. above do. to under side of ground floor .	0 6	under floor.
9 0		468 0	8 6	height of story.
			9 0	
2)5 0 ½	35 0	D^{dt}· openings	5 0	
3 6		17 6	0 4½	upper reveal.
2)5 4½ 2½	45 8	D^{dt}· reveals . Windows	5 4½	
4 3			3 6	
2)4 3 2	29 9	D^{dt}· backs .	0 9	2 side reveals.
3 6		59 6	4 3	
7 6 ½	26 3	D^{dt}· openings		
3 6		13 2 Door.		
7 10½ 2½	33 5	D^{dt}· reveal .		
4 3				

FOUNDATIONS, BRICKWORK, ETC.

ft. in.	bks.	ft. in.			ft. in.	
26 0 13 0	2½	338 0	Add B. W. to ground floor	{	1 0	thickness of floor.
					12 0	height of room to under side of one pair floor.
2)7 6 3 6	½	52 6	D$^{dt.}$ openings 17 6	⎫	13 0	
2)7 10½ 4 3	2	66 11	D$^{dt.}$ reveals 133 10	⎬ Windows.		
2)4 3 2 6	1½	21 3	D$^{dt.}$ backs	⎭		
10 0 3 6	½	35 0	D$^{dt.}$ opening 17 6	⎫		
				⎬ Front door.		
10 4½ 4 3	2	44 1	D$^{dt.}$ reveal 88 2	⎭		
26 0 13 0	2	338 0	Add B. W. to one-pair floor 676 0	⎫⎬⎭	1 0 12 0	thickness of floor. height of room.
					13 0	
3)8 0 3 6	½	84 0	D$^{dt.}$ openings 28 0	⎫		
3)8 4½ 4 3	1½	106 9	D$^{dt.}$ reveals	⎬ Windows. (See Figs. D, E & F.)		
3)4 3 2 3	1	28 8	D$^{dt.}$ backs	⎭		
26 0 9 11	1½	257 10	Two-pair floor, B. W. to under side of tie-beam.			
3)5 6 3 6	½	57 9	D$^{dt.}$ opening 19 3	⎫		
3)5 10½ 4 3	1	74 11	D$^{dt.}$ reveal	⎬ Windows		
3)4 3 2 9	¼	35 1	D$^{dt.}$ backs 17 7	⎭		
26 0 3 2	1	82 4	Add B. W. to parapet to under side of coping.			
			In making deductions for revealed windows, if the wall is only one brick thick, take one reveal in and one out as follows :—			
5 8¼ 3 10½	1	22 0	L$^{dt.}$ upper windows, suppose wall only one brick, and the window openings of the annexed dimensions		5·6 3·6	

MEASUREMENT OF BRICKWORK.

Measuring Chimneys.

The height of the rooms supposed to be 10 feet.
Do. of the chimney-openings, 4 feet.

(See Plate, No. 1.
Figs. I, L, M, N.)

ft. in.	bks.	ft. in.		
10 0 4 6	3	45 0	B. W. to angle chimney. 90 0	
4 0 3 6	2	14 0	Dᵈᵗ· opening. 28 0	ft. in. 101 3 8
10 0 4 6 2 3		101 3	Cube B. W. to angle chimney. I.	9)810 0 90 0
10 0 9 0	3	90 0	B. W. to angle chimney. L. 180 0	
4 0 3 6	2	14 0	Dᵈᵗ· opening. 28 0	
10 0 5 6	1	55 0	B. W. to chimney-breast. M.	
4 0 3 6	2	14 0	Dᵈᵗ· opening. 28 0	
10 0 4 9	5	47 6	B. W. to angle chimney. N. 95 0	
10 0 4 0	2	40 0	Dᵈᵗ· B. W. angle.	
4 0 3 6	2	14 0	Dᵈᵗ· opening. 108 0	
		54 0		

{ This, though taken before, is entered again to show the manner of abstracting cube B. W. red. to 1½ brick.

All gauged work is first measured in with the common brickwork, and afterwards taken at per foot superf. measured as follows:—

3 6
0 4   1 2 soffit to gauged arches.

3 10
1 0   3 10 face of ditto.

(See Plate I., fig. G.)

ON ABSTRACTING.

In abstracting bricklayer's work, although it will be found advantageous, it is not so absolutely requisite to observe a regular rotation as in joiner's work. But particular attention is required in abstracting bricklayer's work, to place the contents of the dimensions, according to their different thicknesses, and the deductions thereon, so that they may be *reduced* to the proper standard or thickness (of one brick and a half, or thirteen and a half inches) in the abstract; which will be perfectly easy after considering the explanation given, and seeing the form of the following abstract:—

Place the cube brickwork in the first columns.
{ One column for one brick thick.
One do. for one and a half do. } Add.
One do. for one brick thick.
One do. for one and a half do. } D^{dt}.

By which method you may abstract brickwork to any thickness. Thus:—

If half a brick thick, one-half the quantity may be placed under the head of one brick, or one-third the quantity under the head of 1½ brick.

If two bricks in thickness, twice the quantity may be placed under the head of one brick.

If two and a half bricks in thickness, the same quantity must be placed under the head of one brick, and also under 1½ brick.

If three bricks in thickness, twice the quantity must be placed under the head of 1½ brick.

In this manner brick walls of all thicknesses may be abstracted under two heads, and thereby avoid having a column for every thickness of wall in the building.

Next proceed with the different descriptions of tilings, and all other work measured by the square of 100 feet.

Next, the pavings, brick-nogging, and other work measured by the yard square of nine feet.

Next, the work measured by the foot superficial; and next with the work measured by the foot run, as shown in the following abstract.

ABSTRACTING BRICKWORK.

The following are imaginary dimensions, to explain the manner in which walls of any number of bricks in thickness may be abstracted under the two heads of one brick and one brick and a half. These being the general thicknesses of walls, it very seldom occurs that the walls are of the thicknesses here stated, which are only given to make the principle understood.

ft. in.	bks.	ft. in.	To be abstracted as ft. in. bks.	ft. in.	bks.	ft. in.	To be abstracted as ft. in. bks.
5 6 2 3	½	12 4	6 2 . 1	8 0 6 4	3½	50 8	{ 50 8 . 1½ { 101 4 . 1
7 6 5 8	1	42 6	42 6 . 1	7 6 3 9	4	23 1	112 4 . 1
10 3 6 9	1½	69 2	69 2 . 1½	12 6 3 8	4½	45 10	137 6 . 1½
8 6 6 2	2	52 5	104 10 . 1	10 6 5 2	5	54 3	271 3 . 1
10 0 4 6	2½	45 0	{ 45 0 . 1½ { 45 0 . 1	8 4 3 9	5½	31 3	{ 31 3 . 1½ { 125 0 . 1
9 0 5 3	3	47 3	94 6 . 1½	10 10 5 6	6	59 7	238 4 . 1½

Walls one brick thick are *reduced* to 1½ brick by multiplying their area by 2 and dividing by 3.

Abstract arranged as before stated. ABSTRACT OF BRICKLAYERS' WORK done for A. B. by X. Y.
Abstracted with Z. January 1, 1870.

	RODS.							SQUARES.			YARDS.				FEET.		
Cube B. W.		Superficial B. W.						Tiling.		Paving.							
Add.	Dt.	Add.		Dt.				Plain.	Pan.	Grey Stock.	Dutch Clinkers.	10-inch Tiles.	Brick-nogging.		Super.	Runs.	Nos.
		1 Bk.	1½ Bk.	1 Bk.	1½ Bk.												
ft. in.	ft. in.	ft. in.	ft. in.	ft. in.	ft. in.												
161 3		27 6	27 6	17 6	45 8												
		80 4	468 0	45 8	33 5												
		338 0	338 0	59 6	17 6												
		676 0	257 10	13 2	21 3												
		82 0	4 90	33 5	28 0					Com'on Bricks.							
		55 0	0 180	133 10	106 9												
		95 0	0 95	17 6	19 3												
		6 0	2 69	88 2													
		42 6	6 45	28 8							12-inch Tiles.						
		104 10	94 6	74 11													
		45 0	50 8	17 7													
		101 4	137 6	28 0													
		112 4	31 3	28 0													
		271 3	238 4	28 0													
		125 0		108 0													

These columns are to be added up, subtracting the deductions from the additions, and the remainder is to be reduced to the standard thickness of 1½ brick, and brought into rods of 272 ft. 3 in. superficial.

If different sorts of bricks are used, separate heads must be formed in the Abstract, each detailing the various proportions and descriptions of the work.

ROTATION.

To be attended to in bringing the quantities into Bill.

BRICKLAYER.

Yds.	ft.	in.	
			Cube of concrete to foundations, composed as specified . . .
			Supl. do. to floors, &c., less than 12 in. thick . . .
			(State thickness.)

Rods	ft.	in.	
			Supl. reduced brickwork, if stock bricks, if part with other bricks, their proportions, &c. . . .
			Do., do., to garden walls . .
			Or whatever way the work may be done at per rod.

Sqrs.	ft.	in.	
			Supl. pan-tiling, if dry or pointed inside or out. . . .
			Do., plain-tiling, if double fir laths and wrought nails, &c. . .
			Or other articles by the square.

Yds.	ft.	in.	
			Supl. brick-nogging, flat or on edge .
			Do., brick paving, do. . . .
			Do., 10 in. or 12 in. tile paving .
			Do., pebble paving . . .
			Do., tuck pointing . . .
			Or other articles by the yard superficial.

Ft.	in.	
		Supl. gauged arches . . .
		Do., malm facings, either as best or seconds . . . .
		Do., extra only, in cement . .
		Do., do., to arches in cement . .
		Do., half-brick trimmers . . .
		Do., cutting splays, &c. . .
		Do., asphalting, or slates in cement to form damp-proof course . .
		Do., extra to battering face of wall .
		And all other articles at per foot superficial.
		Run of cutting to narrow splays, or birds'-mouths, &c. . . .
		Do., do., and pinning into wall .
		Do., pipe drains jointed in cement or clay . . . . .
		And all other articles at per foot run.

Nos.	
	Terra-cotta chimney moulds, and setting in cement . . . . .
	Door and window-frames, bedded and pointed . . . . .
	Flues cored . . . . .
	Traps, junctions, bends, &c., to drains
	And all other articles that are numbered.
	Labour and materials to setting stoves, ranges, coppers, &c. . .
	Attending on other Trades . . .

VALUATION OF BRICKLAYER'S WORK.

SIZE AND WEIGHT OF VARIOUS ARTICLES.

	Length.		Breadth.		Thickness.		Weight.	
	ft.	in.	ft.	in.	ft.	in.	lbs.	oz.
Stock bricks . . . each	0	8¾	0	4¼	0	2½	5	0
Paving do. . . . . do.	0	9	0	4½	0	1¾	4	0
Dutch clinkers . . do.	0	6¼	0	3	0	1½	1	8
12-inch paving tiles . . do.	0	11¾	0	11¾	0	1½	13	0
10-inch do. . . do.	0	9¾	0	9¾	0	1	8	9
Pan tiles . . . . do.	1	1½	0	9½	0	0½	5	4
Plain tiles . . . do.	0	10½	0	6½	0	0⅜	2	5
Pantile laths, per 10 ft. bundle .	120	0	0	1½	0	1	4	6
Ditto, per 12 ft. bundle . .	144	0	0	1½	0	1	5	0
A bundle contains 12 laths.								
Plain tile laths, per bundle . .	500	0	0	1	0	0¼	3	0
Thirty bundles of laths make a load.								

A bricklayer's hod measures 1 ft. 4 in. × 9 in. × 9 in., and contains 20 bricks.

A single load of sand is 27 cubic feet, or one cubic yard.

A double load of sand is 54 cubic feet, or two cubic yards.

A measure of lime is 27 cubic feet, or one cubic yard, and contains from 16 to 18 bushels.

QUANTITIES, ETC.

A rod of brickwork measures 16 ft. 6 in. × 16 ft. 6 in., or 272 ft. 3 in. superf., 1½ brick, or 13½ in. thick, called the standard thickness, or 306 cubic feet, or 11⅓ cubic yards.

A rod of brickwork laid to a 12-in. gauge, i.e., four courses to measure one foot in height, requires 4356 stock bricks.

Ditto, laid to 11½-in. gauge, requires 4538 stock bricks.

A foot superficial of reduced brickwork requires 16 bricks.

These calculations are made without allowance for waste; and indeed there is very little, as nearly every part is worked in, and much space is occupied by timbers, flues, &c., for which no deduction is made in measurement; and therefore in the erection of dwelling-houses containing flues and bond timbers, 4300 stocks is quite sufficient, and this is the usual number allowed for a rod of brickwork.

5370 stocks to the rod, if laid dry.

4900 do. in wells and circular cesspools.

A rod of brickwork, laid four courses to gauge 12 in., contains 235 ft. cube of bricks, and 71 ft. cube of mortar; and the average weight is about 15 tons.

A rod of brickwork requires 1½ cubic yard of chalk lime and three loads of sand; or one cubic yard of stone lime, and 3½ loads of sand; or 36 bushels of cement, and 36 bushels of sharp sand.

A cubic yard or load of mortar requires nine bushels of lime and one load of sand.

Facing requires 7 bricks per foot superficial, headers and stretchers being used alternately.

Gauged arches require 10 bricks per foot superficial.

Brick-nogging per yard superficial, requires 80 bricks on edge, or 45 laid flat.

PAVING.

Description.		Number required.
Stock bricks, laid flat.	per yard	36
Ditto on edge	do.	52
Paving bricks laid flat	do.	36
Ditto on edge	do.	82
Dutch clinkers do.	do.	140
12-inch paving tiles	do.	9
10-inch ditto	do.	13

TILING.

	Gauge.	Number required.
	Inches.	
Pan tiles, per square	12	150
Do. do.	11	164
Do. do.	10	180
A square of pan-tiling requires one bundle of laths and 1¼ hundred of 6d. nails.		
Plain tiles, per square	4	600
Do. do.	3½	700
Do. do.	3	800
Do. do.	laid flat	210
A square of plain-tiling requires one bundle of laths and nails, one peck of tile pins, and three hods of mortar.		

CALCULATION OF LABOUR.

The following table, although far from complete, contains constants for some of the principal descriptions of bricklayer's work:—

	CONSTANT.
	To be multiplied by the rate of wages for a labourer per hour
Concrete in Foundations. — Labour in mixing, wheeling, throwing-in, per yard cube	2·35

	To be multiplied by the rate of wages for a bricklayer and labourer per hour.
Brickwork, per rod	42·5
Pointing, per foot super	·09

	To be multiplied by the rate of wages for a bricklayer per hour.
Extra labour to malm facings, per foot super	·07

		To be multiplied by the rate of wages for a bricklayer and labourer per hour.
Paving, including levelling ground.		
Brick paving laid flat in sand	per yard	·40
Do. laid on edge in sand	do.	·52
Do. laid flat in mortar	do.	·61
Do. laid on edge in mortar	do.	·84
Paving-brick paving laid flat in sand	do.	·46
Do. laid on edge in sand	do.	1·06

		To be multiplied by the rate of wages for a bricklayer and labourer per hour.
Paving-brick paving laid flat in mortar	per yard	·75
Do. on edge in mortar	do.	1·21
Clinker paving on edge in sand	do.	1·32
10 or 12 inch tile paving	do.	·10
Tiling.		
Pan-tiling laid dry	per square	2·75
Do. pointed outside	do.	5·00
Do. pointed inside and outside	do.	7·00
Plain-tiling laid to a 4-inch gauge	do.	5·00
Do. to a 3½-inch gauge	do.	6·20
Do. to a 3-inch gauge	do.	7·20

It would be impossible to give examples for every case that might occur; but the following will show the method of valuing the principal descriptions of bricklayers' work.

Ex. 1.—To find the value of a cubic yard of concrete, made in the proportion of six parts of gravel to one of lime.

£ s. d.

1·1 yard of gravel, at per yard, prime cost
Carriage of above to the works.
Three bushels of lime, at per bushel

—— per cent. profit
Labour on the above, found by multiplying the rate of wages per hour for a labourer by 2·35

Value per cubic yard . . . £

Ex. 2.—To find the value of a rod of brickwork.

£ s. d.

4300 stocks, at per thousand
1½ yards of lime, at per yard
Three loads of sand, at per load

—— per cent. profit
Scaffolding
Labour per rod, found by multiplying the rate of wages per hour for a bricklayer and labourer by 42·5 .

Value per rod . . . . . £

Ex. 3.—To find the value of a foot of malm facing.

	£	s.	d.
No. 7 best malms (or seconds, as the case may be), at —— each			
D⁰· the value of seven bricks, according to the quality with which the walls are built, the facing having been measured with the wall —— at —— each . .			
Extra value of the malm bricks			
Extra labour on the malm bricks, found by multiplying the rate of wages per hour for a bricklayer by ·07 .			
	£		

Ex. 4.—To find the value of a yard of paving,—say with stock bricks laid flat in sand.

	£	s.	d.
36 stocks, at —— each			
Sand			
—— per cent. profit			
Labour, found by multiplying the rate of wages per hour for a bricklayer and labourer by ·4 . . .			
Per yard	£		

Ex. 5.—To find the value of a square of plain tiling, laid to a 4-inch gauge.

	£	s.	d.
600 plain tiles, at per thousand			
One bundle of laths and nails			
One peck of tile pins			
Three hods of mortar			
—— per cent. profit			
Labour, found by multiplying the rate of wages for a bricklayer and labourer per hour by ·5 . . .			
Per square	£		

VALUE OF BRICKWORK.—SLATING.

TABLE.

Showing the value of reduced brickwork per rod, calculated at the several prices of £4, £4 5s., £4 10s., £4 15s., £5, £5 5s., per rod, for mortar, labour, and scaffolding; and of bricks from 30s. to 60s. per thousand; allowing 4500 bricks to a rod.

Bricks per M.	Mortar, Labour, and Scaffolding, per rod.											
	£4		£4 5s.		£4 10s.		£4 15s.		£5.		£5 5s.	
s.	£	s.	£	s.	£	s.	£	s.	£	s.	£	s.
30	10	15	11	0	11	5	11	10	11	15	12	0
32	11	4	11	9	11	14	11	19	12	4	12	9
34	11	13	11	18	12	3	12	8	12	13	12	18
36	12	2	12	7	12	12	12	17	13	2	13	7
38	12	11	12	16	13	1	13	6	13	11	13	16
40	13	0	13	5	13	10	13	15	14	0	14	5
42	13	9	13	14	13	19	14	4	14	9	14	14
44	13	18	14	3	14	8	14	13	14	18	15	3
46	14	7	14	12	14	17	15	2	15	7	15	12
48	14	16	15	1	15	6	15	11	15	16	16	1
50	15	5	15	10	15	15	16	0	16	5	16	10
52	15	14	15	19	16	4	16	9	16	14	16	19
54	16	3	16	8	16	13	16	18	17	3	17	8
56	16	12	16	17	17	2	17	7	17	12	17	17
58	17	1	17	6	17	11	17	16	18	1	18	6
60	17	10	17	15	18	0	18	5	18	10	18	15

SLATING.

SLATE is a natural material, which splits readily into thin slabs, and is largely used for building purposes, more especially as a covering to roofs, being lighter and more inpervious to wet than tiles. There are various qualities of roofing-slates, those obtained from North Wales being most highly esteemed, on account of their great hardness and capability of being split into very thin plates. The weight of this slate is 180 lbs. per cubic foot. The Welsh slates are made into a great variety of sizes, but when used in roofing, one size only is used throughout the same roof; so that the *gauge* to which they are laid, or the distance from the bottom of one slate to the bottom of the next above or below, is the same all over the roof. They are usually laid with a *lap* of 2½ in.; the bottom of the next course above the eaves course extending 1¼ in. below the middle line of the first or eaves course, and the bottom of the third course extending 1¼ in. below the middle of the second course, and consequently 2½ in. (the *lap*) beyond the top of the first or eaves course ; and so on all up to the top of the roof. The *gauge* is therefore the half length of the slate less half the lap.

Thus, in fig. 17 we have 3 courses of slates, in which the distance from *a* to *b* represents the *lap*, and the distance from *b* to *c* represents the *gauge*.

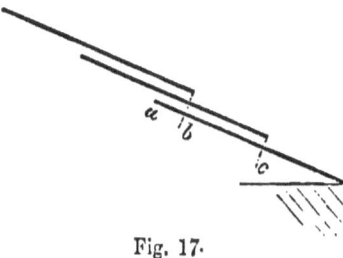

Fig. 17.

Westmoreland slates are coarser and thicker than the Welsh, although their weight per cubic foot is less, being about 173 lbs. They are used of all sizes in the same roof, the largest being selected for the lower course, and gradually diminishing in size towards the top.

Slating is sometimes laid upon close boarding nailed to the rafters ; but is more frequently laid upon fir battens or slate-laths nailed to the rafters at distances from centre to centre equal to the gauge to be given to the slates.

Thus for *countess* slating with slates 20 in. long, in order to give a 2½ in. lap, the battens must be 8¾ in. from centre to centre. When battening is used for Westmoreland slates, the distance apart of the battens diminishes as they get nearer the top. As the battens are often put on by the slater, they may be included in the measurement and price of the slater's work. Slates are fastened to the battens with copper or zinc nails, two to each slate, except in the very small sizes, which have one nail each. Holes are punched in the slates near the middle, for the nails to pass through. When slating is laid upon close boarding it is usual to put a layer of *asphalted felt* upon the boarding before laying the slates.

Slating is measured superficial, and charged per square of 100 ft. In measuring, allow for the eaves, whatever the bottom course measures, and for the hips and valleys measure their length by 12 in., namely, 6 in. on each side ; also the length of all irregular angles, as chimneys, dormers, &c., by 6 in. wide, as a fair allowance for cutting and waste. For circular slating allow one-third extra.

Slate is also sawn into coverings for the hips and ridges, which are measured by the foot run, their thickness and width or girt being described. Slate steps are measured by the foot run. Slate slabs, shelves, &c., are measured by the foot superficial; so also are slate cisterns and tanks, including iron bolts.

In the following Table the slating is all supposed to be laid with a 2½ in. lap.

VALUATION OF SLATER'S WORK.
TABLE OF MATERIALS AND LABOUR.

Sizes and Description of Slates.	No. of square ft. covered by 1 M. of 1200.	Wt. per M. of 1200 in cwt.		No. of slates to cover 1 sqre. of 100 ft.	Wt. of copper nails to 1 sqr.	CONSTANT. To be multiplied by the rate of wages for a slater and labourer per hour.	
		1st	2nd			Laying only.	Preparing and laying.
inches.					lbs.	per sqr.	per sqr.
Singles . . 12 × 6	240	14	14	550	2¼	1·80	7·0
Doubles . 13 × 7	306	18	21	400	3½	1·73	6·6
Ladies, small 14 × 8	385	22	26	312	3	1·60	6·3
Do. large . 16 × 8	450	25	33	266	2½	1·50	6·0
Viscountess . 18 × 9	580	31	40	210	2½	1·45	5·8
Countess . 20 × 10	730	40	54	164	2	1·37	5·5
Marchioness. 22 × 11	900	48	60	133	2	1·32	5·3
Duchess . 24 × 12	1000	60	80	120	1¾	1·20	5·0
Princess . . 24 × 14	1254	70	90	96	1½	1·20	5·0
Imperials . 30 × 24	A ton will cover 225 ft.						7·5
Rags and Queens } 36 × 24	to 250 ft.	...	...	...	...		
Westmorland, various	225	...	...	...	...	...	8·0

Ex.—To find the value of a square of duchess slating, copper nailed.

	£	s.	d.
No. 120 duchesses, at per thousand . . .			
1¾ lbs. copper nails, at per lb.			
—— per cent. profit			
Labour in preparing and laying, at per hour .			
Value per square	£		

MEASUREMENT OF SLATER'S WORK done for A. B. Esq.

2)26 0			Countess slating, 2½" lap, 2—1½" copper nails to each slate.
16 6	858·0		
2)22 0			Add for cutting to hips and valleys.
1 0	44·0		
12 6			Do., do. round chimney.
6	6·3		
	908·3		
5 6			D⁰. do. for chimney.
1 2	6·5		
	901·10		
26 0	26·0		Sawn slate ridge.
2)26 0			Felt laid under slates.
16 6	858·0		
10 6			1" sawn slate sides to cistern.
2 0	21·0		
3 6			1¼" do. bottom do
2 6	8·9		

No. 4.—½" iron bolts 2'·6" long with nuts and screws.

ABSTRACT.

Super.	Super.	Run.
Countess slating. 2½ lap, 2—1½ in. copper nails. 100)901 ft. 10"	1" sawn slate cistern. 21 ft. 0'	Sawn slate ridge. 26 ft. 0".
9 sqrs. 2 ft.		No.
Felting. 100)858·0	1¼" ditto. 8 ft. 9".	½" iron bolts, 2'·6" long, with nuts and screws.
8 sqrs. 58 ft.		4

BILL OF QUANTITIES.

sqrs.	ft.	in.		£	s.	d.
9	2	0	Super Countess slating, laid with a 2½" lap, and 2—1½" copper nails to each			
8	58	0	Do., felting laid under slating			
	21	0	Do., 1" sawn slate cistern			
	8	9	Do., 1¼" do. do.			
	26	0	Run sawn slate ridge			
No. 4			½" iron bolts, 2'·6" long, with nuts and screws			

£

CHAPTER IV

CARPENTRY AND JOINERY.

DEFINITIONS.

CARPENTRY is the framing together of timbers, as in forming roofs, floors, partitions, and all other work in which rough timber is employed.

JOINERY is the framing of wood together, for internal and external finishings of houses; thus the linings of walls, the coverings of timbers, laying of floor-boards, the construction of doors, windows, stairs, &c. are included in the joiner's work.

CARPENTRY.

NAKED FLOORING is the term applied to the timbers used in forming the floor of a room. There are three different kinds of naked flooring; namely, single-joisted floors, double floors, and framed floors.

SINGLE-JOISTED floors consist of one series of timbers called *joists*, which are laid across the room from wall to wall with their broadest side vertical; they are usually placed about 11 in. apart, and the floor boards are nailed on the top edge, the laths for the ceiling of the room below being nailed to their bottom edge. Their *scantling* or sectional dimensions, varies according to the span of the room, or length of bearing as it is termed, and the load which they are required to carry; and as their strength increases with the *square* of their depth, but only *as* the breadth, a doubling of their depth produces the same effect on the strength as quadrupling their breadth, but with half the quantity of timber. Hence the advantage of deep and narrow joists, to which there is no limit but in the tendency to lateral bending, for which transverse struts afford a remedy. These are usually placed in pairs crossing like an ⋈, and are then called *herring-bone strutting;* a row of which should be repeated every 5 or 6 ft. length of joist. The ends of joists are laid in the walls upon *wall-plates,* or pieces of timber built in with the brickwork and generally the thickness and breadth of one course of bricks; by means of the wall-plates the pressure of the joists is uniformly distributed over the whole length of the wall. When the joists are not all of exactly the same depth, the upper edges must be brought up to a perfect level by slips of wood laid under their ends upon the wall-plates; this is called *furring-up* the ends of the joists.

In forming the lowest floor of a building, it is usual to lay the joists upon *sleepers,* which are timbers laid upon brick piers or *sleeper-walls* at frequent intervals; by which means the carpenter is enabled to use joists of smaller scantling and to save timber.

The following is about the usual scantling for fir joists according to their bearings:—For 5 ft. bearing $4\frac{1}{2}$ in. deep × 2 in. wide; for 10 ft. bearing, 7 in. × $2\frac{1}{4}$ in.; for 15 ft. bearing, 10 in. × $2\frac{1}{2}$ in.; for 20 ft. bearing, 12 in. × 3 in. Where the floors have to carry extraordinary loads, the above scantlings must be increased in proportion.

On account of flues, fire-places, openings for staircases, &c. it often happens that some of the joists cannot have a bearing on the wall. In such cases a piece of timber, called a *trimmer,* is framed between two of the nearest joists that have a bearing on the wall; and into this trimmer the ends of the joists requiring support are mortised. This mode of framing joists is therefore called *trimming.* Trimmers should always be of the same depth as the joists, but 1 inch thicker. The joists into which the trimmers are framed are called *trimming-joists;* these run in the same direction as the other joists, but must be stouter because they are weakened by having to receive the ends of the trimmers which are tenoned into them.

When one piece of timber is framed into another piece, the part cut out of the latter is called a *mortice,* and the part of the former which goes into the mortice is called a *tenon*.

DOUBLE FLOORS are formed of three tiers of joists; namely, binding-joists, bridging-joists, and ceiling-joists; the *binding-joists* or *binders* have their ends resting on the wall and support the *bridging-joists* which are notched upon the top of them, and on these last the floor boards are laid. The *ceiling-joists* are merely timbers of small scantling spiked to the underside of the binders to receive the ceiling-laths of the room below. A section of such a floor is shown in fig. 18, in which *a* is the flooring, *b* the

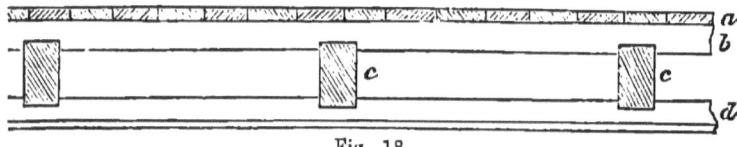

Fig. 18.

bridging-joists, *c* the binders, *d* the ceiling-joists. Fig. 19 shows a transverse section of the same floor.

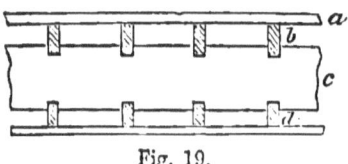

Fig. 19.

The scantling of the bridging-joists depends on the distance apart of the binders, and will be found as in single-joisted floors. The scantling of the binders must also depend upon their distance apart, and the load they have to carry; but for common floors, where the distance apart is 6 ft., their scantling, when of fir, may be as follows:—When the bearing is 10 ft., the scantling may be 8 in. × 8 in., or 10 in. deep × 5 in. wide; when it is 15 ft., the scantling may be 10 in. deep × 11 in. wide, or 13 in. deep × 7 in. wide; when it is 20 ft., the scantling should be 14 in. × 11 in. The ceiling-joists should be at least 3 in. × 2 in. when the binders are 6 to 8 feet apart.

FRAMED FLOORS have the binders framed into large baulks of timber called *girders*, which last rest upon the walls or upon piers. In fig. 20 is shown a section of a floor of this description, *a*, *a* being the girders, *b* the binders, *c* the bridging or floor joists, *d* the ceiling-joists.

The scantling of fir girders 10 ft. apart for ordinary floors, should be as follows. When the bearing is 15 ft., the scantling should be 12 in. deep × 10 in. wide; when

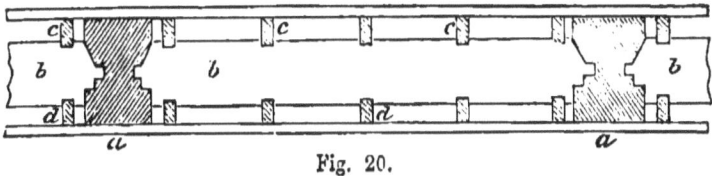

Fig. 20.

it is 20 ft., the scantling should be 14 in. × 13 in., or 15 in. × 11 in. When it is 25 ft., the scantling should be 18 in. × 14 in., or 17 in. × 16 in.; when it is 30 ft. the scantling should be 20 in. × 16 in. When baulks of large scantling are used for the girders, they should be always sawn down the middle lengthwise, reversed, and the two parts bolted together; since timbers when used in large baulks are liable to twist.

LINTELS are pieces of timber laid across the openings in walls to carry the weight of the superstructure; their thickness or depth must depend upon the bearing or span of the opening and the weight above, but for ordinary windows and doors they are usually from 4 to 6 in. deep, and rest 9 in. at each end on the wall.

BREAST-SUMMERS, or BRESSUMMERS, are lintels placed over wide openings in walls, as for shop-windows, cart-ways, or openings for bay-windows; they are generally sawn down, reversed, and bolted together; and a truss of iron is sometimes introduced between the two halves or *flitches*. A plate of wrought iron, the exact depth of the beam, is sometimes introduced in the middle between the two flitches, and the beam is then termed a *flitch-girder*: the thickness of the iron plate should be about one-twentieth of the thickness of the wood. If we multiply the breadth of the beam in inches by the square of the depth, and divide the product by the span or bearing in inches, the result multiplied by 2300 will give the number of lbs. avoirdupois that may with perfect safety be distributed uniformly over the *whole length* of the girder, fir being the timber supposed to be used.

QUARTERED PARTITION is a framework of timber for dividing the internal parts of a house into rooms, where there is no cross wall on the lowest story to carry a brick partition. This kind of partition should never be allowed to rest any considerable part of its weight upon the floor-joists, but should be trussed in such a manner as to throw the weight as much as possible on the walls at the two ends. Quartered partitions are formed of longitudinal timbers at top and bottom, called *head* and *sill*, into which are framed *braces*, or timbers placed diagonally across the partition; they are filled in with upright pieces about 11 in. apart, called *quarters*, to which the plaster-laths are nailed.

The principal timbers of a trussed partition should be 4 in. × 3 in. for a bearing under 25 feet; 5 in. × 3½ in. for a bearing between 25 and 30 ft.; 6 in. × 4 in. when the bearing is between 30 and 40 ft. The quarters need not exceed 2 in. in thickness, and may be stiffened when very long by short struts fixed between them. When the partition has to carry the weight of a floor above, the main timbers must be strengthened accordingly, and a vertical iron bolt may be employed to form a truss as in a king-post roof (p. 85).

Roofs in carpentry, are the timbers framed together for the purpose of receiving the slates, tiles, lead, or other

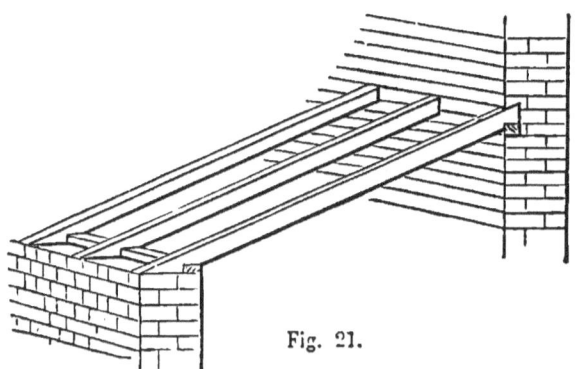

Fig. 21.

material by which the top of the building is to be covered in. The simplest kind of roof is that called a *lean-to* or *shed-roof,* in which a number of timbers called *rafters* rest upon wall plates laid on two walls, one of which is higher than the other, as shown in figure 21, and consequently the rafters have a slope or *fall* towards the lower wall.

A very common form of roof in town houses is the *V-roof,* or double lean-to, as shown in fig. 22.

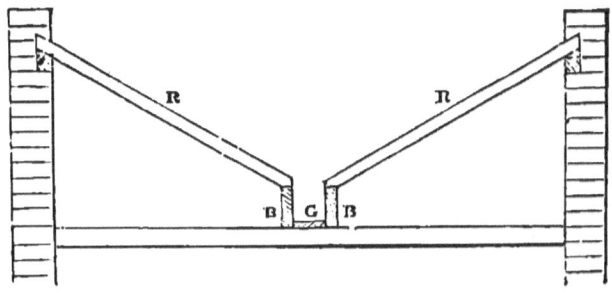

Fig. 22.

In this roof the rafters (R) rest at their feet upon two bearers (B) carried from back to front of the house, and forming a trough-gutter (G) along the middle. The upper ends of the rafters are supported by the party-walls.

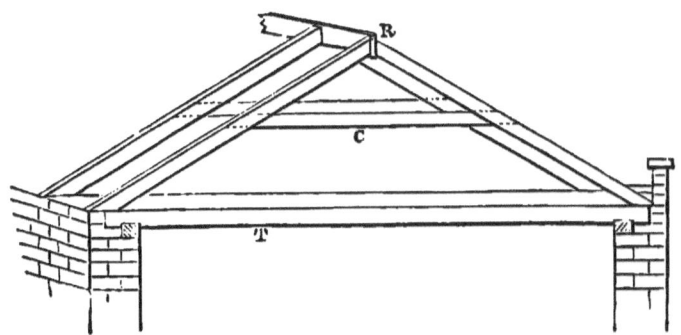

Fig. 23.

When the walls are both of one height the rafters are generally put together in pairs, sloping upwards from each wall to a *ridge-piece* (marked R) in the centre, as shown in fig. 23.

G 2

The following are the minimum scantlings of rafters for a slated roof having a pitch of 30°; when the horizontal distance from foot to head of rafter is 5 ft., the rafter should be $3\frac{1}{2}$ in. deep × 2 in. wide; for 6 ft., $4\frac{1}{2}$ in. × 2 in.; for 7 ft., 5 in. × $2\frac{1}{2}$ in.; for 8 ft., $5\frac{1}{2}$ in. × $2\frac{1}{2}$ in.

Hip-Roofs are those whose ends rise immediately from the wall, having the same inclination to the horizon as the sides have; a hipped roof is of a pyramidal form, and the angles made by the meeting of the planes which form the pyramid are called the *hips*. *Jack-rafters* are the short rafters rising from the walls and framing into the hip-rafters. The length of the hip rafter is found by dropping a plumb line from its vertex to meet a horizontal line from its foot, then adding together the squares of the lengths of those two lines, and taking the square-root of their sum.

Valleys are the opposite of hips, being the internal angles formed by the two planes of a roof. *Valley-boards* are boards laid on each side of the angle to receive the lead.

Collars are introduced in order to prevent the rafters of a roof from thrusting out at the feet, and are horizontal pieces of timber (marked c, fig. 23) nailed across each pair of rafters, at any convenient height, and *halved*

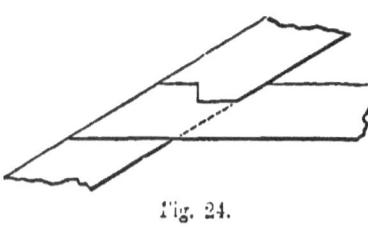

Fig. 24.

on to them, as shown on fig. 24, the collar having the same scantling as the rafters. When this piece is placed at the *feet* of the rafters it is called a *tie-beam* (marked T), and in that case the roof has no outward thrust on the wall. When the span of the roof is considerable, the tie-beam will have a tendency to bend in the middle; to obviate which a piece of timber called a *king-post* (marked K, fig. 25) is introduced between the heads of the rafters and the centre of the tie-beam; into the head of this post the rafters are framed and thus hold up the post, *shoulders* being formed in it for that purpose, the king-post holding up the centre of the tie-beam by means of a strap which is passed under it. Such a combination of timbers is

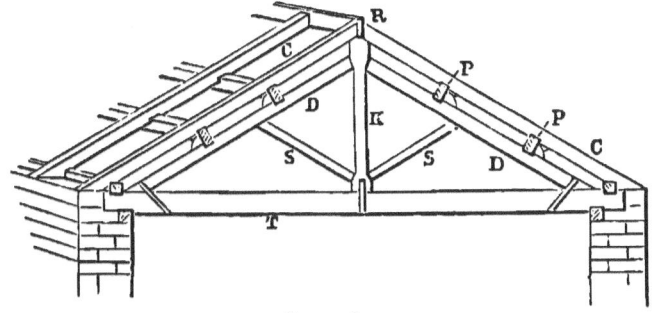

Fig. 25.

called a *truss* or *principal*. When two upright pieces (fig. 26) are introduced to hold up the tie-beam, they are called *queen-posts* (marked Q), and the horizontal piece between their heads is called the *straining-beam* (marked B).

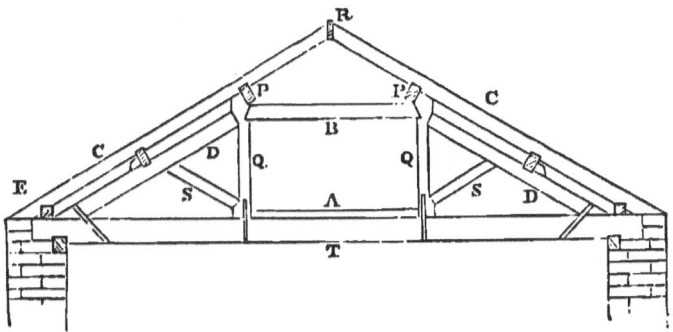

Fig. 26.

In order to stiffen the main rafters, pieces of wood called *struts* (marked S) are framed into the feet of the king or queen posts and also into the centre of the rafters. In the king-post roof the opposite thrusts of the struts counterbalance each other on the foot of the king-post; but in the queen-post roof their thrusts have to be conveyed along a *straining sill* (A) placed between the feet of the queen-posts upon the top of the tie-beam. The mode of framing the feet of the rafters into the tie-beam is shown in fig. 27.

When a roof is framed in either of the foregoing methods, the trusses do not themselves directly carry the slates or other covering, but are placed about 10 ft. apart and receive longitudinal beams called *purlins* (marked P), laid upon the *principal rafters* (D) of each truss, about 5 ft. apart; upon these purlins are laid the *common rafters* (marked C) about 11 in. apart, on which the covering of slates, &c. is laid. The feet of the common rafters rest upon a piece of timber laid upon the wall or upon the ends of the tie-beam, which is called the *pole-plate* (marked E).

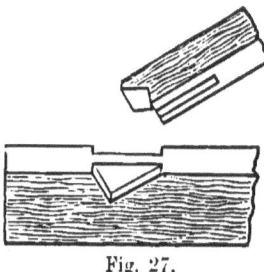

Fig. 27.

When the covering for the roof is to be lead, the rafters must be laid over with *close-boarding*, on which the lead is secured by means of rolls of wood placed every 2 ft. or 3 ft. apart and fixed from bottom to top, and over which the lead is dressed. If slate or tile is the material of the covering, battens or laths are nailed horizontally along the rafters at distances apart regulated by the gauge of the slates or tiles (see SLATING and TILING): at the eaves of a slated roof an *eaves-board* is generally laid, to give solidity to the slating at that part; and in order to check the rush of water into the gutter at the eaves, the slates are tilted up there by means of a strip of wood called a *tilting-fillet;* similar fillets are also laid along the edges of valleys, and wherever the slating abuts against a wall.

When the ridge or hips are to be covered with lead or zinc, a rounded roll of wood is spiked to the whole length of hip-rafter or ridge piece, and is called a *ridge-roll*. When there is a parapet wall at the eaves of the roof, a *gutter* has to be formed by means of horizontal pieces called *bearers* spiked to the feet of the rafters, and on which the *gutter-boards* are laid to receive the lead (fig. c, Plate III.).

The mode of forming a gutter between two roofs having eaves-boards, is shown on fig. D, Plate III.; the eaves-boards being feather-edged in order to allow the slating to lie evenly upon them. These gutters are generally laid with a fall of 2 inches in every 10 feet, consequently they are wider at one end than at the other. The bearers are spiked to the feet of each rafter.

Ashlering is the name given to upright quarters fixed between the joists and rafters of an attic or room in the roof, to cut off the acute angles formed by the feet of the rafters with the joists.

CURB-ROOF, or MANSARDE, is one in which the rafters on each side are in two separate lengths, and form an external angle (A) at their junction, as in Fig. 28. A

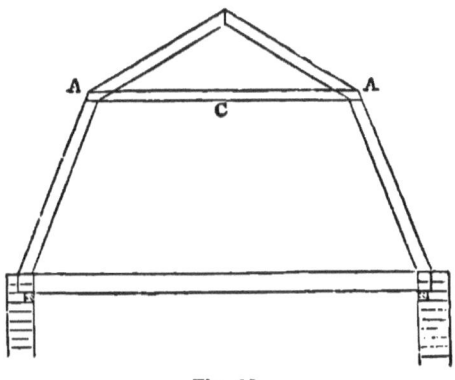

Fig. 28.

collar-beam (c) is introduced at the junction of the two sets of rafters, and acts as a ceiling-joist for the rooms in the roof.

The *pitch* of a roof is the angle which the feet of the rafters make with the tie-beam; and the pitch is said to be *high* or *low*, according as this angle is large or small. When the covering of the roof is lead or zinc, the pitch may be as low as 4°, or just sufficient to allow the water to run off; for slating and tiling, it should never be less than 25° to 30°.

Roofs may have the feet of their rafters prevented from

thrusting outwards without employing a *horizontal* tie-beam, as shown in figs. 29, 30, 31.

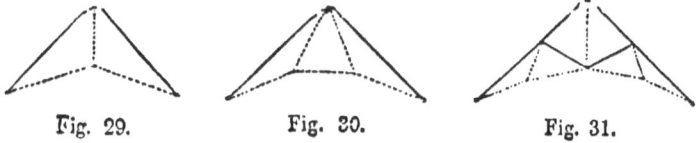

Fig. 29. Fig. 30. Fig. 31.

This form of truss is however more suitable for iron than for wood.

DOMICAL or CYLINDRICAL roofs may be constructed of timber, on the principle suggested by Philibert de Lorme, as shown in fig. 32. In this method a series of curved ribs are placed so that their lower ends stand upon a curb at the base and the upper ends meet at the top, diagonal struts being introduced between them. These ribs are formed of planks put together in thicknesses, with the joints crossed and well bolted together; there should be at least three thicknesses in each rib, not bent, but applied flat together in a vertical plane, and their edges cut to the proper curvature; the layers of the ribs may be held together without bolts, by merely the horizontal rings or purlins, which pass through a mortice hole in the middle and have themselves a slit into which a wooden key is driven on each side of the rib, as shown in the figure. For a dome 24 ft. diameter the ribs should be 8″ × 1″; for 36 ft., 10″ × 1½″; for 60 ft., 13″ × 2″; for 90 ft., 13″ × 2½″; for 108 ft., 13″ × 3″; the ribs being formed in two thicknesses, and placed 2 ft. apart at the base. They are stiffened by having the rafters notched upon them to receive the boarding, and also horizontal ties on the inside.

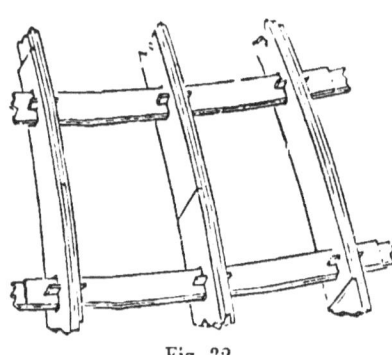

Fig. 32.

COLLAR-ROOFS.

COLLAR-ROOFS are frequently used over Gothic buildings of moderate span, as shown in fig. 33. In

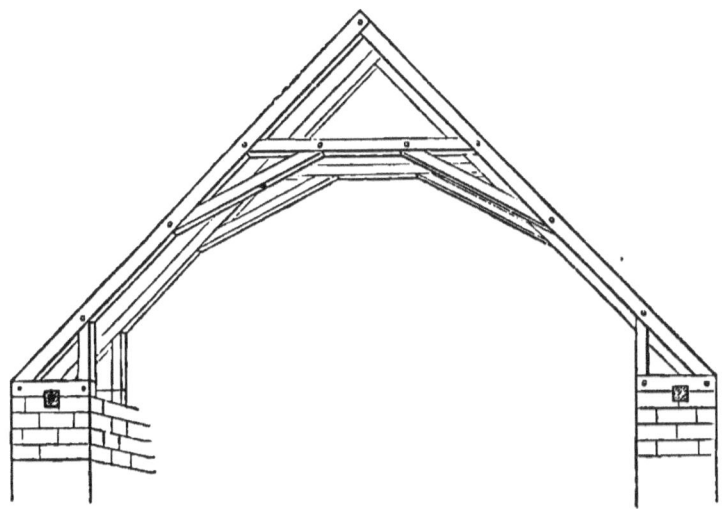

Fig. 33.

this form of roof the *collar* is placed high up, tenoned into the rafters, and secured thereto with oak pins. Diagonal pieces, called *braces*, are also tenoned into both the collar and the rafters, and secured with pins. The foot of each rafter is framed into a horizontal *wall-piece*, which is notched upon the wall-plate and lies across the whole thickness of the wall; into the inner end of this wall-piece a vertical strut is framed, and also into the rafter itself. By this arrangement the outward thrust on the wall is greatly counteracted, and the weight thrown nearly vertically upon it.

HAMMER-BEAM roofs are sometimes found over old Gothic buildings, and their form is shown in fig. 34, the timbers being fastened together with oak pins.

In this kind of roof we may suppose that the feet of the rafters are first prevented from spreading by being framed into a tie-beam; the middle part of the tie-beam is afterwards cut away, and the remaining parts (marked H) are called *hammer-beams*. To prevent these beams from thrusting outwards, a diagonal strut (marked s) is framed into its inner end, and also into a vertical wall-

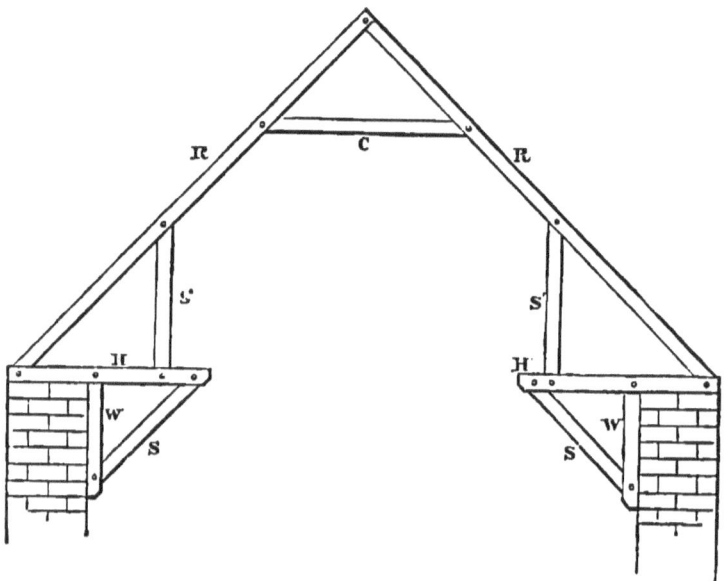

Fig. 34.

piece (w), which is itself framed into the underside of the hammer-beam. A vertical strut (marked S') is also placed between the rafter and the end of the hammer-beam. By this means a considerable amount of the thrust of the rafters is thrown vertically down the walls. There will, however, always remain sufficient horizontal thrust to push out the walls, if they are not built very strong, or supported by external buttresses.

BARGE-BOARDS are boards of an ornamental character, used on gables where the covering of the roof extends beyond the face of the wall. The *barge* projects from the wall a few inches and either covers the rafters or occupies their place. The slating or tiling projects over the face of the board.

When a wall-plate is carried along two adjoining sides

of a wall, as in the case of a hipped-roof, it is usual to *halve* the two ends upon each other, so as to prevent them from separating at the angle, as shown on fig. 35.

It frequently happens that a very long straight piece of timber is required, as in the tie-beam of a roof of large span: it is then usual to make the beam in two lengths, which are joined in the middle. The simplest method of joining beams longitudinally is by means of a *fished-joint*,

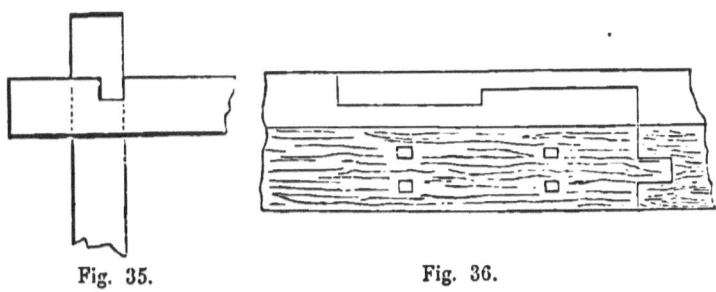

Fig. 35.   Fig. 36.

the two ends being made to abut against each other, and a plate of iron bolted on each side, so as to cover the joint and prevent it from opening. Another mode is called *scarfing*, as shown on fig. 36, in which each piece is toothed out to receive the other, and the two are fastened together by keyed wedges driven through the timber.

CENTERING is a framework of timber temporarily erected to support the arch-stones or *voussoirs* of an arch during its construction, and until the mortar in which the stones are bedded has become sufficiently set to allow the arch to stand without any intermediate support. Centerings are composed of several separate vertical trusses connected by horizontal ties, and stiffened by braces, the nature of such trusses depending upon the span of the arch. In small arches formed over openings in the wall of a house, a centering is made by forming two *turning-pieces* cut to the shape of the arch, to which boards are nailed for the voussoirs to rest upon. In large arches great care has to be taken in the construction of the framing for the centering, in order that it may not lose its shape by the pressure of the voussoirs, the laying of which must always be proceeded with uniformly on the two sides of the arch. In designing the centering of a large arch, it must be borne in mind, that the voussoirs near the crown press with greater force upon the framework than those near the springing; hence it is necessary that the centering be made stronger and stiffer in the upper parts than in the lower. On account of the adhesion of the mortar to the bricks or stones forming the arch, there is little or no pressure on the centering until the joints make an angle of 30° with the horizon: and when they make an angle of 45° with the horizon the pressure amounts to only one-fourth of their actual weights; at an angle of 60° their pressure on the centering is over one-half their actual weights; but it may be taken as a general rule, that any voussoir in which a plumbline from the centre of gravity falls outside its lower joint, presses with its full weight upon the centering until the whole arch is completed or keyed up. For a full description of the modes of constructing centering for large arches, see "Carpentry," (Weale's series, No. 182).

Fences for Parks or Gardens are formed with oak posts let into the ground 9 ft. apart, to which are framed horizontal bars called *arris-rails*, against which the *feather-edged* (thicker on one edge than the other) oak palings are nailed. A 1¼ inch oak plank 12 in. wide is fixed at the bottom. Fences are measured by the rod run of 16 ft. 6 in. or 5½ yards, and valued according to the height of the palings. If oak capping is placed on the top, it is measured extra; the plank at bottom is also measured extra.

Timber used in carpentry being cut out of trees which have grown by annual deposits of new layers of wood on the outside, it follows that the different layers of which it is composed possess different qualities of hardness and strength; the old or inner layers being the hardest, and the outer or new layers being the softest. If we suppose fig. 37 to represent the section, taken transversely, of a timber tree, the inner circle *a* will be the boundary of the *wood* proper; the next circle *b* that of the last deposit, and the space between these two circles is called the *sap-wood*; the outer circle *c* represents the *bark*, which is stripped off before the tree is cut up into timber. If, then, a *baulk* of timber is cut out of the tree as shown by the square *d e f g*, the four corners will consist of sapwood, and will be useless for purposes of construction. To obtain a baulk entirely free from sap-wood it is necessary to cut the tree as shown by the square *h i k l*. The sapwood possesses very little strength, and is also very liable to decay; its presence in timber can

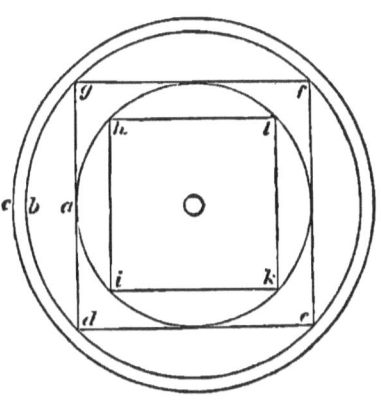

Fig. 37.

easily be detected by its totally different appearance to the inner wood. In old trees the part in the centre (O) called the *pith* is the first of the inner wood that decays, and the *dead-knots* found in wood are the dried-up or partly decayed pith of the branches. The knots in a piece of timber weaken it very much, and prevent it from bearing so great a strain as one without knots, fracture being generally found to take place at a knot. Baulks of timber are also frequently found, when cut down the middle, to be rotten in the centre; hence it is never advisable to use them in their natural size and shape; but when employed as girders or bressummers, they should be sawn down, reversed, and bolted together, by which means any tendency to twist is prevented, as well as any internal decay detected.

STRENGTH OF TIMBER as represented by the weight acting transversely, which a beam, when supported at each end, will bear in the centre, is found by multiplying its breadth by the square of its depth, and dividing the product by the length of the bearing, all the dimensions being in inches; the quotient is then to be multiplied by 4300 for Riga fir, and by 6600 for English oak, and the result is the *breaking-weight* in lbs. avoirdupois. If the weight is uniformly distributed over the entire length of the beam, the *breaking-weight* will be twice as great as when it is all concentrated at the centre. The *permanent-load* which a beam of wood is made to bear, should not be greater than one-sixth of the breaking-weight. If a beam whose transverse section is a parallelogram of unequal sides is laid with its broader side horizontal, it will bear much less load than if laid with its broader side vertical, the strength being proportional to the *square* of the depth, but simply *as* the breadth. Thus, if one side of the section is double the other side, the strength of the beam, when laid with the broader side vertical, is double what it is when laid with the same side horizontal.

Timber when used as struts, pillars, or shores, has to sustain a load acting in the direction of its length, which tends partly to crush the fibres and partly to bend them. When the length of a pillar is more than twenty-five times the breadth or diameter, the resistance to crushing does not come in play, but the timber yields by bending only. In this case, the *breaking-weight* in lbs. avoirdupois is found by dividing the *fourth* power (or square of the square) of the diameter in inches by the square of the length in feet, and multiplying the quotient by 24,542 for oak, and by 17,511 for red deal. The *permanent* load ought not to exceed one-tenth of the breaking-weight, so that it may be found at once by using 2454 and 1751 respectively as the multipliers. In this case the weight is supposed to act directly down the axis of the pillar; but if it acts otherwise, as down the diagonal, the resisting power will be reduced to one-third of the above.

When the length of pillars or struts is less than twenty-five times and more than ten times their diameter, they yield partly by crushing and partly by bending. To obtain approximately their strength, first find the breaking-weight, as above described for long pillars, and multiply that quantity by the area of the section of the pillar in inches, and then by 6000 for fir, and by 10,000 for oak; then take three-fourths of the area of section multiplied by either of the above numbers (according to the material), and add it to the breaking-weight first found; divide the first found breaking-weight by the last-named sum, and the quotient is the true breaking-weight in lbs. avoirdupois.

For pillars whose length is less than ten times the diameter, the resistance to crushing need only be considered, and it is found by multiplying the area of section in square inches by one or other of the above-mentioned numbers according to the material used.

The following rule will enable us to find with sufficient accuracy the breaking-weight of a square pillar of oak; in which the length is more than 5 times the diameter or side of the square: take the *ratio of length to diameter*, square it, and add it to 350; then divide 34,000 by the sum, and the result is the breaking-weight in cwts. per *square inch of section;* this multiplied by the number of square inches in the cross section gives the strength of any particular pillar. If the pillar has a circular section, first multiply the ratio of length to diameter by $\frac{6}{7}$ before applying the rule. The safe permanent load may be taken as $\frac{1}{10}$th of the breaking-weight, as found by this rule. The ends of the pillars are supposed perfectly flat and parallel, and the load pressing directly down the axis of the pillars.

These rules for calculation of the strength of timbers can only be considered as approximations, as much will depend on the *part* of the tree from which the beam is cut; the wood cut from the inner part being stronger than that cut from the outer part. The strength of timber also varies very much according to the amount of water it contains: that of very wet timber being only half as great as of that which has been well dried.

When timber is used in tie-beams, braces, or collars, it

has to resist a *stretching-force*, which is directly proportional to the area of its transverse section. The force (in tons) that will tear asunder the fibres of a piece of timber by stretching in the direction of its length is found by multiplying the sectional area in square inches by from $3\frac{1}{2}$ to 6 for fir, and by from 4 to 8 for oak. For the strength of other kinds of wood, see the above-named treatise on Carpentry.

FRAMED TRUSSES have the several timbers which compose them subjected to every variety of strain that has been above mentioned. The rafters have to bear a compressing-force in the direction of their length, and also a transverse strain perpendicular to their length. The tie-beam is subjected to a stretching-force acting in the direction of its length, and also to a transverse strain arising from its own weight and that of the ceiling attached to it. The collars are strained by a stretching-force in the direction of their length, and also transversely by the action of their own weight, which may be considered as a load uniformly distributed over their whole length. The king and queen posts are strained by a stretching-force only acting in the direction of their length; the struts by a compressing-force acting down their length, and also by their own weight acting transversely when they are not placed vertical. Braces are always subjected to a stretching-force acting longitudinally, and also to a transverse strain from their own weight, increasing in magnitude as they are more and more inclined from a vertical position. Straining-beams are subjected to a longitudinal compressing-force, and also to the transverse strain from their own weight. Purlins are only strained by a transverse force arising from their own weight and the load of the roof covering which is laid upon them. Hammer-beams are subjected to a longitudinal extending force. The strains on the timbers of floors, lintels, and bressummers, may be considered as entirely transverse or perpendicular to their length. The action, however, of all transverse strains is to stretch the fibres on one side of the beam, and to compress them on the opposite side.

The scantlings of the several timbers of a *king-post* roof vary according to the span or bearing. For a span of 20 ft., the tie-beam should be 7 in. by 3 in.; for 24 ft., 8 in. × 4 in.; for 30 ft., 9 in. by 5 in. The scantling of the king-post for the above spans should be—3 in. by 3 in.; 4 in. by 4 in.; 4 in. by 5 in. The principal rafters —4½ in. by 3 in.; 5 in. by 4 in.; 5 in. by 5 in. The struts—4 in. by 2 in.; 4 in. by 3 in.; 5 in. by 3 in. The purlins, when the trusses are 10 ft. apart—7 in. by 3 in.; 8 in. by 3 in.; 9 in. by 5 in. Common rafters, 4 in. by 2 in. to 5 in. by 3 in.

In a queen-post roof of 36 ft. span, the tie-beam should be 8 in. by 5 in.; of 40 ft. span, 9 in. by 5 in.; of 44 ft. span, 9 in. by 6 in. The queen-posts for the above spans should be—4 in. by 4 in.; 5 in. by 4 in.; 6 in. by 4 in. The principal rafters—5 in. by 5 in.; 6 in. by 5 in.; 6 in. by 6 in.; the straining-beam—7 in. by 5 in.; 8 in. by 5 in.; 9 in. by 6 in. The struts, 4 in. × 3 in. The purlins, where the trusses or principals are 10 ft. apart, 9 in. by 4 in. The common rafters, 4 in. by 2 in. to 5 in. by 3 in.

The pitch of the roof is supposed here to be about 30°, the covering slate, and the timber the best Memel or Riga fir. If a heavier material, as tiling or lead, is employed as a covering, greater strength must be given to the timbers.

The tie-beams whose dimensions are given above are supposed to have to carry the weight of a ceiling in addition to their own weight. Where there is no ceiling, a light iron tie-rod may be substituted, with economy, for the heavy tie-beam.

SHORING is a framing of rough timbers fixed against an old wall in order to render it secure during the process of alteration or partial rebuilding. The tendency of the wall being to fall outwards when the lateral supports are removed, a piece of timber, called a *raking-shore*, is fixed firmly against it in an inclined position, as D and E (fig. 38).

A plank or *waling-piece*, A B, is fixed in an upright position against the wall, and through it is driven a horizontal piece of wood or *needle*, as C, which passes completely through the wall. Another plank or *template*, is firmly bedded in an inclined position, at some distance from the wall, as H; and the raking-shore, D or E, is tightly fixed by means of wedges between H and C, so as to act as a temporary buttress. The weight of the wall above the needle C presses down the shore and prevents it from lifting, which it would tend to do if the wall moved outwards about its base. In lofty walls a raking-shore is generally fixed about half-way up the shore E, as F, resting on a piece G; such a shore, however, is of very little use unless there is a considerable weight of wall above the top of it to keep it from overturning about its base. In order to stiffen the shores and prevent them from *sagging*, or bending in the middle, the struts, K, are usually introduced. For an investigation of the mechanical principles involved in shoring, see Tarn's "Science of Building."

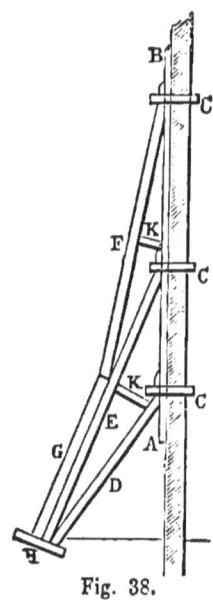

Fig. 38.

NEEDLING is a shoring by means of upright shores, and is employed when the lower part of a wall has to be removed or rebuilt without disturbing the upper part. A hole is cut through the wall and a large baulk of timber, or *needle*, is firmly fixed therein, being held up by two vertical shores firmly wedged on a template laid horizontally on the ground.

The timbers used in shoring and needle are all unsawn, and are measured as *cube fir framed rough;* and as the same timbers can be used several times over for similar purposes, they are not charged their full value, but are taken as *use and waste*.

WEATHER-BOARDING consists of boards nailed lapping one over the other, so as to keep out weather; the boards are generally made thicker at one edge than the other, and are termed *feather-edged boards;* the thick edge of one board is lapped about an inch over the thin edge of that next below it, and the nails driven through the lap. This is measured by the *square*.

SOUND-BOARDING consists of short boards placed transversely between the joists, on fillets nailed half way down the joists, for the purpose of receiving a coarse plaster, called pugging, to prevent the transmission of sound from one story to another. This is measured by the square, not deducting the joists.

SCAFFOLDING or staging erected of whole timber for the erection of buildings in which there is heavy masonry, is taken as *cube fir rough*, and valued as *use and waste*. All bolts, straps and other iron work being charged by weight. The term GANTRY is often applied to this kind of scaffolding.

WOOD BRICKS are pieces of deal cut to the size of an ordinary brick, and fixed into a brick wall as the work proceeds, wherever there is any joiner's work to be fixed in the finishing of the house. These are numbered in the Carpenter's Bill.

ON MEASURING CARPENTER'S WORK.

There are two methods of measuring carpenter's work: one by taking the superficial contents of roofs, floors, partitions, &c. at per square of 100 feet for the labour and nails, and then the cube contents of the timber without labour. The other method is, by measuring the cube contents of the timber as cube fir and labour, framed, &c., &c.

If the scantlings of the timber are small or light, it will pay the carpenter best to measure the roofs, floors, &c., as labour and nails, and the timber as no labour. But if the scantlings of the timber are large and heavy, then it will be more to his advantage to measure the work as timber, with the particular labour thereon, as follows :—

If the work is measured as timber and labour, the scantling of each piece is taken as cube fir or oak and labour, and entered accordingly; as

Cube fir, or oak, in ground joists, bonds, lintels, plates, &c., labour and nails included.
Do. framed in roofs, partitions, naked floors, &c., labour and nails included.
Do.   do.   truss framed   do.
Do. wrought and framed, labour and nails included.
Do. wrought, framed, and rebated,   do.
Do. wrought, framed, rebated, and beaded,   do.
Do. in door-cases.
Oak trusses put into girders, per foot run, stating their size, as 4 in. square, &c.

In measuring for labour and nails to roofs, naked framed floors, ceiling joists, quarter partitions, or any other rough framed work, the dimensions should be taken from the extreme ends of the timber each way, to ascertain the superficial contents thereof, as labour and nails at per square of 100 superficial feet. The openings to chimneys, staircases, &c., are not to be deducted, as the trouble of framing the trimmers and the joists into those openings is fully equivalent to running the joists through them. The same rule must be observed in taking the labour and nails in quarter partitions, as doors, &c., which must be entered in the measuring-book, and valued according to the description of the work, as follows :—

For Roofs.

Labour and nails to common shed roofing.
    Do.   do.   with purlins.
    Do.   do.   with purlins and struts.
    Do.   do.   common span or valley with purlins and two orders of rafters.
    Do.   do.   span with collars, dovetailed into sides of principal rafters, and these notched to receive purlins, filled in with common rafters.
    Do.   do.   framed with principals, king-posts, two struts and purlins, filled in with common rafters.
    Do.   do.   do.   with king and queen posts.
    Do.   do.   common Mansarde, or curb roof.

For Floors.

Labour and nails to fir ground joists, bedded and not framed.
Do. do. pinned down on plates and framed to chimneys.
Do. do. single framed floors, trimmed to chimneys and stairs.
Do. do. with girders and cased bays.
Do. do. framed floors with girders, binding, bridging, and ceiling joists.
Do. do. to common framed ceiling floors, with binding and ceiling joists.

Quarter Partitions.

Labour and nails to common 4 in. quarter partitions.
Do. do. 5 in. do.
Do. do. 6 in. do.
Do. do. truss framed with king-posts.
Do. do. do. with king and queen posts.
If oak is used, describe it.

Having taken the labour and nails, you must then proceed to take the timber therein, which must be entered as cube fir, or oak, without labour.

In roofs, it is customary to take the highest timbers first, as the ridge piece, hips, &c., next the rafters, and so proceed downwards.

In partitions, floors, &c., begin with the timbers of the largest scantlings. Wherever a tenon is made, the length must be taken from the ends of the tenon, and not from the shoulders. Likewise the length of joists, including the part in the wall.

In measuring king and queen posts, take the whole length by the scantling of the shoulders. The parallel pieces sawn out for the abutment of the principal rafters must be deducted, should they exceed two feet in length and $2\frac{1}{2}$ inches in thickness; but taken five or six inches short of the length between the shoulders as the saw cannot enter with much less waste. But if the pieces are less than $2\frac{1}{2}$ inches thick, no deduction must be made, they not being worth more than the labour of cutting them out.

ROOFS.

Hips and valleys to be taken at per foot run, for cutting and waste.

All plates, lintels, discharging pieces, to be taken as bond timber.

Gutter plates, diagonal ties, dragon-pieces or braces, struts, and tie-beams, as fir framed.

Deduct half the length of bond timbers running through openings.

Allow the length of dovetails or scarfing in bond timber but only taken as bond timber.

Fixing iron straps, screw bolts, hanging ditto, and all iron work, to be taken and allowed extra.

FLOORS.

Oak trusses, let into breast-summers, to be taken at per foot run.

Oak king or *queen posts*, let into breast-summers, each at ——.

Girders sawn-down, reversed and bolted, per foot run extra.

Letting-in screw-bolts, plates, &c., each extra.

Common or *herring-bone strutting* between the joists, per foot run extra.

Furrings to ceilings, quarter partitions, battenings to walls, &c., are measured by the square, including labour and nails, and valued according to the thickness of the deals used, from ¾ to 3 inches thick. Describe the battening either as framed or nailed only, or if plugged, or if with horizontal backings.

All *wall-hooks* and *holdfasts* to be allowed extra.

Centering to groins, vaults, recesses, &c.—Take the depth by the circumference for the superficial dimensions, which is valued at per square for use and waste, materials and time. If taken in this way, the whole of the vaults or recesses must be taken, although the same centering might have been used. But where there are a number of vaults or recesses of the same size, the fairest way is to allow the whole of the materials and time, or, if any trifling alteration only is wanted, to allow the time expended in doing it.

If to small openings, as windows, recesses, doors, &c., they may be measured at per foot superficial, viz. :—

ft. in.	ft. in.		ft. in.
3 6	0 4	Superf. of centering to apertures, as windows, &c. (Plate III., fig. 1 A.)	
12 10	0 9		8 2 4 1 0 7
			12 10
16 10	1 10	Superf. of semicircular centering to revealed windows. (Plate III., fig. 2 B.)	10 8 5 4 0 10
			16 10

Bracketing to plaster cornices (Plate III.)—To be measured at per foot superficial, according to the girt, viz., 24½ inches by the length, as whole or 1½ inch deal, according to the thickness of deals used. Some allow the bracketing the same girt as the cornice.

inches.
6
1½
9
6½
1½
———
24½

Cradling for entablatures over shop windows, &c., measured and charged per foot superficial, according to their thickness.

All *circular* bracketing, cradlings, &c., to be charged double those of straight work.

Ashlering is taken at per foot superficial, according to the thickness of the deals used.

ft. in.	ft. in.	
20 0		Supposed length } (See Plate III., fig. C.)
2 0		Height. . .

Gutters and *bearers* (Plate III., fig. C).—Measure the length, then the breadth of the bottom and half the eaves-board.

Gutters between the roofs having two eaves-boards, one on each side, take for the width of gutter one of them. (Fig. D.)

ft.	in.	ft.	in.		
20	0			Supposed length	
1	1½			Width (fig. C.)	Enter them as—
					Superf. of whole deal gutters and bearers.
20	0			Do. (fig. D.)	
1	6				

Arris or *fillet* gutters per foot superficial.

Water trunks for carrying off rain-water, per foot run; describe size, and allow for laps and half the length of shoe. These are now seldom used, iron or lead pipes being generally employed in preference.

Doorcases if solid are measured by the foot cube, and described as *proper* if rebated and beaded.

JOINERY.

JOINER'S-WORK requires greater accuracy of workmanship than carpenter's-work, being nearer to the eye and subject to closer inspection; the joints must therefore be accurately fitted, and the exposed surfaces rendered perfectly smooth. The wood used in joinery, called *stuff*, consists of *planks* or *boards*, *deals*, and *battens*, according to their widths; their thickness as imported being 2½ in. or 3 in. *Battens* vary in width from 2 in. to 7 in.; *deals* are generally 9 in. wide; *planks* are 11 in. wide.

GROOVING, or PLOUGHING, is the forming a channel or *groove* of uniform width in a piece of wood as in fig. 39, marked G.

Fig. 39.

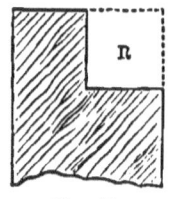
Fig. 40.

REBATING is the cutting a rectangular strip or *rebate* out of one side of a piece of wood, as in fig. 40, marked R.

MORTISING is the cutting a rectangular cavity or *mortise* within the surface of a piece of wood, in order to receive a corresponding rectangular piece, called a *tenon*, projecting from another piece of wood, as in fig. 41, M being the mortise, and T the tenon; the parts (s) from which the tenon projects are called the *shoulders*. Mortise holes are cut in the stile of a door to receive the lock.

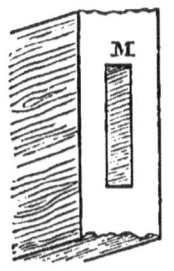 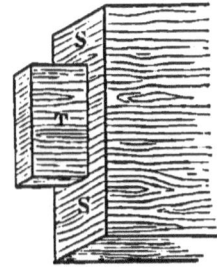

Fig. 41.

ARRIS is the sharp edge or external angle formed by the meeting of two plane surfaces.

CHAMFER, or BEVIL is the cutting-off of an arris by a plane making obtuse angles with the sides, as at M, fig. 43.

TONGUEING is the insertion of a thin slip of wood or iron, called a *tongue*, into grooves cut or *ploughed* in two pieces of wood, for the purpose of keeping their faces in one continuous plane, as in fig. 42, T being the tongue; when the tongue is cut across

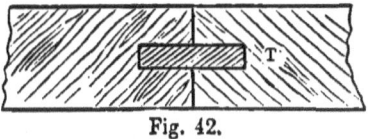

Fig. 42.

the grain of the wood, it is called a *feather-tongue*. The tongue is bedded in glue and the pieces fixed firmly by blows of a mallet.

A MITRE is the diagonal joint which two pieces of wood make with each other when meeting at an angle, as M in fig. 43. A *mitred border* is often put round the hearth of a fire place, to prevent the ends of the floor boards from abutting against the stone.

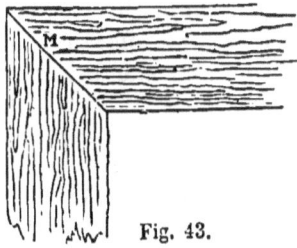

Fig. 43.

SHOOTING is making straight edges to boards which

are then said to be *shot,* in order to fit them closely together.

DOVETAILING is the joining two boards by indenting them one into the other, the sections of the projection and hollows being in the form of a *dovetail,* as shown in fig. 44. It is executed in different ways according to the kind of work.

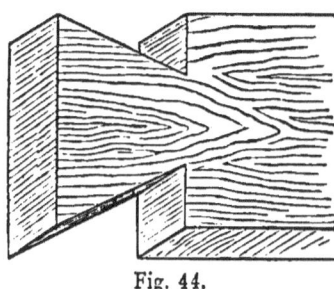

Fig. 44.

CLAMPING is when several boards are fastened together to form one plane surface, by means of another board, or *clamp,* fixed transversely at each end by means of mortise and tenon, or by a groove and tongue. If the clamp forms a mitre at each end with the two outer boards, it is then said to be *mitre-clamped.*

BLOCKINGS are small pieces of wood fitted and glued to the interior angle of two boards, in order to strengthen the joint, as in the junction between the treads and risers of stairs.

HOUSING is the cutting a piece out of one board for the insertion of the end of another, in order to fasten the two together, as in letting the ends of treads and risers of stairs into the strings.

STAIRCASES are flights of steps for passing from one floor to another, and are formed with *treads* and *risers,* the riser being mortised into the tread which overhangs it with a *nosing,* which is generally rounded or moulded (Plate V.) The ends of the treads and risers are supported on *string-boards,* one placed against the wall and called the *wall-string,* and the other at the outer ends called the *outer-string.* The treads and risers are housed into the wall-string, and also into the outer-string which is then called a *close string;* or the outer string-board is cut away to receive them, in which case it is mitred to the treads and risers, and the nosings of the steps are continued round their outer ends. When the stairs are formed in two flights, with a landing or resting-place between, the term *quarter-space* or *half-space* is applied to the resting-place, according as a quadrant or semi-circle is described in passing from one flight to the other.

The stairs which are all the same width and are all parallel to each other are called *flyers*, and those which turn round a solid *newel-post*, or a circular *well-hole*, are called *winders*, the treads being wider at the inner end next the wall than at the outer end. The *newel-post* is an upright piece of wood into which the string-boards are framed. The term *dog-legged* stairs is applied to those which have no well-hole, the steps being fixed to newels and string boards. The *carriages* are pieces of timber placed underneath the steps and following the pitch of the staircase, to give firmness to the flight. The *hand-rail* is a bar or rail raised upon slender posts called *balusters*, to protect the outer edge of the staircase and to assist persons in ascending and descending. When the handrail is made continuous round a well-hole staircase, the part where it turns round the well is called the *writhe* or *wreath*. A *ramp* is where the handrail makes a sudden rise without turning round, as in framing into a newel-post. Handrails are jointed together in lengths and fastened with *handrail-screws*. A *curtail-step* is the first step by which a staircase is ascended, finished at the end in the form of a scroll. The *apron-piece*, or *pitching-piece*, is a horizontal timber let into the walls on each side of the landing or half-space of a staircase to which the carriages are fastened. The *apron-lining* is the thin deal facing which covers the apron-piece.

When there is a circular well-hole to a staircase the string-board is glued up in vertical slips to form the curved part of the well. The making of the *writhe* of the handrail round the well is a very nice piece of joinery, and is always executed by men especially trained to this work, moulds having to be got out by geometrical methods before the handrail itself can be cut. The writhe is elliptical in form, and is jointed at the middle of the turn, as the two parts twist in opposite directions from that point. For simple methods of getting out the moulds the reader is referred to Collings' "Handrailing" (Weale's Series).

The ballusters which support the handrail are housed into the treads of the stairs.

SKIRTING is a narrow board or margin placed against the wall of a room next the floor, and is either moulded or plain. It is fastened to a narrow slip of wood called the *skirting-ground*, which is nailed to plugs or wood bricks let into the wall.

If the skirting has no moulding on the top, it is called *square*; if it has a simple bead, it is called *torus*; moulded skirtings usually have the moulding planted on the top. Skirtings are mitred and tongued at their angles. They are measured by the foot super, unless they are very narrow when they are taken by the foot run. When fixed on a slope, as up the side of a staircase, they are termed *raking*.

DADO is a piece of framing fixed round the base of the walls of a room, generally about 2 ft 6 to 3 feet in height, and consists of three parts, namely, the *plinth* or *skirting* next the floor, the *surbase* moulding at the top, and the plain surface or *dado* proper between them. The construction is fully described on p. 122, and shown on Plate III.

GROUNDS are pieces of wood which are placed round openings, and wherever the plaster of a room is to stop abruptly, for the plaster to finish against, and are made the exact thickness of the plaster; they are nailed to plugs or wood bricks let into the wall (D, figs. 3, 4 and 5, Plate IV.).

ARCHITRAVES are the mouldings which are planted round door, window and other openings (M, fig. 8, Plate IV.).

BOXINGS are the cases into which shutters (E, fig. 8, Plate IV.) are folded when hung in more than two widths.

Back-flaps F and G are the parts of the shutters which fold into the boxings. *Window-back* K, fig. 7, is the framed lining below the window, between the window-sill and the floor; *elbows* H, fig. 8, are the corresponding linings under the boxings of the shutters (fig. 8, Plate IV.).

ANGLE-STAFFS are slips of wood fastened to plugs at the corners of chimney-breasts for the purpose of protecting the plaster, which would be broken off if brought to a sharp arris. Angle-staffs are either square or beaded: if the walls are intended to be papered, square staffs are used; but if they are to be painted, beaded staffs are fixed, and the plaster is *quirked* against them.

MATCHED-BOARDING is lining with battens of equal width, having a bead and tongue worked on one edge and a groove on the other edge, and fitted together to form one uniform surface.

SASH is the part of a window intended to receive the glass, and is made to open and shut (figs. 6, 7, 8, 9, and 10, Plate IV.). *Hung-sashes* are formed in two parts or *sheets*, called the *top* and the *bottom* sash; these work in two grooves in the sash-frame or *case*, and are hung by cords or chains passing over pullies and attached to balance weights concealed at the back of the case (figs. 6, 7, 8, Plate IV.); the grooves in which the sashes run are called *pulley-pieces*, and are separated by a *parting-slip*, which keeps the two sashes apart, and enables them to pass each other without touching.

The bottom of the sash-frame, called the *sill*, which is generally of oak, is sunk, and *weathered* outwards to carry off the rain. The *sashes* consist of upright pieces or *stiles*, top, bottom, and meeting *rails*, and sash-bars. These are rebated on the outside to receive the glass, and moulded on the inside in various forms, the most common of which are the plain chamfer bar, the ovolo, the lamb's-tongue, and the astragal and hollow. The sashes and frames are always taken in one measurement, the thickness of the sash and character of moulding being described, and also the quality of axle-pullies, whether iron or brass, and the kind of cord or chain used for hanging. When only one sheet of the sash is made to open, it is said to be *single-hung;* when both sheets open, the sashes are *double-hung*.

CASEMENTS are sashes which are hung to a solid rebated frame by means of hinges; they are also called *French-windows* when hung *double*, or in two parts folding together.

In casements which are made to open inwards the stiles

are grooved and a small projecting fillet is fixed on the side of the frame which fits into the groove in the sash when closed so as to render it weather-tight. India-rubber tubing will also serve the purpose of making a tight joint between sash and frame.

In order to prevent the rain from driving under the sill, a projecting piece of wood or *weather-board* is fixed to the bottom rail of the casement; or the plan shown in fig. 45 may be employed with advantage, the *throating* T being made to stop the water from driving under, and holes are drilled in the sill to carry off the water. Folding casements are fastened with an *epagnolette-bolt* which is made the whole length of the sashes and has a hook at top and bottom by which they are firmly held together. Another mode of fastening is by a metal tongue which fits into a groove the whole height of the casement and helps to keep out the weather.

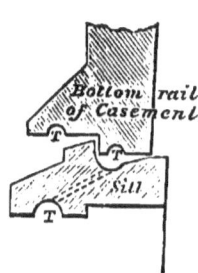

Fig. 45.

VENETIAN-WINDOWS are divided into three parts by mullions, the centre sashes being generally much wider than the side ones.

FANLIGHT is a sash fixed over an outside door, the door frame being rebated to receive it. It is generally made to open by turning on centres, or on hinges as a casement.

SASH-BARS are made of various sections, those in common use being shown on figs. 46, 47, 48, 49, and 50.

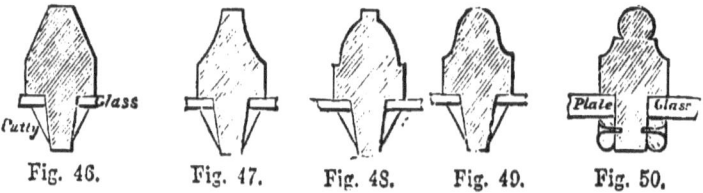

Fig. 46.   Fig. 47.   Fig. 48.   Fig. 49.   Fig. 50.

Fig. 46, is a plain *chamfer;* fig. 47, a *hollow;* fig. 48, an *ovolo;* fig. 49, *lamb's-tongue;* fig. 50, *astragal and hollow.*

DORMER is a window made upon the slope of a roof, the frame being set up perpendicularly; the sides next the slating are termed the *cheeks*.

SKYLIGHT is a window placed over an opening made in a roof, and generally follows the slope of the roof, being raised a few inches above the slating to keep out the rain-water.

BRASS SASH-BARS for shop windows are made with a core of wood, covered over by a thin sheet of brass stamped or rolled to various forms and mouldings; they are measured by the foot run, the price varying with the size and character of the moulding.

STALL-BOARD is the framing which forms the part of the shop-front which is below the glass of the window.

FACIA is the board placed above the glass of a shop window, and covers up the bressummer.

DISHING is cutting away to a bevel the edges of a hole cut in a piece of wood, as in the seat of a water-closet; the term *dishing-out* is also applied to the formation of *coves* to a ceiling.

CANTING is a term used for cutting off the corner of an angular piece of wood or framework.

FLOORS. (PLATE II.)

FLOOR-BOARDS vary in thickness up to 3 in.; and being always cut out of $2\frac{1}{2}$ in. or 3 in. stuff, they are often described according to the number of saw-cuts required to be made in the original stuff. Thus, when 3 in. stuff is cut into floor-boards by one saw-cut, the floor-boards so obtained are said to be *one cut*, or $1\frac{1}{2}$ in. thick; if by two saw-cuts, they are termed *two-cut*, or 1 in. thick; if by three saw-cuts, they are termed *three-cut*, or $\frac{3}{4}$ in. thick. The actual thickness of floor-boards, when laid, is always about one-eighth of an inch less than that by which they are called, on account of the loss by the saw-cut and the planing of one side. Floor-boards are fixed to the joists by means of *floor-brads* or nails driven through from the surface; but in thick floors *edge-nailing* is sometimes adopted, the nails being driven at the edges of the boards in an oblique direction so that the heads are not visible on the flooring; the next board is then tongued into the edge that is nailed. Floor-boards are cut out of planks, deals, or battens; but the best floors are made out of battens, which should be as narrow as possible, since wide boards are liable to shrink and warp and become hollow in the middle.

FOLDING-FLOORS (fig. 1, Plate II.) are laid four boards together, which are shot as nearly as possible to fit a given space, and forced downwards *folding* into their places. These are the commonest and cheapest kind of floors.

STRAIGHT-JOINT FLOORS (fig. 2).—The boards are carefully laid the length of the room in regular straight joints, and their *heading-joints* should be either splayed (fig. 6), ploughed and tongued (fig. 7), or grooved and tongued (as fig. 5), taking care to break them at proper distances. Sometimes the edges are also ploughed, and fastened together with tongues of wood or iron; or else mortised and tenoned by a mortise-groove being cut on one edge, and a tenon on the other edge of each board; these are called *tongued-floors*.

DOWELLED-FLOORS (fig. 3).— The boards are laid straight, joined with wood or iron *dowels*, or pegs let into the edges to confine them down, instead of nails from the face of floors, having them only on the edges of the boards.

Wainscot-floors should have iron dowels, but deal-floors may have dowels made of beech, as the dowel should certainly be made of a material much stronger than the floor. If beech, they should be formed as at A, and cut square; and being driven into round holes in the battens makes them draw.

In dowelled-floors the dowels are set from 6 in. to 8 in. apart, and the heading-joints ploughed and tongued; and no heading-joint of any two boards ought to be allowed to meet that of two other boards, or to form a straight line equal to the width of two boards.

Wainscot-floors and first-class deal batten floors should never be laid directly on the joists, but a commoner kind of floor should be first formed, and the finishing one laid in the opposite direction upon it.

PARQUETRY-FLOORS are made with pieces of variously coloured woods, laid in geometrical patterns, and tongued together, a common deal-floor being first laid upon the joists to receive the parquetry.

DOORS, SHUTTERS, AND FRAMING.

LEDGED-DOORS are the commonest kind of doors, and are formed by placing boards side by side, tongued or tenoned one into the other, and fastening them together at the back by horizontal boards called *ledges*. Coach-house doors are also made in this manner, but with upright pieces at each edge framed into the ledges, and cross-braced in addition.

PANEL-DOORS and PANEL-FRAMING consist of vertical and horizontal pieces mortised and tenoned into one another, so as to form rectangular compartments. (Fig. 5, Plate VIII.)

The upright and horizontal pieces of any framing are called respectively *stiles* and *rails*, and the thin filling-pieces *panels*. The grooves to receive panels are one-third the thickness of the framing; but the panels themselves may vary from this up to two-thirds thereof, or twice the groove. In this latter case they are *flat* on one side, coming no more forward than the groove; but on the other side they come out to the same plane with the face of the framing, and are called *flush* panels. To obtain the strength of these without their bald flat appearance, the continental artists invented the mode of sloping off a margin all round them, so that the framing, though not more prominent than their centre part, might project and cast shadow on this margin; and these are called *raised* panels. In the richest variety, one or more mouldings separate the margin from the centre, and these are *moulded raised* panels. Panels may also be as thick as the framing, and either *raised on both sides*, or *flush and raised*.

The chief modes of finishing or marking the separation of the framing and panels are these :—

BEAD-BUTT, or BEAD AND BUTT, when the panel's face is flush, and two small round mouldings, or *beads*, are struck along its two lateral edges.

BEAD-FLUSH, or BEAD AND FLUSH, when similar beads are struck, not on the panel itself, but on the edges of the stiles and rails, so that they form a bead border all round it, mitring at the corners.

SQUARE, when the panel's face is flat, and the framing simply projects before it with a square edge.

CHAMFERED, when the square edge is merely pared off to a narrow diagonal face; and the chamfering is said to be *stopped* when this face does not mitre round the corners, but finishes a little short of them, to avoid weakening the junctions. Stopped chamfers are common in Gothic work, both for edges of panelling and for angles of timbers or posts.

BEAD AND FLAT is when a flat panel is surrounded by a bead, or three-quarter cylinder, wrought on the projecting stiles and rails.

OVOLO AND FLAT, a convex moulding, whose section is a quarter circle or quarter oval, surrounding a flat panel.

OVOLO AND RAISED, the same round a raised panel.

OGEE FLAT, or OGEE AND RAISED *Panel*, a moulding of contrary flexure, instead of one only convex.

QUIRK OVOLO, or QUIRK OGEE, &c., when the curve of either of these mouldings returns inward, after its greatest projection forward.

MOULDED is a general term that will apply to the four last finishings, or to any in which these mouldings are combined, to whatever extent.

The description of any piece of panelled framing, as a door, &c., may thus be given by describing each side separately, as in the abbreviations about to be proposed. See p. 118.

The panelled construction is properly only adapted to internal doors, for external ones should present, outwards to the weather, as little lodgment, and therefore as smooth a face, as possible; and the natural ornament or relief to the baldness of such a face is obviously furnished by the hinge attachments and other metal work.

In doors moulded on both sides, the grooves for panels must be ploughed deeper than the moulding, to prevent light showing through the mitres, should the deals shrink; but if framed with a square back, there is no necessity for ploughing so deep.

As the mouldings, however, are now seldom really part of the work, but imitated on strips of wood stuck on afterwards, the above rule seldom applies.

The joints of panels should be ploughed and tongued.

All tongues should be cut across the grain of the wood.

LININGS *of a door* are the internal facings which surround the *jambs* or sides, and the *soffit* or top, of a door opening; they are *rebated* to receive the door; when rebated on both edges for the sake of uniformity, they are said to be *double-rebated*.

SHUTTERS to house windows are usually framed with panels, stiles and rails, as above described for doors and other framing. If hung *folding*, they are in two or more widths on each side of the window, and when open are generally concealed in a *box*, the front shutter only showing on the window jamb. This mode of hanging is shown on plan at fig. 8, Plate IV., where L is the front of the *boxing*, E the shutter, F and G the *back-flaps*, the whole being hung together with hinges, so that when closed they extend half-way across the window, and meet the shutter from the other side with a rebated joint.

Lifting-shutters are generally employed for bay-windows, where there is a difficulty in fitting boxing-shutters. These are hung in a similar manner to sashes, with cords, pulleys and balance-weights, sliding up and down in a pulley-stile, either in one or two heights; when open the top is concealed by a flap which falls down and covers them up from view.

Shop-shutters are made in separate pieces of framing so as to put up and take down by hand, and are rebated at each edge. They are fixed in a horizontal groove at the top of the window, and secured below by bolts passing through the sill of the sash to the inside. *Coiling-shutters* do not form part of the joiner's work, except so far as relates to preparing for them; and the reader will find these described under the Smith's work.

WINDOW-BLINDS, whether external or internal, form an important feature in the finishing of houses, the object aimed at in using them being to enable the inmates to regulate the quantity of light to be admitted by the windows. In situations which are not exposed to a strong sunlight, *inside* blinds are sufficient for the above purpose; but where the sun shines for many hours together upon a window, *outside* blinds become necessary in addition, in order to prevent the rays of the sun from falling upon the glass, and thereby producing a large extent of heating surface, which raises the temperature of the air of the room by contact with the glass.

For *inside* blinds the simplest and cheapest are Holland *roller-blinds*, placed on wooden rollers with brass ends, which turn in brass or wooden holders; these have a pulley at one end with an endless cord passing over it, the cord also passing round a pulley below attached to a *rack*, whereby it can be tightened at pleasure. Some are fitted with self-acting spring rollers, which draw up the blinds when the catch which holds them is released. Roller-blinds are charged by the square foot, and the price depends on whether they have plain or spring rollers.

In sunny situations *inside Venetian-blinds* are mostly used; these consist of a number of thin laths attached to webbing, so as to lap one over the other, the laths being capable of being set to any angle by means of cords attached to them. These blinds are drawn up and down by means of cords passing over pulleys at the top, or else by a patent action, in which a spring performs the work of drawing up the blinds, and can be stopped at any height. Venetian-blinds are valued by the square foot, measuring the opening of the window, the patent action being charged extra according to its character.

Outside-blinds are various in character, the simplest and cheapest being the Venetian-blinds, as above described for the inside, fitted into an outside case fixed between the reveals of the window. Then there are what are termed "jalousies," or Venetian shutter-blinds, consisting of a number of laths or louvres fixed in a frame hung upon hinges, so as to open back against the face of the wall. These blinds are also made to slide upon horizontal iron rods.

The *Florentine* and *Spanish* blinds are made of striped linen tick, working up and down in cases, and when let down they project forward so as to admit air to the window at the bottom only. The *Heliocene* is made of similar material, but being formed in a series of hoods one above the other, it affords a shelter from the sun's rays without concealing the view from persons within; it also admits the air to every part of the window, and may be drawn up or down at pleasure.

All these blinds are valued by the foot superficial, according to the measurement of window covered by them; if the window is less than 16 feet in area, the blind is charged as 16 feet.

Water-closet fittings, consist of seat, riser, clamped flap, beaded frame and narrow skirting, fixed on framed bearers; a *dished hole* being cut in the seat, and a beaded hand-hole for handle of apparatus. These may be taken as *each* with full description and dimensions; or every part may be measured separately, the area of the seat and riser being taken as one item, that of the flap and frame as another, the run of the skirting round the seat, and the number of holes cut in seat; the hinges to be numbered; the run of the rounded or moulded edge to be taken. An allowance is also made for attendance on plumber in fitting the apparatus.

Sketches should be made in the margin of the bill of quantities of any work that cannot be accurately described, such as mouldings, enrichments, trusses, balusters, newels, &c.

ABBREVIATIONS.

The same observations respecting abbreviations will hold good, but to a greater extent, with the carpenter and joiner than any of the other trades; and even the most complicated, as sashes and frames, which may appear at first unintelligible, will very soon be read with as great facility and equal accuracy, with all their varieties, as they could possibly be if written at full length : viz. :—

For Timber.

L N O	Labour and nails only.	Ro and L	Rough and Labour.
L^r to Q^r P^{ns}	Labour to Quarter Partitions.	W	Wrought.
		F	Framed.
Fir or ⎰ Ro	Cube Fir rough.	B	Beaded.
Oak ⎱ Bnd	Bond.		

Example.

C Fir, W, F, R, & B . . Cube Fir, wrought, framed, rebated, and beaded.

For Deals, after describing their thickness.

Inch deal R	Inch deal rough.	D	Dovetailed.
E S	Edges shot.	F	Framed.
W 1 S	Wrought one side.	K	Keyed.
W 2 S	Wrought two sides.	M & C	Mitred and chamfered.
G	Grooved.	S	Sunk.
B	Beaded.	P	Plugged.
P T	Ploughed and tongued.	L	Ledged.

Example.

Inch deal, W 2 S, F & B. Inch deal, wrought two sides, framed and beaded.
Whole deal, W 1 S, P T. Whole deal, wrought one side, ploughed and tongued.
1½ deal, W 2 S, M & C . 1½ deal, wrought two sides, mitred and chamfered.

Doors.

R, G & L	Rough, grooved, and ledged.	O R P	Ovolo raised panel.
		Q O B	Quirk ovolo and bead.
W, L, R, B	Wrought, ledged, rebated, and beaded.	O G	Ogee.
		Q^k O G	Quirk ogee.
S	Square.	Q^k O G B	Quirk ogee and bead.
B, B & S	Bead, butt, and square.	O G F	Ogee flat.
B F^h	Bead flush.	O G R P	Ogee raised panel.
B F^t	Bead flat.	D M	Double margin.
B S	Both sides.	B M	Broad margin.
O F^t	Ovolo flat.		

Example.

1¼ D, W L, R & B door . . .	One and a quarter inch deal, wrought ledged, rebated, and beaded door.
1½ D¹, 4 P, Q O G & B, & B F⁴ door.	One and a half inch deal, four panel, quirk ogee and bead, and bead flush door.

FLOORS.

Inch W D Floor F	Inch white deal floor, laid folding.	Ro, E S	Rough, edges shot.
		W F	Wrought, laid folding.
1½ Y D, R F Floor	Inch and half yellow deal, rough folding floor.	W S J	Wrought, straight joint.
		D	Dowelled.

Example.

1¼ inch Y deal, W S J floor, H J P T, E N. } 1¼ inch yellow deal, wrought straight joint floor, heading joints ploughed and tongued, edge-nailed.

SASHES AND FRAMES.

D C frames, O D S sills, W P P, B & P S. 1½ W a & h Sashes, D h, B P, P L & L weights. } Deal case frames, oak double sunk sills, wainscot pulley pieces, beads and parting slips, 1½ inch wainscot astragal and hollow sashes, double hung, brass pulleys, patent lines, and lead weights.

Any variation from this description may be made with ease; viz., if

I B P	Iron box pulleys.	M P P, B & P S	Mahogany pulley pieces, beads and parting slips.
B A P	Brass axle pulleys.		
S H	Single hung.		
C W L	Common white line.	D P P, B & P S	Deal pulley pieces, beads and parting slips.
I P	Iron pulleys.		
I W	Iron weights.		

MEASURING JOINER'S WORK.

In the measurement of joiner's work, there is very little cubing as in carpentry, most of the work being taken superficial by the foot, or square of 100 feet. Narrow pieces are measured by the foot run, and articles that cannot be measured are numbered.

IRONMONGERY which is fixed by the joiner is included in the valuation, the several articles being numbered and described. There are two methods that may be adopted in measuring joiner's work; one is to take all works of the same kind together throughout the house, such as floors, framing, skirtings, sashes, &c. The other method is to take each floor by itself, keeping all the work of each separate until you come to the abstract and bringing into bill. Whichever method is adopted must be kept to throughout in order to prevent confusion; in large buildings the latter system will generally be found preferable, as the several items can more easily be referred to, and any errors or omissions detected; while the former system answers better for small works, and involves a smaller amount of calculation. In taking the superf. of any work, no allowance is to be made for tenons, or any parts that are concealed, the nett area only being measured.

Flooring (Plate II.).—In measuring boarded flooring, the dimensions must be taken, allowing the thickness of the skirting, and valued at per square.

Enter them in your book according to their thickness, and if yellow or white deal, if common or second-best or clean deal; if laid *folding, straight-joint* or *dowelled*.

The slabs are not generally deducted if they have *mitred borders;* if they have not mitred borders, deduct the opening or slab from the flooring. If the deduction is made when there are borders, the *borders* must be taken at per foot run, which will amount to as much as the deduction made on the floor.

Mouldings, such as architraves, round doors, windows, &c., base, sur-base, &c., &c., are to be measured round the mitres and girt with a fine tape, and entered as moulded architrave, base, &c., as the case may be. But in the abstract they must be all classed under the same head as mouldings.

Single mouldings, as Q^k O G and bead, or Q^k ovolo and bead, &c., may be taken at per foot run, but their girth must be described, as they will be valued accordingly.

Doors, linings, &c., &c. (Plate IV.).—*Doors* are measured and valued at per foot superficial, according to their description. *Door linings, grounds,* &c., at per foot superficial; *Solid doorcases* at per foot cube; as follows:—

MEASURING JOINER'S WORK.

Solid doorcases and doors.

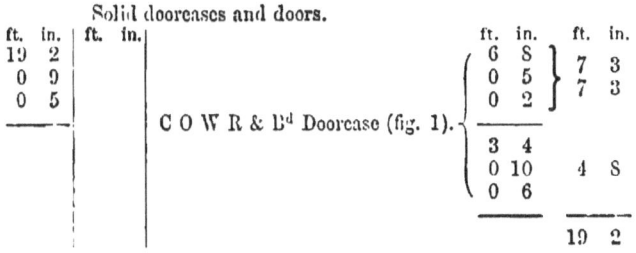

ft.	in.		ft.	in.
19	2			
0	9			
0	5			

C O W R & B^d Doorcase (fig. 1).

ft.	in.		ft.	in.
6	8			
0	5		7	3
0	2		7	3
3	4			
0	10		4	8
0	6			
			19	2

If there is a sill, take it the same as the head, viz., by making an allowance for its passing under and beyond the jambs, as may be; and also allow the additional length of jambs for framing into ditto. If a stone sill, iron shoes should be secured to the bottom of jambs, which must be numbered.

ft.	in.	ft.	in.
6	8½		
3	5		

W^h Deal, 2 S L, R & B^d Door, size including the rebates. (Fig. 1.)

Number the bolts, and enter the hinges per pair.

Framed partitions are measured same as doors.

Spandril framing is measured separately, length by half the height.

ft.	in.	ft.	in.		ft.	in.	
6	8			Whole deal 4 P, Q O G & b and B F	6	8	
3	1			Doors (fig. 2, and A, fig. 3, 4, 5), or as it may be. But the door must be taken first between and including the rebates.	6	8	
					3	1	their thickness.
16	7				0	2	
0	6½			2nd. The linings by calculation . .	16	7	Twice the width of grounds.
					0	9	
17	4			W^h deal, P F B & b lining (as B, fig. 3 and 4).			
0	4½				17	4	One face of architrave for mitres.
				3rd. The grounds, viz.—Inch deal framed grounds. (D, fig. 3, 4, 5.)	0	8	
18	0			4th. Architraves.—Superf. moulded architraves. (C, fig. 3, 4, 5.)	18	0	
0	9						

If mitred and block plinths, number them, but observe to take the architraves short.

Number the locks, hinges, bolts, &c., describing them.

Fig. 3 and 4, the common methods for doors in partitions: No. 4 has the preference. Fig. 5, for doors Q O & b^d b s in walls, consequently wide linings framed in panels to answer them.

Skirtings, either *plain* or *raking,* taken at per foot superficial; if very narrow, by the foot run.

If *raking,* to be taken for the width as per sketch. (Plate V., fig. B.)

If on narrow *grounds,* take the grounds by the foot run.

If *plugged* to the walls, allow extra for plugging.

Moulded plinths.—Measure the square part by the length and width, and enter it—Whole deal, wrought one side, rebated and backed plinth. Girt the moulding, and allow half an inch behind the plinth. (See Dado, Plate III., fig. n.)

When skirtings are *tongued* or *mitred* at the angles, the *number* of the angles is taken separately.—*Housings* into architraves are also numbered.

Scribing is fitting the edge of a board to any surface, as the bottom edge of a skirting to a floor, or the floor boards to the line of wall; it is measured by the foot run.

Pilasters.—Girt and enter them thus:—

ft.	in.	ft.	in.			ft.	in.
7	6			1½ inch deal, glued and blocked pilasters, framed Q^k O G, or ovolo and bead, as may be. (Plate 3, fig. 3.)			
2	0						
2	6					0	6
0	9			Moulded impost . .	}	0	3
						0	9

The plinth may be measured in with the pilaster.

Dado (see Plate III.). Elevation and section, showing base and surbase moulding, plinth, &c., and that the heading-joints should be broken, as they are in a straight-joint floor. By the narrow grounds K, tongues I, and keys G, the dado hangs unconfined, the joints being also secured by slips ploughed and glued into the back, as at H, and dovetailed pieces inserted at regular distances, as at M, the top and bottom of dado not being confined, and the joints thus secured, there will be no danger of the joints opening, even should the deal shrink. The tongues, I, through the grounds, K, should be about three feet asunder, as also the keys, G: these must be about three inches wide at the bottom. The heading-joints should be ploughed and tongued.

B, the common, though bad method of rebating the dado into the grounds.

E, fillet in floor to secure plinth.

F, the best method, by grooving the plinth into floor. The angles of all dados must be grooved.

Measure the height of dado within half an inch of the top of surbase, as it will do for dado and grounds; then take superf. of moulded base and surbase mouldings: girt the surbase from plastering to face of dado, and the base from dado to top of plain plinths; then add half an inch for rebate. Enter the dado according to its description, viz.:—

As inch deal keyed dado.
Do., dovetailed at the back, with grooved rail, or as the case may be.
Do., do., raking.
Do., do., circular on the plan, grooved and backed on the cylinder.
Do., do., wreathed.
Number each external mitre.

Sashes and frames, shutters, and fitting up to windows (see Plate IV.).—Take the dimensions from the beads of sashes on the inside, and allow seven inches additional height for head and sill, and eight inches in width for frames in common sashes; but nine inches for large sashes.

ft. in.	ft. in.		ft. in.
9 1			8 6
4 10		(Fig. 7) D C F, O S sills, W P P, B & P S .	0 7
			9 1
			4 2
			0 8
		2 inch W a & h sashes, D h, B A P, P L, L W .	4 10

French sashes, or casements, hung on hinges, or

sashes hung on centres in solid frames.—Take the sashes separate, and the frames as directed for doorcases. If Venetian frames, describe them as such.

If mouldings up mullion, take them per foot run.

If circular heads, take the sash by itself, and the frames as run of circular frames, as per description; viz., with beads, parting slips, &c., &c., as may be.

Window shutters are taken per foot superficial, allowing for the rebates. If *lifting*, include frames in measurement.

Number the sash fastenings, locking bars, spring latches, hinges, &c., &c.

The framed grounds, rebated and beaded boxings, linings, moulded architraves, &c., are taken per foot superficial, similar to the doors, viz. :—

ft. in.	ft. in.		ft. in.	
2)8 8		(Fig. 8.)	8 6	top and bottom bds.
0 11		1¼ D$^l.$ 4 Pan$^l.$ Q O b & b b shutters, (E) hung in two heights.	0 2	
			8 8	
2)8 8		Do., Back flaps (F).		
0 9½				
2)8 8		Inch deal, do, do., (G).		
0 6½				
		N. 4 pair 2½ butts.		
		8 pair back flap hinges.		
		1 locking bar.		
		2 brass knob spring latches.		
		1 patent sash fastening.		
2)8 10		1¼ deal, 4 panel, b b, back lining (H).	8 8	
0 10½			0 2	
			8 10	
4 10		1¼ deal, Q O & B soffit (fig. 6, I)	4 2	
0 11			0 8	
			4 10	
6 2		1¼ deal, 3 panel, Q O b backs and elbows (fig. 6 and 7, K)	4 4	
2 6			0 11	
			0 11	
			6 2	
4 4		Run of slit deal, beaded capping to back.		
		No. 2. Caps and elbows.		

MEASURING JOINER'S WORK.

ft. in.	ft. in.		ft. in.	
2)11 2		1¼ deal splayed and framed boxings (fig. 8, L).	8 8 / 2 6	
0 5			11 2	
5 2		1¼ deal framed grounds (fig. 6, N).	4 4 / 0 10	
0 5			5 2	
27 5		Moulded architrave, M. 1 pair of mitred and blocked plinths 8½ in. high.	11 2 / 0 6½	width of architrave.
0 9			11 8½ / 0 8¼	D⁰ h⁰ of plinth.
			11 0 / 11 0	
			4 4 / 1 1	width of architrave.
			27 5	

If boxings are executed, as shown at L (fig. 10), they must be taken as splayed, framed, rebated, and beaded boxings, per foot superf., and the mouldings forming the architrave at per foot run.

In *lifting* shutters, the flap to cover same is taken by foot run.

Shop-windows may be measured by the foot superficial, or, if in very large squares, the sash-bars, head and sill may be taken at the foot run.

Staircases (Plate V.) are taken per foot superficial, by girting the riser and tread by the length of the step, allowing extra for the thickness of the skirting, which is entered in the measuring-book according to their thickness and description, viz., inch deal common steps, risers, and carriage.

1¼ inch deal second-best, steps, risers, and carriage, with moulded nosings, close or cut string; or,

1¼ inch deal second-best, S R & C, M nosings, mitred to receive brackets or string boards and return nosings, and dovetailed to receive balusters.

1¼ inch clean deal, do., do.

1¼ inch clean deal, S R & C, to geometrical stairs on a circular plan, the risers mitred to the string board.

METHOD OF MEASURING STEPS, RISERS, AND CARRIAGE.

ft.	in.	ft.	in.		ft.	in.	
3	6			Length of tread.	0	10	
1	5			Supr. 1¼ deal, S R & C to fliers (fig. B & C)	0	7	
					1	5	

If geometrical winders (as plan A), consequently wrought and blocked carriages (as fig. F and G), they must be taken thus, and described as such :—

				Winders with circular ends. (Enter description.)			
7	2				ft.	in.	
3	9				0	7	Project.
27	10			Risers, the lengths collected . . .	0	1	of nosing.
0	8				0	8	

1	2			Dd⁰ opening.
1	6			

9	6			Whole deal framed string.
0	10¾			

4	4			Whole deal apron, 2 sides (fig. D) } Return landing.
0	9			¾ do., ploughed in (fig. E) . .

4	4		
0	4½		

N.B. All winders must be taken as before described.
Fig. F shows a single wrought and blocked carriage for a geometrical winder; G, a set of do. as fixed; the dotted lines show the fronts of steps.
If moulded return nosings, or brackets, either straight or circular, number them.
Iron balusters, do.
Block steps, do.
Veneered curtails, do. (Plan of do., fig. H, showing the manner of veneering it; I, section of wedge.)
Turnings to newels, do.
Pendent drops, do.
Handrails, either straight, ramped, circular, or wreathed, per foot run, the different parts being kept separate.
Planceers, newels, bar balusters, &c., do., but if the balusters are carved, they should be numbered and described.

ROTATION.

In measuring the carpenter's work of a building, it is usual and customary to begin with taking the roof; then the plates, bond timbers, &c., next the quarter partitions; then the naked floors under ditto.

If it is determined to take the timber in the above without labour, then the labour and nails at per square must be measured as such before the cube timber is taken.

In measuring joiner's work, on entering each room, first take the boarded floors, then the dado or skirting, next the battening or bracketing, if any, then the chimney grounds and chimney pieces, next the windows, as sashes and frames, linings, boxings, grounds, architraves, &c., and last the doors, linings, grounds, architraves to ditto, &c., &c.

ABSTRACTING.

In abstracting carpenter's and joiner's work, the greatest possible care must be taken to prevent confusion, for when several thousand dimensions have to be entered under their respective heads, unless a regular rule be observed in drawing out the abstract, and placing every description of work in the situation usually allotted to it, much time would be consumed in referring to the different heads.

Proper attention to the form here given, for abstracting the quantities and bringing the different articles into bill according to their regular rotation, will prevent the student from experiencing this inconvenience.

The abstract for carpenter's and joiner's work should be made on very large paper, and care taken to give sufficient length in each column for all the dimensions that it may be requisite to enter in them. The deals, as shown in the lower range, should be put on the other side, or on another sheet of paper, under their respective thicknesses. The partitions, backs and elbows, soffits, dados, columns, pilasters, stairs, strings, gutters and bearers, &c., &c., should be placed. It is also better, in abstracting the work of a large building, to keep the ironmongery on another paper, as every care should be taken to keep all the articles and entries separate and distinct.

The abstract of the carpenter's work should be made separately from that of the joiner's work, as shown on the following pages.

ABSTRACT OF CARPENTER'S WORK.

PER SQUARE.					PER FT. CUBE.				
LABOUR AND NAILS.		FURRINGS AND BATTENINGS.		BOARDING.	OAK.			FIR.	
Qr. P. Roofs.	Floors	¾ in.	2 in.	Ro. Wea-ther.	Cube No lab'ur	Plank 1 in.	2½ in.	Cube No lab'ur	W & F
		1 in.	2½ in.	W 1 S					
				Sound	Do. lab'ur	1½ in.	3 in.		W F R&Bd
								And lab'ur	
		1½ in.	3 in.	W 2 S					
					Wrot.	2 in.	4 in.		Ppr. Door Cases
					Wrot. and fram'd			Wrot	

ABSTRACT OF JOINER'S WORK.

PER SQUARE.		SUPERS IN DEAL.							SUPER IN MAH'Y OR WAINSCOT.		PER FT. RUN.				NUMBERS.
FLOORING.	SKIRTINGS.	DADO.	FRAMED PARTITION.	WINDOW SHUTTERS.	DOORS.	SASHES.	FRAMES.	SASHES AND FRAMES.	W.C. FITTINGS.	DOORS & FRAMING.	DEAL.	MAH'Y.	MOULDINGS. DEAL. MAH'Y.		DEAL. MAH'Y.
White Inch. Yell'w Inch. Ro. Inch Wrot. Fold'g.	Inch. Sqre.					1¼ in Dl. ovolo.	D C F O S sill W P P B & P S								
			Spandril framing.												
1¼ in. 1½ in. 1¾ in. Wrot. S J	1¼ in. torus.	Plinth	Mouldings.			2 W ovolo.			Mo'ldings.				SUNDRIES. DEAL. MAH'Y.		
	1¼ in. mold.			Staircase.											
SLIT DEAL	¾ IN. DEAL.	INCH DEAL.	1¼ IN. DEAL.	1½ IN. DEAL.	1¾ IN. DEAL.	2 IN. DEAL.	2½ IN. DEAL.	IN. DEAL							
Ro. 1 S 2 S	Ro. 1 S 2 S	Ro. 1 S 2 S	Ro. 1 S 2 S	Ro. 1 S 2 S	Ro. 1 S 2 S	Ro. 1 S 2 S	Ro. 1 S 2 S	Ro. 1 S 2 S							

ABSTRACT OF IRONMONGERY.

SCREWS.		BOLTS.		ESPAGNIO-LETTE BOLTS.	HINGES.		LOCKS.	SASH BARS.	RINGS	SUNDRIES.
Iron.	Brass.	Iron.	Brass.		Iron.	Brass.				
		Barrel	Barrel		Butt.	Butt.	Cupboard.			
					Back flap.		Rim.			
	Bright rod.	flush.				Spring.	Mortice.			
					Cross garnet.		Drawback.			

ROTATION.

To be attended to in bringing the quantities into Bill.

Sqrs. ft. in.

CARPENTER'S BILL.

Labour and nails to roofs, according to description . . . .
Do. do. to floors, naked framed do. .
Do. do. to quarter partitions . .
Inch deal furrings, according to description . . . . .
Do. battenings . . . . . do.
Do. rough boarding . . . do.
Do. wrought do. . . . . do.
Do. weather do. . . . . do.

Ft. in. pa.

Then the cubes, as—
Cube oak, no labour. . . .
Do. bond . . . . . .
Do. wrought, &c., &c. . . .
Cube fir, no labour . . .
Do. bond . . . . . .
Do. wrought and framed, &c., &c. .
Cube fir, wrought, framed, and rebated
Do. proper doorcases, or any other, according to the work thereon . .

JOINER'S BILL.

	Ft.	in.

Superf. of inch oak plank, *then the other thickness of oak plank, with the labour,* &c. . . . . .
Superf. of ½ in. deal rough, labour and nails . . . . . .
Superf. of do. wrought one side . .
Superf. of ¾ in. deal, *and proceed to the thicker deals, with their labour, as the case may be, commencing with the thinnest, and proceeding in regular succession, according to their thickness and the labour thereon* . . . .

Then the framed work, as—

Inch deal square framed partitions .

Next the doors, as—

1¼ in. 4 panel bead flush and square doors . . . . . .

Then the windows, viz. :—

Inch deal bead butt back linings, quirk ogee and bead backs, elbows, and soffits . . . . . .

Shutters—

Bead butt back flaps, quirk ogee and bead shutters, &c.. . . . .

W. C. fittings—

Inch mahogany seat and riser . .
Do. clamped flap and frame . . .

Sashes and frames—

1½ in. deal ovolo sashes . . .
Deal cased frames and sashes, according to their descriptions . . . .
Staircases, per foot super.
Superf. of mouldings . . . .
The work per foot run . . . .
Do. numbered . . . . .
Attending on other Trades . . .

VALUATION OF CARPENTER'S AND JOINER'S WORK.

MEMORANDA.

50 cubic feet of timber equal one load.
100 feet superficial equal one square.
120 deals are called one hundred.

A reduced deal is $1\frac{1}{2}$ inch thick, 11 inches wide, and 12 feet long.

120 12-feet 3-inch deals equal $5\frac{2}{5}$ loads of timber.

400 feet superficial of $1\frac{1}{2}$-inch plank or deals equal one load.

Planks are 11 inches wide; deals, 9 inches; and battens, 7 inches or less.

A square of flooring requires:—

	Number of 12 ft. boards.
Laid rough	$12\frac{1}{4}$
Do. edges shot	$12\frac{1}{2}$
Wrought and laid folding	13
Do. straight joint	$13\frac{1}{2}$
Do. do. and ploughed and tongued	14

	Number of 12 ft. battens.
One square of wrought folding floor requires	17
Do. straight joint	18

WEIGHT OF TIMBER.

39 cubic ft. of oak	equal	1 ton.		51 cubit feet of beach equal			1 ton.
65 „	fir	„	do.	45	„	ash „	do.
66 „	deals	„	do.	34	„	mahogany „	do.
60 „	elm	„	do.				

CALCULATION, showing the method of ascertaining the VALUE of a CUBIC FOOT of FIR or other Timber from the prime cost prices:—

	£	s.	
Fir timber, at per load, say	5	0	0
Carriage (according to distance)	0	5	0
Sawing, on an average	0	10	0
	5	15	0
Waste in converting $\frac{1}{10}$	0	11	6
	6	6	6
20 per cent. profit	1	5	$3\frac{1}{2}$
	7	11	$9\frac{1}{2}$

$$\frac{£\ s.\ d.}{\ 7\ 11\ 9\frac{1}{2}}{50}$$ or $2s.\ 0\frac{1}{4}d.$ per foot cube.

The constants in the following tables are to be multiplied by the rate of wages for a carpenter per hour.

LABOUR AND NAILS* TO ROOFS.

At per square of 100 superficial feet.

	Labour. Hours.	Nails. s. d.
To common shed roofs	5·2	2 3
Do., do., with purlins	5·8	2 3
Common span or gabled roof, with purlins, and principal and secondary rafters, two stories high	11·6	3 6
If three stories, add	·84	
Framed roofs, with collars dovetailed into sides of principal rafters, and these notched to receive purlins, and filled in with common rafters	15·5	3 6
Roofs framed with principals, king-posts, purlins, braces, and common rafters	20·0	4 0
Do., do., with king and queen posts	21·0	4 0
Mansarde or curb roofs on one side	7·7	2 9
If two sides, add	·8	
If three sides, add	1·7	
If above two stories, add	1·0	

LABOUR AND NAILS TO NAKED FLOORS.

At per square of 100 superficial feet.

	Labour. Hours.	Nails. s. d.
Ceiling floors, joists only	5·2	2 3
Do. framed with tie-beams	7·6	2 0
Do. with binding and ceiling joists	9·0	2 3
Ground joists, bedded but not framed	5·2	1 2
Do. pinned down on plates and framed to chimneys	8·4	1 6
Single framed floors trimmed to chimneys and stairs	8·8	1 2
If above 9 in. deep, add	1·7	
Framed with girders and cased bays	17·0	3 6
Framed with girders, binding, bridging, and ceiling joists	25·0	5 6

LABOUR AND NAILS TO QUARTER PARTITIONS.

At per square of 100 superficial feet.

	Labour. Hours.	Nails. s. d.
Common 4 in. partitions	7·0	1 8
Do. 5 in. do.	9·0	1 9
Do. 6 in. do.	9·5	2 3
Truss framed with king-posts	12·0	1 9
Do. with king and queen posts	14·0	2 6
If oak, one-third extra.		

* Labour and nails include glue, attendance on sawyers, and profit.

LABOUR ON FIR TIMBER.
At per foot cube.

	Hours.
Cube fir bond	·50
Do. framed	·75
Do. truss framed	1·00 to 1·6
Do. framed and chamfered	1·50
Do. wrought and framed	1·75
Do. do. and rebated	2·25
Do. W, F, R, and beaded	2·50
Do. W, F, R, and D beaded	3·00
Do. proper doorcases	3·50
Planing fir, per foot superf.	0·10

Bond timbers, wall plates, wood bricks, pole and curb, &c., are all to be under the head of *bond*.

CALCULATION, showing the method of finding the VALUE of DEALS or BATTENS from the prime cost prices.

	£	s.	d.
Prime cost per hundred of 12 ft. 3 in. deals, say	35	0	0
Carriage, according to distance	0	10	0
	35	10	0
20 per cent. profit	7	2	0
	£42	12	0

£ s. d.
42 12 0
――――― or 7s. 1d. to be allowed in day-bills for each
120

	£	s.	d.
3 in. deal	0	7	1
In measured work, allow for waste $\frac{1}{10}$	0	0	8½
	0	7	9½

In calculating the value of deals in thicknesses, add the value of the sawing, according to the number of cuts.

Every rise and fall of 9*l.* per hundred, will increase or diminish the price of deals as nearly as possible, per foot superficial, 1*d.* per inch in thickness. This rule will be found sufficiently correct for practice where the quantities are not large: where they are, the exact calculation should be made.

LABOUR ON DEALS, AT PER FOOT SUPERFICIAL.

In order to facilitate the fixing of proper prices for the labour on deals, at per foot superficial, the different descriptions of work which have always been considered of equal value are classed together, by which the system adopted for valuing the various sorts of labour on deals will be rendered more simple and easy. Over the column in which is inserted each kind of work of equal value, is placed the decimal which, multiplied by the rate per hour allowed for a carpenter or joiner at the time and place where the work is performed, will show the fair and equitable price to be allowed.

	No. 1.	No. 2.	No. 3.	No. 4.
For deals from ½ to 1½ in. thick	·05	·2	·3	·4
For deals from 2 to 3 in. thick	·08	·35	·45	·5
	Edges shot Plugged Jacked Rounded	Labour and nails Planing on each side Grooved Rebated Ploughed and tongued Framed Battened Mitred Scribed Backed Throated Clamped Beaded	Cut circular	Cut standards Sunk shelves Scolloped Ledged Dovetailed

BATTENING, PER SQUARE.

	Labour. Hours.	Nails. s. d.
¾ in. to 1¼ in. placed 12 in. from centre to centre	3·0	1　0
If plugged to walls, add	2·0	0　6
Extra for wall hooks.		

WEATHER BOARDING, PER SQUARE.

Rough	5·2	2　9
Do. splayed edges	6·8	2　9
Wrought	9·0	2　6
Do. and beaded	11·0	3　0

ROUGH BOARDING, PER SQUARE.

	Labour. Hours.	Nails. s. d.
¾ in. deal, rough	4·8	2 4
Do. edges shot	6·0	2 0
Do. ploughed and tongued	8·0	3 0
Inch deal, rough	5·0	2 6
Do. edges shot	6·5	2 0
Do. ploughed and tongued	8·5	3 0
Whole deal, rough	5·2	2 9
Do. edges shot	7·0	3 0
Do. ploughed and tongued	10·4	3 0
1½ in. deal, rough	5·4	3 0
Do. edges shot	7·5	3 0
Do. ploughed and tongued	11·0	3 0

DEAL FLOORS, PER SQUARE.

Inch, rough edges shot	7·0	1 9
Do. wrought folding	8·0	1 9
Do. do. straight joint	10·0	1 9
Whole deal, rough edges shot	7·5	2 6
Do. wrought folding	8·5	2 9
Do. do. straight joint, splayed headings	11·0	2 9
Do. do. dowelled	24·0	6 0
1½ in. deal, rough edges shot	8·0	3 0
Do. wrought folding	9·2	3 0
Do. do. straight joint, splayed heading	12·0	3 0
If ploughed and tongued headings, add	1·5	
If ploughed and tongued edges, add	2·5	
For tongues to edges of boards, add	4·0	

BATTEN FLOORS, PER SQUARE.

Inch, wrought folding	8·5	2 3
Do. straight joint, splayed headings	10·5	2 3
1¼ in. wrought folding	9·0	2 8
Do. straight joint, splayed headings	11·3	2 9
Do. dowelled	27·0	4 6
If ploughed and tongued headings, add	4·3	
If ploughed and tongued edges, add	6·5	
For tongues to edges of boards, add	6·0	
If battens less than 5 in., add	3·3	

FRAMED GROUNDS, PER FOOT SUPERFICIAL.

	Labour and Nails.
Common framed grounds	hours ·50
1 in. do. ploughed for plastering	·58
1¼ in. do. do. do.	·60
1½ in. do. do. do.	·65

SKIRTINGS, PER FOOT SUPERFICIAL.

	Labour and Nails.
Plain skirting	hours ·33
Do. raking cut to steps	·52
Torus skirting	·45
Do. raking cut to steps	·66

GUTTERS AND BEARERS, PER FOOT SUPERFICIAL.

Inch or whole deal	·50

DOOR LININGS, PER FOOT SUPERFICIAL.

Plain single rebated	·56
Do. and beaded	·63
Do. double rebated	·70
Do. do. and double beaded	·77
Square framed jambs, each in 2 panels and soffit in 1 panel	1·05
If bead butt, or moulded, add	·13
Bead flush, or quirk moulded	·27
Raised panel and moulded	·42
For every extra panel if square	·21
Do. flush or moulded	·27
If double rebated	·21
If double beaded	·13

LEDGED DOORS, PER FOOT SUPERFICIAL.

1¼ in. rough edges shot	·40
Add,	
If ploughed and tongued	·13
If ploughed and beaded	·21
If wrought each side	·13
If braced	·27
If hung folding	·21
If 1 in. thick	·13

FRAMED PARTITIONS, PER FOOT SUPERFICIAL.

1½ in. square framed	·55
2 in. do.	·60
Add,	
If B B or moulded	·27
If B F or quirk moulded	·42

DEAL MOULDINGS, FIXED COMPLETE.

Common mouldings	1·25
Add, if quirked	0·28

The materials for mouldings in deal will be found as

near as possible of the same value as the labour. Small mouldings may be measured at per foot run, and valued according to the girt and form.

DOORS HUNG COMPLETE, PER FOOT SUPERFICIAL.

	Labour and Nails.
Two-panel square framed	hours ·68
Add, for every additional two panels:	
If framed square,	
For 1½ in. deal	·15
2 in. do.	·17
2½ in. do.	·27
Add, for every two panels:	
If framed B B and square—For 1½ in. deal	·21
2 in. do.	·24
2½ in. do.	·27
If framed B F and square—For 1½ in. do.	·27
2 in. do.	·37
2½ in. do.	·42
If framed Qk. O G and Bd. and square, or Q Ov. and Bd. and square—For 1½ in. deal	·21
2 in. do.	·24
2½ in. do.	·27
If double margins 4½ in. wide	·21
Do. 5¼ or 6 in. wide	·42
Hung folding	·13

WINDOW LININGS, PER FOOT SUPERFICIAL.

Inch deal two-panel square framed back linings	·80
If B B, or moulded, add	·13
B F, or quirk moulded, add	·21
For each panel above two, if square	·21
Do. do. if moulded	·27
If splayed linings, add	·07

WINDOW BACKS, ELBOWS, AND SOFFITS, PER FOOT SUPERFICIAL.

Inch deal, plain keyed or two-panel square backs	·75
Do. two-panel square backs, elbows and soffits	·95
Add for each panel above three,	
If splayed	·10
If bead butt, or moulded	·13
B F, or quirk moulded	·21

BOXING TO WINDOWS, PER FOOT SUPERFICIAL.

Framed, rebated, and beaded boxings	1·01
Splayed F R and beaded boxings	1·20

INSIDE WINDOW SHUTTERS, PER FOOT SUPERFICIAL.	Labour and Nails. hours
¾ in. deal clamped flaps in one height	1·20
Inch do. two-panel square in one height	1·25
For every panel above two add,	
If framed square	0·22
If B B, or moulded	0·22
B F, or Q^k· moulded	0·26
Q O G & b, or Q O & b & square	0·26
For every extra height, add	0·13

SASHES AND FRAMES HUNG COMPLETE, PER FOOT SUPERFICIAL.	
Sashes—	
1½ in. deal ovolo sashes	0·49
Do. wainscot or mahogany	1·05
If 2 in. or 2½ in. sashes deal, add	·15
If do. wainscot or mahogany, add	·15
If astragal and hollow in deal, add	·10
If do. in wainscot or mahogany, add	·10
Frames—	
Deal cased frames O S sills, D P P B & P S, S hung	·45
If prepared for 2 or 2½ sashes, add	·15
If prepared with wainscot or mahogany P P B^{ds}· & P slips, add	·40
If for 2 or 2½ in. sashes, add	·15
If double hung, add	·18

To find the value of sashes and frames, add to the above for labour and nails only, the amount of materials expended.

STAIRCASES, PER FOOT SUPERFICIAL.	Labour and Nails. hours
Common steps and risers and two fir carriages	·80
Do. moulded nosings and close strings	·98
Do., do. mitred to cut string boards (to imitate stone steps) and dovetailed to balusters	1·20
Add, if winders circular at one end	·30
Do. circular at two ends	·55
Do. geometrical, with wrought and blocked carriages	·56
Riser tongued to step bottom edge	·10
Do., do. both edges	·20
Feather tongued joints	·15
Add for each,	
Quarter curtail glued upright	3·00
Do. blocked and veneered	6·30
Proper curtail step and riser	18·00
Returned moulded nosing (to imitate stonework)	·50
Do. circular	·75
Plain cut bracket	·90
Do. circular	1·55

	Labour and Nails.
Housing to step and riser	hours ·40
Do. to winders	·80
Do. to moulded nosings	·50
Do. to do. circular ends	1·00

OUTSIDE STRINGS TO STAIRS, PER FOOT SUPERFICIAL.

Whole deal, plain	·55
Do. sunk	·80
Do. sunk and moulded	·85
Do., do. cut (*i.e.*, to imitate stone steps)	·95
Do., do. mitred to risers	1·00

If ramped (but in one plane), once and a half the above.
If wreathed, four times.

WALL STRINGS, PER FOOT SUPERFICIAL.

Plain and plugging	·70
If moulded, add	·21
If rebated for plastering, add	·28

DADOS, PER FOOT SUPERFICIAL.

Proper dado, with dovetailed keys, joints secured with slips, and dovetails hung to grounds by keys grooved into do. and dado	·70
Add,	
If raking scribed to steps	·10
Do. to moulding nosings	·16
If base grooved into floor	·09
For each external mitre beyond two in the room	2·28

If circular on the plan,—double the above.
If wreathed do.,—treble do.

IMITATIONS OF COLUMNS AND PILASTERS, PER FOOT SUPERFICIAL.

1¼ in. deal plain pilasters, properly glued and blocked	·90
Do., do. diminished	1·00
1¼ in. deal diminished columns, properly glued and blocked, under 14 inches diameter	2·20
Do., do. above do.	2·00
Add for,	
Arris or deep fluting to pilasters, one inch wide	·21
Do. two inches wide	·28
Do. three inches wide	·42
Arris or deep fluting to columns, one inch wide	·27
Do. two inches wide	·42
Do. three inches wide	·56
Straight grooves to columns	·21
Headings to flutes to do.	·70
Straight grooves to pilasters	·13
Headings to flutes to do.	·42

SAWYER'S WORK.

The charges for sawyer's work are often very inconsistent, and differ widely in various parts of the country. They also vary according as the work is done by hand or by machinery, which latter mode is most commonly used, there being less waste in machine-sawing than in hand-sawing.

The proper mode of valuing the labour on sawing fir or any other kind of timber is by the square of 100 superficial feet, the price depending on the usual rate of wages and the hardness of the timber.

Sawing to old timber is usually charged double, on account of the extra labour occasioned by nails, &c.

Small scantlings may be charged by the foot run.

Planks, deals, battens, and flat cuts, according to their length, at per dozen cuts.

And all other descriptions of sawyer's work may be valued in a similar manner, according to the circumstances of the case.

CHAPTER V.

MASONRY.

TECHNICAL TERMS AND EXPLANATIONS.

The term Masonry is generally applied to the art of building with stone laid in mortar or cement; it is also used in a restricted sense as applying only to the parts of a building in which worked or *hewn* stone is employed, as in the formation of arches, piers, and walls, built with solid blocks of stone.

Building Stones, or stones used for ordinary building purposes, may be divided into five classes; namely, Granites, Sandstones, Limestones, Oolites, and Magnesian Limestones.

Granite is one of the hardest, strongest, and most durable of the building stones, having a crystalline character, but varying greatly both in durability and weight, according to its composition. Its chief components are quartz, felspar, and mica; and the durability very much depends on the proportions in which these minerals are combined; those granites which contain a large proportion of felspar being rapidly decomposed by the action of the weather, since that mineral contains a considerable proportion of clay. The red colour which some granites possess is owing to the presence of the oxide of iron.

The granites used in this country are principally obtained from Guernsey, Cornwall, and Scotland; the most durable and useful for building purposes being found in the neighbourhood of Aberdeen, which weighs about 166 lbs. per cubic foot, and has a resistance to crushing of 10,000 lbs. per square inch. The use of granite on a large scale, is confined to the building of walls for river embankments, piers, and arches of bridges, and for other engineering purposes, where great power of resistance to weather, or to the action of water, is required. For ordinary buildings, it is only employed where no other material is obtainable, on account of the great difficulty in working it. As, however, it is capable of taking a beautiful polish, which is not injured by exposure to weather, it is occasionally used in the ornamental features of first-class structures.

Cornwall supplies a large quantity of very excellent granite, that from the Cheesewring quarries near Liskeard being much used for engineering purposes. It is of a light greyish colour, and weighs 166 lbs. per cubic foot. Granites of good quality are also obtained from Devonshire, as that from Hay Tor near Teignmouth, which is fine-grained, hard, and durable. Its colour is a bluish grey, and its weight 165 lbs. per cubic foot. Lundy island, off the north-west coast of Devon, also supplies a good grey granite. An excellent granite for paving purposes is obtained from Caernarvonshire, weighing 160 lbs. per cubic foot. By far the largest quantity and best quality of granites are obtained from Scotland, especially from Aberdeenshire. These are of various tints—grey, blue, red, and pink. Granites of the latter tints are found in the island of Mull on the west coast, and resemble the Aberdeen in quality. On the south-west coast of Scotland good grey granite is obtained from Wigtown Bay, and also from Dalbeattie, near Dumfries. Ireland supplies a considerable quantity, that from the neighbourhood of Wicklow and Dublin being a speckled grey; and from Galway a reddish granite is obtained. Carlow also supplies several varieties of granite. The Dublin stone weighs 169 lbs. per cubic foot.

For curbs the granite obtained from the islands of Guernsey and Jersey, and also from Aberdeen and Devon is generally used; and for paving in blocks or *pitching* to roads subjected to heavy traffic, the Aberdeen granite is found to be most durable, being used in blocks from 12 to 18 inches long, 3 inches thick, and 9 inches deep. For macadamised roads the Guernsey granite answers best.

The following terms are used in Aberdeenshire for worked granite: *Hammer-blocks* are used only for basement stories, foundations and underground work. *Scappled-blocks* are squared with a heavy pick, and employed for engineering purposes. *Rough-picked* is finished better than the last, but used for similar works. *Close-picked* has the beds and arrises worked fair, the face being finely worked by the pick: this is used for ashlar facing. *Single-axed* is finished finer than the last, and used for dressings and moulded works to buildings. *Fine-axing* is the highest finish that can be given to granite by means of the axe. The working of granite can only be properly done by men especially trained to the use of that material.

SANDSTONES are minerals composed almost entirely of silica or sand, with a small proportion of the carbonate of lime as a cement for the siliceous particles. They vary greatly in hardness and durability, as well as in heaviness; some, as the Reigate stone, weighing only 103 lbs. per cubic foot, whilst others found in Scotland, Yorkshire, and the west of England, weigh more than 160 lbs. to the cubic foot. Sandstones are generally more or less *laminated*, that is, the particles appear to have been deposited in layers. These laminations are always parallel to the natural *bed* of the stone, which can be easily detected in most stones of this class; and, as a general rule, sandstones should always be laid in a building upon their natural bed, otherwise they are liable to disintegration. The best stones for building purposes are those which are called *grits*, and are found in Yorkshire, Lancashire, the south of Scotland, Northumberland, and Derbyshire. These are heavy stones, weighing from 140 to 160 lbs. to the cubic foot, and having a resistance to crushing of from 4000 to 8000 lbs. per square inch. The best known of the *grits* in this country are those obtained from the neighbourhoods of Edinburgh, Leeds, Huddersfield, Halifax, Harrogate, Heddon, and Kenton. *Flagstones*, used for paving, sills of windows, steps, &c., are grits of a highly laminated kind, being easily split into thin slices or flags.

A good building sandstone is obtained from Mansfield in Nottinghamshire, of which there are two qualities, known as the *red* and the *white* Mansfield. It contains a considerable proportion of magnesia, and consequently is not so durable as most of the grits or other sandstones in which that material is absent. The white stone is considered to be a rather better weather stone than the red. Mansfield stone weighs 146 lbs. per cubic foot, and has a resistance to crushing of 5000 lbs. per square inch.

Red sandstones are very much in demand for decorative purposes, both external and internal; among the best of these is the red Mansfield, and the stone from Dumfries and Corsehill in the south-west of Scotland.

LIMESTONES are largely employed for masonry, but vary greatly in quality, some being very soft and others too hard to be worked. One of the best limestones found in England is the Chilmark stone, which is very durable, and weighs more than 150 lbs. per cubic foot. It has a resistance to crushing of 6000 lbs. per square inch. That obtained from the Isle of Purbeck is also a very useful material for steps, landings, and paving; it weighs 151 lbs. to the cubic foot. The *Rags*, as they are termed, are limestones of a very hard and unworkable character, fit only for common rubble walling; the Kentish-rag is a well-known material belonging to this class. Limestones which are not suitable for masonry are generally valuable for burning into lime, which forms a principal ingredient in mortar.

OOLITES, or *roe-stones*, are limestones of a peculiar structure, being composed of small particles in appearance like the *roe* of a fish. The *Portland* oolite is highly prized as a building stone, being strong and durable, when obtained tolerably free from fossil remains. That from the Waycroft quarry weighs 136 lbs. to the cubic foot; and its resistance to crushing is about 4000 lbs. per square inch. The *Ancaster* stone is an oolite which also possesses valuable qualities, although inferior in strength and durability to the Portland. Its weight is 140 lbs. per cubic foot, and its resistance to crushing about 2300 lbs. per square inch. The *Bath* stone is of a very soft character, although that which is obtained from the *Box* quarries becomes harder by exposure to the air, and is a serviceable material for external work in localities which are tolerably free from coal smoke. It weighs 123 lbs. per cubic foot, and its resistance to crushing is about 1500 lbs. per square inch. The stone obtained from the other quarries in the neighbourhood of Bath is only fit for internal use.

The stone obtained from the Doulting quarries near Shepton Mallet, Somerset, is a good building oolite, being harder, heavier, and more durable than that from the Bath quarries; it contains a small proportion of silica, which mineral is entirely absent from the Bath stone; and weighs 134 lbs. to the cubic foot, its crushing strength being 2 tons per square inch. A somewhat similar stone is obtained at Painswick in Gloucestershire, and is much used for steps on account of its hardness. The stone brought from *Caen* in Normandy is largely used in this country for internal masonry, but is too soft to be safely employed for external work; its weight is 125 lbs. per cubic foot, and its resistance to crushing 2000 lbs. per square inch. *Aubigny* stone, imported from the neighbourhood of Falaise in Normandy, is harder and more durable than that from Caen; its weight is 150 lbs. per cubic foot.

MAGNESIAN-LIMESTONES, which are composed chiefly of the carbonates of lime and magnesia, are not found in such large quantities as the other classes of stone, and are seldom very durable when used externally, especially when exposed to a smoky atmosphere. The hardest and most durable is that from Bolsover, in Derbyshire, the weight of which is 152 lbs. per cubic foot, and its resistance to crushing 8000 lbs. per square inch. The stone from *Anston* in Yorkshire is another of this class, and weighs 144 lbs. per cubic foot. Magnesian-limestones are obtained from other places in Yorkshire, as Roche-abbey, near Bawtry, the stone from which weighs 139 lbs. per cubic foot; also Parknook, near Doncaster, the stone from which weighs 137 lbs. per cubic foot.

PRESERVATION OF STONE.—Since it is found that some

stones are very liable to disintegrate when exposed to the action of weather, or to the acids found in a smoky atmosphere, several methods of checking such destroying influences have been tried, and one of the most effectual of these appears to be that of giving the stone a coating of a material called the *silicate of lime*. This is done by first washing the surface with a solution of silicate of soda, or flints boiled, until they dissolve, in caustic soda. The surface is afterwards played upon with a solution of the chloride of calcium, or lime dissolved in muriatic acid. By the chemical change which immediately ensues, the silicate of soda becomes silicate of lime, and the chloride of calcium is changed into common salt, which is easily washed off with water, leaving the face of the stone coated with silicate of lime, which is insoluble in water and is unaffected by weak acids.

Another method is by brushing over the surface of the stone with a solution or varnish made of the following materials : 85·5 per cent. by weight of benzoline or other similar spirit, 10 per cent of gum dammar, 2 of wax, 2 of sugar of lead, and ·5 per cent of corrosive sublimate. The surface must be thoroughly cleansed before the varnish is applied.

FOUNDATIONS of stone walls are formed by laying large flat stones or rough landings extending some considerable distance on each side of the basement wall, and of sufficient thickness to bear, without breaking, the superincumbent weight; these stones are called *footings* or *bottoms*, and are measured and valued separately from the walling, being taken by the superficial foot or yard, describing their thickness.

RUBBLE-WALLS are those that are built of unhewn stone, either with or without mortar. When no mortar is used, as in fences, it is called *dry-walling*. *Coursed-rubble* walling has the stones gauged and dressed with a hammer, and laid in parallel courses varying in thickness. *Uncoursed-rubble* walling has the stones laid without attention to placing them in courses. This kind of walling is measured either by the superficial yard, describing the thickness, and whether *hammer-dressed* or otherwise finished or *dressed* on the face; or else the walling is taken by the cubic yard, and the work to the face by the superficial yard. Rubble-walling is seldom built less than 18 in. in thickness, and consists of an inner and outer casing, connected together at intervals by *bond* or *through-stones* which are blocks of stone laid the full thickness of the wall, and tilted slightly towards the outer end to prevent the weather from driving through; the middle space between the two casings is filled in with broken stones, and grouted with liquid mortar. In situations exposed to driving rains, it is advisable to omit the grouting, to prevent the wet from passing through to the inside.

FLINT-WALLING is executed with the *flints* found in large quantities in the chalk formation, which are often cut and dressed so as to form very ornamental features. Walling with flint always requires the mortar to be used *hot*, otherwise it will not adhere to the stone.

BATTER is the term applied to walls when the face slopes back from the base towards the top. An extra price is allowed for walls that batter, or the walling may be taken in the ordinary way, and the battered surface measured separately.

ASHLAR-FACING is stone carefully dressed and laid in courses of equal width to form the outer casing of a wall; it is generally backed up with rubble-walling or brick-work, and the whole tied together by *through-stones* or *bond-stones;* where the backing is of brick, the connection is sometimes made by iron *ties* or *cramps* let into the edges of the ashlar and carried through the wall. In this kind of wall the ashlar is measured separately from the backing; the ashlar is taken by the cubic foot, and the labour on the face and beds by the superficial foot, measuring one plain bed to each stone, and the sawing, if any, to the back. The rubble-backing is measured by the cubic yard; bond-stones by the cubic foot.

Walls built of solid hewn-stone are measured by the cubic foot, and the labour of working the faces and beds taken as above described for ashlar.

QUOIN-STONES are the blocks of hewn-stone placed at

the outer corners of walls which are built with brickwork or rubble masonry; they are measured in the same way as ashlar, but no deduction is made for them in measuring the rubble-walling.

SAWING is measured by the foot superficial to all concealed parts of hewn-stone which are not left rough.

MOULDED-WORK, as in cornices, string-courses, &c., is girt and taken by the foot superficial. Moulding to arches or other circular work is taken in the same way, and described as *circular-moulded*.

HEWN-STONE employed in a building is measured by the cubic foot as in the rough before being worked.

JOGGLED-JOINT is a method of fastening the edge of one stone into another by a projecting piece in one edge entering a groove in the other, and answers to ploughing and tongueing in Joinery: this method is sometimes used in forming balconies or landings of thin slabs of stone, which are made in several pieces. Joggled-jointing is measured by the foot run, or it may be girt and taken as sunk work per foot superficial. Joggles are considered by some masons to weaken the stone, as a very moderate strain will break them off.

THROATING is cutting a groove on the underside of a projecting sill, to prevent the rain-water from returning and running down the wall; it is taken at per foot run.

SPLAY is where one surface of a stone makes an oblique angle with an adjacent side, and the splayed surface is said to be *sunk* from the original square of the stone; it is taken at per foot superficial, as sunk work.

CHAMFER is a small splay which just takes off the edge or *arris* of a stone, and is measured by the foot run.

ARCHES are combinations of wedge-shaped stones, or *voussoirs*, built together for the purpose of carrying walls over openings whose width is too great to be spanned by single stones or lintels. The voussoirs of an arch are worked so as to fit accurately, and are laid upon a centering of timber until the mortar is set and the superstructure erected. When the centering is *struck* or taken down, the voussoirs are kept in their places by their mutual pressure, by the weight of the wall above, and by the resistance of the piers or *abutments* on which they stand. The inner line of the face of an arch is called the *intrados*, and the outer line of the voussoirs the *extrados*. The *crown* or topmost stone of an arch is called the *keystone*, the base is called the *springing*, and the parts between the bases and the crown are called the *haunches*, which are the weakest parts of all arches. In every arch there is, from its very nature, a certain amount of *horizontal-thrust* tending to push over the abutments, which must be sufficiently strong and heavy to resist it.

In a semicircular-arch having all the voussoirs of equal depth, the weakest joint is that which makes an angle of 30° with the horizon, or at one-third of the distance from the springing to the crown: it is at this point that the horizontal thrust of the arch has its greatest effect on the abutments; it increases with the size of the arch, being generally proportional to the square of the span. If the height of the abutment from the ground to the springing of the arch is equal to the radius or half-span of the arch, its thickness ought not to be less than one-fourth of the span; for a greater height of abutment a greater thickness will be necessary. (For investigations on the thrust of arches, see Tarn's "Science of Building.")

In *segmental-arches*, or those which have less than half a circle for their intrados, the springing-joint is called a *skew-back*.

GOTHIC-ARCHES are those which are formed of two segments of a circle meeting in a point at the crown. An *equilateral* Gothic-arch is one in which the *chords* of the two segments are each equal to the span, the radius at the springing being horizontal: in this arch the weakest joint is that which makes an angle of 16° with the horizon, and the horizontal-thrust is greatest at this point. If the abutment equals in height the span of the arch, its thickness must not be less than three-tenths of the span.

GOTHIC-VAULTING is formed by main *ribs* or arches, thrown across from wall to wall of a building, with *diagonal-ribs* springing from the same point and intersecting each other in the middle. The space between the ribs is filled in with light masonry, also arched and springing from the ribs.

Arch stones are measured by the cubic foot according to their original dimensions before being worked; one bed is taken as *sunk* work by the foot superficial; the two ends as circular work, if plain; or circular moulded, if any mouldings are worked upon them.

HOISTING stone above 30 ft. of height, is charged extra, according to the number of cubic feet in the stone, and the height to which it is raised.

Stones above 5 ft. long have their cubical contents measured and described as *scantling lengths*, their value being greater per foot than that of smaller stones.

FLAGS are thin paving stones from 2 in. to 3 in. thick, and not exceeding a square yard in area; this kind of stone is generally termed "York," being mostly obtained from the county of Yorkshire. Paving is measured by the superficial yard, and described as *self-faced flagging, quarry-worked* or *boasted, tooled* or *rubbed*. If the edges are worked as well, they are measured by the foot run extra.

LANDINGS are flags of larger size than 1 superficial yard, and above 3 inches in thickness. They are measured by the superficial foot, and described as tooled or rubbed. The edges are measured by the foot run, and described as plain or joggled. The sizes of the stones must be described, as the larger landings are more valuable in proportion than the smaller.

CUTTING AND PINNING is letting the ends of stone steps or landings into a wall already built; and is measured by the foot run.

COPING is the name given to the stones laid on the top of a wall to protect it from weather, or for the purpose of receiving an iron balustrade. When the coping is of equal thickness all over, it is termed parallel coping; when thinner at one edge than the other, it is *feather-edged* or *weathered*; when thinner at the edges than in the middle, it is *saddle-backed*. If the coping projects beyond the face of the wall, the under edge is usually *throated*.

Coping is measured by the foot run, and its form described; unless of extra large size, when it must be measured as other hewn-stone.

Cutting holes for rails in coping are numbered and charged at so much each.

SILLS are the stones laid outside the bottom of windows. When not more than 3 inches thick, they are taken at per foot run, and described as plain, sunk or weathered, throated, &c. Thicker sills are taken as hewn-stone.

SINKS are measured by the superficial foot, their thickness and depth of sinking being described, holes and rounded corners being numbered.

CRAMPS are pieces of iron or copper turned down at each end, and used to hold two stones together; they are generally run with lead.

DOWELS are straight pieces of stone or metal let into mortice-holes cut in two abutting stones, for the purpose of preventing them from moving laterally; they are either fixed in cement or run with lead. Cramps, dowels, &c., are numbered, as well as the holes made to receive them.

The LEWIS is a contrivance for temporarily attaching large blocks of stone to the chain by which they are to be hoisted. It consists of two pieces of iron or steel cut in dovetail form, which are let into a hole of similar shape cut in the middle of the stone, and forced into their places by a straight piece of metal fitting in between them. The three pieces are connected at top by a ring attached to the hoisting chain, and are drawn tight by the strain put upon the chain. Very soft and friable stones of large size cannot generally be hoisted by the lewis with safety, as the edges of the hole in the stone are liable to break away.

BANKER is the name given to the block of stone or bench upon which the mason works hewn-stone.

BEAM-FILLING is a term used in some stone districts for the filling with rubble-walling the interstices on the top of a wall between the feet of the rafters, up to the underside of the slates.

ON MEASURING STONE-MASON'S WORK.

There is a variety of opinions respecting the manner of measuring stone-mason's work, both in taking the dimensions for the stone, and also for the labour. It certainly requires more practical knowledge of the operative or working part of the business than any other trade, to determine correctly between these conflicting opinions. The following rules may be considered sufficiently explanatory of the principle on which the practice is governed or founded.

In measuring cube Portland or other stone, all stones that are worked square should be taken accurately as they come from the saw to the banker, of course including the parts laid on or pinned into the walls. But as bevelled or irregularly formed stones cannot be converted without more waste than square ones, the dimensions should be taken so as to make a fair allowance for such additional waste, particularly as the solid contents of all the different descriptions of Portland stone, whatever shape the stones may be worked to, are abstracted under the same head (viz., cube Portland), and therefore should be of the same value; but which cannot be the case unless the extra waste in the bevelled stone, &c., be allowed for in taking the dimensions. When this is done, it is only requisite in estimating the prime cost, to calculate for the waste as if all the stones in the building were cut and worked square. If this method were not adopted, it would be requisite, in ascertaining its real value, to make as many different heads in the abstract for cube Portland, as there are different shaped or bevelled stones, accurately describing each, when the calculations for waste, and of course the price, must vary according to each particular form, the trouble of which would be endless, and without any advantage. Indeed, it would come to the same thing, viz., making the necessary allowances for waste, according to the form of the stone. Bevelled or arch stones should be taken about one-sixth above the mean dimension, to allow for waste.

In measuring the cubic contents of *spandril steps*, some

difference of opinion exists as to the best method of taking the requisite dimensions. The following three methods are in common use, viz. :—

1st. Take the length of the step by its extreme width, and by the whole height of the riser measured from tread to tread.

2nd. Take the length of the step by the extreme width from the nosing of the tread to the acute angle, and by half the height of the riser taken from the top of the tread to the acute angle downwards.

3rd. Take the length of the step by its extreme width, and by three-fifths of the depth of the riser taken from the top of the tread to the acute angle downwards.

To illustrate these different methods, a diagram is given, Plate 6, fig. 3, showing the method of sawing two spandril steps out of the same block, by which it will be seen that, allowing half an inch only in each step for waste in sawing and taking them out of winding, the original block must not be less than twelve inches deep; and supposing the extreme length of the step, including the part pinned into the wall, to be five feet, the size of the block will be

```
       5  0
       1  3
       1  0
      ——— 6  3 ;
```

and each step will therefore contain

```
       5  0
       1  3
       0  6
      ——— 3  1  6
```

By method 1st, we have

```
       5  0         length of step.
       1  3         extreme width.
       0  6         whole height of riser.
      ——— 3  1  6   which is correct.
```

It should, however, be observed, that if the steps, instead of having moulded nosings, were worked plain, the block would only require to be 11 inches deep, as shown by the dotted lines, or one-twelfth less than for moulded steps; whilst the rule gives the same content as before, and consequently it should only be applied for the latter description of step.

By method 2nd, we have
5 0 0   length of step.
1 6 0   extreme width of do. to acute angle.
4 3   { half height of riser from top of tread to
            { acute angle downwards.
——— 2 7 10 which is about one-twelfth less than the real content.

By method 3rd, we have
5 0 0   length of step.
1 3 0   extreme width of do.
5 8   { three fifths of height of riser from top of
            { tread to acute angle downwards.
——— 2 11 5 which is nearly correct.

A better way than either of the above methods is to take the length of the step by a dimension found as follows, allowing half an inch on each step for waste, thus:—

        1 3 width of step.
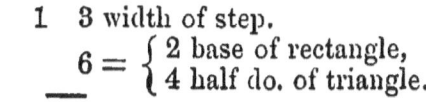

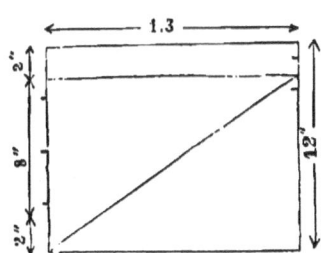

This gives
    5 0
    1 3
      6
  ——— 3 1 6 as before.

In measuring winders, the content may be found in the same way, taking the extreme length of the step by the mean sectional area, making due allowance for waste.

The labour on the under side to be taken as circular sunk work.

In parts of the country where stone is obtained near at hand, and is therefore comparatively cheap, it is usual to work each spandril step out of the solid rectangular block, the whole of the under portion being cut or hacked off with the chisel, so as to be entirely wasted. The stone must in this case be first measured as cube stone, the same as in solid rectangular or square steps, and the labour to the soffit taken as sunk work.

All stone exceeding three inches thick should be taken as cube measure, with the labour, &c., on ditto.

All stones three inches thick, and under, should be taken as slab, at per foot superficial.

The usual custom has been to measure-in such edges as are worked, and show fair. Objections have been made to this practice. and with some degree of justice; but it will make very little difference, if the edges of thin slabs are measured separately, and a fair price allowed for the labour; and for cutting into narrow pieces for mantles, jambs, &c., it would be nearly equal to the value of the stone; but in thick slabs the same argument will not hold good; and, therefore, as the object in measuring work should be to ascertain its real value, and allow only a fair remunerating price, it appears more correct to measure the labour on the edges at per foot run, offering a fair price, according to their thickness, instead of entering it as stone. An extra price should be allowed for very large scantlings, also for hoisting stones on exceedingly high buildings, according to circumstances.

LABOUR ON PORTLAND OR OTHER STONE.

In measuring the labour of working Portland stone, the principal difference of opinion arises in determining what faces or beds should be taken as plain work. Examples are given, showing the methods of taking the labour on different kinds of common work : but in the measurement of superior work, a plain face must be taken previous to measuring the sunk, moulded, or other work, when the mould could not be applied without first making that plain face. There cannot be much difference of opinion in taking the other labour, such as sunk-work, moulded-work, circular-sunk, or circular-moulded work, &c., which must be girt as it appears when the work is finished, but which is not always the case with the plain work: and therefore it is requisite to know the manner in which the work is executed, to form an accurate conclusion, and to do justice to the workman in its measurement.

ABBREVIATIONS RECOMMENDED.

In measuring stone-mason's work, the same rules must be observed in entering the dimensions in the book as directed for the other trades; and the following abbreviations are recommended, for the reason stated under that head :—

C P   Cube Portland.
P W   Plain work.
S W   Sunk work.
M W   Moulded work.

C W   Circular work.
C C W   Circular circular work.
C S W   Circular sunk work.
C M W   Circular moulded work.

MEASUREMENT.

STAIRCASES. (Fig. 3, Plate VI.)

ft.	in.	ft.	in.	
5	0			Cube Portland steps, the 5 ft. including that part of the step
1	3	3	1½	that is pinned into the wall, and also the projection of nosing.
0	6			

                                ft. in.

ft.	in.		
5	0	P W top .	⎰ 1 1 tread.
1	2¼		⎱ 0 1¼ under the next riser.
			⎯⎯⎯⎯⎯⎯
			1 2¼

4	6	M W front. Girt of moulding, nosing and riser
0	7	

1	3	M W end. Taken or girt at the average width.
0	6	

0	6	P W to front the part pinned into the wall.
0	6	

5	0	P W to soffit.
1	1	

[Or the whole flight may be taken in one dimension.]

158   MASONRY.

ft. n.	ft. in.			
			ft. in.	
			0 1¼	rebate to front of step.
5 0		S W rebate	0 1¾	
0 4¾			0 1¾	do. to back of step.
			0 4¾	

No. of steps pinned into wall.
No. of holes cut for ballusters.

5 0			
1 3	6 3	Block of stone required to cut two steps out of.
1 0		

LANDINGS. (Plate VI., fig. 1.)

			ft. in.	
			12 0	
			0 6	in wall.
13 3			0 6	
4 6		C P landing	0 1½	
0 6			0 1½	joggles.
			13 3	

2)13 3	P W top and bottom. Here is more plain work than appears, but the plain faces must be made before the joggles are worked.
4 6	
12 0	M W front.
0 7	

2)0 6	P W to front of landing in the walls.
0 6	
		ft. in.	
		0 9	girt of the joggle.
2)4 6	S W joggles	0 4½	do. of the groove for do.
1 1½	(Fig. 2.)	1 1½	

		13 3	Cut for and pinning landing
		4 6	into wall, which is allowed to
22 3	Run of cutting	4 6	be taken through the doorway,
			&c., for the extra trouble of
		22 3	pinning up the quoins, &c.

SQUARE STEPS TO ENTRANCE DOORS, ETC. (Plate VI., fig. 4.)

2)6 9	C P supposing two steps.
1 1	
0 7	

| 6 9 | P W to bottom step. |
| 1 8 | |

| 6 9 | P W to top step. |
| 1 7½ | |

MEASURING COPING.

ft.	in.	ft.	in.	
6	9			S W rebate for landing.
0	2½			
6	9			2 in. Portland landing. { Portland steps worked to an exact length, and fitted between spandrils, allow one end as plain work.
4	1			
				No. of plugs.

COPING. (Plate VII., fig. 1.)

ft.	in.			
			ft. in.	Nothing extra is allowed
3	6		0 3	for being cut or worked
1	9	C P feather-edged coping.	0 1½	bevel on the face, as it
0	2¼		———	may be done without
			0 4½	extra waste.

			1 9	top.
			0 3	
			0 1½	edges.
3	6	P W	0 1½	
2	4½		0 1½	projection over wall.
			———	
			2 4½	

1	9	P W to Jts. Allow P W to one joint of each stone, which should average 3 ft. in. length.	
0	2½		

			ft. in.	
1	9	P W to return of angles where they occur . .	0 3	edge.
0	4½		0 1½	projection.
			———	
			0 4½	

3	6	Run of throat, or may be taken at . . .	3 6	
			0 1	M W throat.

The more common way of taking *thin* coping is by the foot run, the width, thickness, and work being described. If more than 3 in. thick it must be taken as described above.

Angle quoins may be numbered as extras; or measure the coping through both ways as common coping, which gives an extra length the width of the coping for the extra thickness, and the trouble of sunk work on the top. Or they may be measured thus:—

1	11		No. of cramps.
1	11	C P quoin, fig. 2 . . .	Pairs of plugs.
0	3		Lead for running ditto.

1	11	S W top.
1	11	

160   MASONRY.

ft. in.	ft. in.			
*1 11		P W joint.		
0 3				ft. in.
				0 3
*3 10		P W outside edge and projection	{	0 1¼
0 4½				―――
				0 4½
				0 1½ inside edge.
				0 1½ projection.
0 4		S W inside angle notched . .	{	0 1 throat.
0 4				―――
				0 4

STRING-COURSES.   (Plate VII., fig. 6.)

3 6	C P string-course.	
1 0		
0 8		
3 6	S W top.	
0 2		
3 6	P W . . .	{ 0 7½
0 9½		0 2 projection.
		0 9½
3 6	Throat S W, or run of throat.	
0 1		
1 0	P W to one joint of each stone average 3 ft. in length.	
0 8		

SQUARE PLINTHS WORKED ALL ROUND.   (Plate VII., fig. 9.)

2 0	C P plinth.	
0 11		
0 6		
		0 11
2 0	P W sides	0 6
2 10		2)1 5
		2 10
0 11	P W top.	
0 6		
1 2	S W rebate.	
0 4		
	No. of mortice holes.	

* In taking the angle quoins of coping, some will allow the plain top to be taken first; but this is incorrect, as there is no occasion to make it previous to sinking the top; it being only necessary to bring the stone to its thickness and out of winding, as if for plain work.

MEASURING MASONRY.

WINDOW SILLS. (Plate VII., fig. 4.)

```
ft. in.| ft. in.
 4  2 |         C P window sill.
 0  8 |                                                ft. in.
 0  6 |                                                 0  2
 ─────                                               ⎧  0  6
 4  2 |         P W top, front, and projection   . ⎨  0  2
 0 10 |                                               ⎩ ─────
 ─────                                                  0 10
 0  8 |         P W to one end.*
 0  6 |                                               ⎧  0  6 top.
 ─────                                                 ⎨  0  1 throat.
 4  2 |         S W top and throat .  .  .  .         ⎩ ─────
 0  7 |                                                  0  7
 ─────
```

CURBS. (Plate VII., fig. 5.)

```
 6  0 |         C P curbs.
 0  7 |
 0  6 |                                               ⎧  0  6
 ─────                                                ⎪  0  7
 6  0 |         P W including projection .  . . ⎨  0  6
 1  8 |                                               ⎪  0  1
 ─────                                                ⎩ ─────
                                                        1  8
 0  7 |         P W to one end of each stone, which should not be less on
 0  6 |            an average than 3 ft. in length.
                Take the quoin ends that show fair as P W.

 2 11 |         C P circular curb. (Plate VII., fig. 7.)
 0  9 |
 0  6 |
 ─────
 2 11 |         P W.
 0  9 |
 ─────
2)2 11|         C P W.
 0  6 |
 ─────
2)0  6|         S W to arch joints.
 0  6 |         Plugs per pair, with lead ; or allow the lead per lb.
 ─────         Holes, each.
```

COLUMNS. (Plate VII., fig. 8.)

```
 5  5 |         C P ⎫
 1  5 |              ⎬ shaft.
 1  5 |              ⎭
 ─────
 5  3 |         C P ⎱
 1  3 |
 1  3 |
 ─────
```

---

* This is what is usually allowed. Some claim both ends, others measure them thus :—

```
  2)0  6 |    P W to projection of ends.
    0  2 |
```

ft. in.	ft. in.	
1 11		C P base.
1 11		
0 8		
1 11		C P cap.
1 11		
0 8		
2)5 5		P W ⎫
1 5		⎬ shaft taken two sides.
2)5 3		P W ⎭
1 3		
5 5		Circular work ⎫
4 6½		⎬ shaft.
5 3		Circular work ⎭
3 11		
1 7½		S W to bed for joggle in lower stone.
1 7½		
1 5		P W top bed of upper stone in shaft.
1 5		
1 11		P W top ⎫
1 11		
2)1 11		P W rims ⎬ base.
0 8		
6 0		Circular M work ⎭
0 10½		
1 11		P W top ⎫
1 11		
2)1 11		P W sides ⎬ cap.
0 8		
6 0		Circular M work ⎭
0 8½		

In measuring the circular M work to cap, it should be taken at the average between the angle of abacus and the front.

If the neck moulding is worked in the shaft, the same dimensions may be taken for C P and labour as the bottom stone of the shaft.

If there is an entasis to the column the surface must be taken as super circular-circular work in addition to the above.

MEASURING MASONRY.

ARCHITRAVES OVER COLUMNS. (Plate VII., fig. 10.)

ft.	in.	ft.	in.	
3	0			C P.
1	7			
1	7			
3	0			P W bottom bed.
1	7			
2)3	0			M W to fronts.
2	0			
1	7			P W to end.
1	7			
1	4			S W to the joggle.
1	0			

					ft.	in.
					1	7
					0	2
1	7			S W to end, including the joggle	0	2
1	11				1	11

BLOCKINGS AND CORNICES. (Plate VII., fig. 3.)

				ft.	in
				0	8
				0	4
3	6	C P blocking			
1	0			4)1	0
0	6½				
				0	6 ½ for bevel.

				1	6
				1	7¾
3	6	P W		0	4
3	5¾				
				3	5¾

1	6	P W joint, average size.
0	6	
0	9	Run of groove for plugs.
		No. pairs of plugs, and running with lead, per pair.

		If the plain work to bed of cornice, on which the blocking stands, is not taken, it would be allowed to take the bottom bed, which would make it 4 ft. 1¾ in. for the P W	3	5¾
			0	8
			4	1¾

3	6	C P top member of cornice.
2	4	
0	8	

ft.	in.	ft.	in.	
3	0			C P bottom member of cornice.
1	3			
0	5			
3	6			P W beds . . . . . . $\left\{\begin{array}{cc} 1 & 3 \\ 1 & 3 \\ \hline 2 & 6 \end{array}\right.$
2	6			
3	0			P W under blocking.
0	9			
3	6			Sunk and moulded work . . . $\left\{\begin{array}{cc} 1 & 1 \\ 0 & 10 \\ 0 & 6 \\ 0 & 7 \\ \hline 3 & 0 \end{array}\right.$
3	0			
1	2			Groove run to joints with lead.

NICHES. (Plate VII., fig. 11.)

6)1	0 high			C P ⎫
1	3			⎬ Stones in body.
0	9			
12)2	6			C P ⎭
1	0			
0	9			
0	9			C P head centre stone.
0	9			
0	4			
3)2	9			C P arch-stones taken the whole width, on account of trouble in getting them out.
1	6			
1	3			
3)2	9			P W face of do.
1	6			
6)2	9			S W to arch-joints of do.
1	3			
5	0			Circular work to body . . $\left\{\begin{array}{ll} & \text{ft. in. ft. in.} \\ 3\times 3 & 0=9\ \ 0 \\ & \tfrac{1}{2}=0\ \ 5 \\ \hline & \tfrac{1}{2})\ 9\ \ 5 \\ \hline & 4\ \ 8\tfrac{1}{2} \end{array}\right.$
4	8¼			
1·6				
4	8½			Circular-circular work to spherical head.
·12)2	6			P W to bed of stones in body.
0	9			

ft. in.	ft. in.	
6)1 3		P W to bed of stones in body.
0 9		
2)13)1 0		S W to arch joints.
1 0		
14 8½		S W to front A . . . $\left\{\begin{array}{cc} 5 & 0 \\ 5 & 0 \\ 4 & 8\frac{1}{2} \\ \hline 14 & 8\frac{1}{2} \end{array}\right.$
0 7½		
10 0		Run of bead and double quirk.
4 8½		Circular do.
		No. of cramps.
		No. of plugs.

Stone facings to fronts of houses, if more than three inches thick, should be taken as cube stone, and the face and one bed and joint taken as P W. Bond stones taken one face bed and joint. If not more than three inches, take them as slab, and one bed and joint as P W. If to circular-headed windows, take the arch joints as sunk work, and the soffits as circular plain work, and the straight reveals as P W. If rusticated describe them as such, or take them as S W. If stone facings are taken to a parallel thickness, as for old brick fronts, they may be taken as slab even to four inches thick, but the P W to beds and joints must not then be taken.

In abstracting mason's work, the paper must be ruled in columns, as before described, observing to place the C P in the first column, and leaving sufficient space in the following columns for the different sorts of labour on ditto, as P W, S W, M W, &c.; the next columns for Portland slabs, keeping each thickness in a separate column; the next columns for vein, statuary, and other marble; the next for Yorkshire and Purbeck pavings, and other articles of different descriptions; the following columns for articles taken as running measure, and the last columns for those numbered.

The following Table gives the weight of some of the stones most commonly used; for a larger list, see the Appendix to Tarn's "Science of Building."

WEIGHT OF STONE.

Purbeck stone	14 cubic feet weigh one ton.	
Portland	16½ " "	do.
Bath	18 " "	do.
Yorkshire	15 " "	do.
Granite	13½ " "	do.
Marble	13 " "	do.
Purbeck paving	50 ft. superf. "	do.
Do. step 13 by 6½	25 ft. run "	do.

VALUATION OF LABOUR.

TABLE OF CONSTANTS FOR THE DIFFERENT DESCRIPTIONS OF MASON'S WORK.

N.B.—The factor to be applied is the rate of wages for a mason per hour.

	hours.
Labour, squaring and laying new York or Purbeck paving per foot superficial	·21
If in courses, add	·10
Labour on Portland or similar stone per foot superficial. N.B.—Sawing to be taken as half plain work.	
Plain work to bond stones . . per foot superf.	1·40
Do. to beds and joints . . . do.	1·60
Do. rubbed face . . . do.	1·85
Do. do. circular . . . do.	2·60
Sunk work rubbed . . . do.	2·70
Do. do. circular . . . do.	3·00
Moulded work rubbed . . . do.	3·10
Do. do. circular on plan . . do.	4·20
Circular work to shafts of columns having the neck moulding or part of the base worked in the same stone . . . do.	3·00
Circular-circular or spherical work to domes, balls, or entasis of columns . . . do.	4·30
If rubbed, add extra . . . . do.	·50
Taking up, squaring, and relaying old paving do.	·42
Add if in courses . . . . . do.	·15

LABOUR ON STATUARY OR VEIN MARBLE,

INCLUDING SAWING, WORKING, AND POLISHING.

Plain work . . . per foot superf.	3·50
Circular work . . . do.	5·80
Sunk work . . . do.	8·00
Moulded work . . . do.	11·00
Circular sunk work . . . do.	16·00
Circular moulded work . . . do.	22·00

ON OLD WORK.

	hours.
Old vein marble chimney-piece cleaned and reset, per ft. superf.	1·20
Do. do. squared and reset . . . . . do. .	1·40
Do. do. sanded, grounded and squared . . . do. . .	1·80
Do. do. and reset . . . . . . do. .	2·00
Do. do. sanded, polished, and reset . . . do. . .	3·5
Do. do. sawed, sanded, polished, squared, and reset . . . . . . do. .	3·8

In the west of England, and all the counties in which stone is abundant, it is usual and customary to build with the rough stone of the country, and the practice generally is to measure the walls by the perch of 18 superficial feet, supposing them 24 inches thick, to which thickness all the walls, whether more or less, are reduced by multiplying the superficial contents by the thickness in inches, and dividing them by 24; or they may be reduced from cubic feet to the perch of 36: but some regulate the prices per perch, according to the thickness of the walls. In other parts the walling is measured by the *rod* of 7 superficial yards.

In measuring the work, some contend that the quoins and all projections should be girt, to pay them, as they say, for the extra trouble in working and setting the stones; but this should not be allowed, except for labour only; and even then it is much fairer to measure the quantity of walling as it is, and make a proper allowance for the extra labour, either in quoins, chimney-breasts, flues, reveals, &c.

In taking out quantities or measuring mason's work in stone districts, it is desirable that the surveyor should make himself acquainted with the methods of working the stone employed in that particular district, as well as the terms used in designating the work; since nearly every county has its own peculiarities and technicalities in masonry, as well as different modes of estimating its value; and what might be well understood in one part of the country would in another be quite incomprehensible. Where slabs of stone are used for footings to walls, or *bottoms* as they are sometimes termed, they must be measured separately from the foundations themselves.

ROTATION

To be attended to in bringing the quantities into Bill.

Yds. ft. in.	MASON.	
	Rough stone walling foundations in random courses, well bonded and flushed with mortar, and grouted with hot lime and sand every two courses	
	Do. do. above foundations, levelled every two feet or height of two quoins, well bonded and flushed with mortar every course . . .	
	Superficial of extra labour to external quoins	
	Do. do. to internal quoins, &c. . .	
	Cube Portland, *or any other stone valued per foot cube*	
	Superficial of plain work . . .	
	Do. of sunk work, *or such other labour, as the case may be*	
	Superficial of 1¼ Portland slab . .	
	Do. 2 do.	
	Do. 2½ do.	
	Do. of 1 in. vein marble slab in chimneys, &c.	
	Do. of 1 in. statuary marble slabs, in do., &c.	
	Do. of Purbeck paving	
	Do. of Yorkshire paving, &c., *then the* runs, as run of Purbeck steps, &c. ; *then the Nos.* as number of holes cut, &c.	
	Attending on other Trades . . .	

TRANSVERSE STRENGTH OF STONE.

When a piece of stone is supported at both ends and loaded in the middle, its breaking weight W in *cwts.* can be ascertained by means of the following formula, in which the breadth b, and depth d, are expressed in *inches*, and the length l between the supports is in *feet*. The constant S is found by experiment upon specimens of the various kinds of materials. If the stone is fixed firmly at one end and loaded at the other, the breaking-weight will be one-fourth of that found in the above case. If the load is uniformly distributed over the whole length, the beam will bear twice as much as it will if all placed at one point.

$$W = S\frac{b \cdot d^2}{l}$$

	Value of S.
Roman cement	·047
Stock brick	·110
Mansfield red sandstone	·470
Yorkshire stone	1·000
Sicilian marble	1·000
Caithness stone	2·550
Welsh slate	5·700
Fir	2·000 to 4·000
Wrought iron	22·000

This rule may be expressed in words; multiply the breadth by the square of the depth, all in inches, and divide by the length in feet; multiply the quotient by the value of S, where the beam is supported at two ends and loaded at the middle, or by twice S if the load is uniformly distributed; the result is the breaking weight in cwts.

CHAPTER VI.

PLASTERING.

TECHNICAL TERMS AND EXPLANATIONS.

PLASTERING consists in covering the inner face of the walls and ceilings of a building with a composition, of which the groundwork is lime, sand, and hair, finished with a coating of finer materials, for the purpose of producing a smoother, less absorbent, and more sightly surface than that of the rough walls. The work of the plasterer also extends to the outside of buildings which have to be *stuccoed* or covered with a coating of cement, and decorated with moulded cornices, string-courses, architraves, and other ornamental features.

Before ceilings or quartered-partitions can be coated with plaster, it is necessary to cover the timbers with *laths* nailed thereon, so as to form a hold or *key* for the plaster. When walls are to be plastered the joints of the brick or stone-work are left quite rough, so that the plaster may have something to hold to. Outside walls in exposed or damp situations are sometimes *battened* with narrow slips of wood fastened to the inside of the walls at intervals of 12 inches, and to these the laths have to be nailed before the walls can be plastered, as in timber partitions and ceilings.

CORNICES and other plaster mouldings are *run* by means of moulds cut in zinc, the exact form of the moulding intended to be used, and fastened to a wooden frame. When a cornice is to receive ornaments, the plasterer leaves sinkings to lay them in. When large or heavy cornices are to be run round the ceiling of a room, *bracketing* must be fixed by the carpenter before the lathing is nailed to the joists; this consists of pieces of wood fixed at intervals of 12 inches apart all round the portion of the ceiling intended to receive the cornice, and is covered with laths like the rest of the ceiling (fig. 1, Plate 8.) Ornaments are usually cast in plaster-of-paris or papier-maché, from models previously made in clay, and are attached to the cornice or ceiling with plaster-of-paris.

When a very fine surface is required to the walls of a room, it is usual to finish the plaster with MARBLE-CEMENT, which is capable of taking a brilliant polish, and is therefore suitable for internal decorations, as columns, pilasters, architraves, dados, &c. When it is required to paper or paint the walls shortly after they are plastered, PARIAN-CEMENT is employed as a finishing coat, which also produces a very hard and smooth surface.

When ceilings are ornamented with *centre-flowers* or other decorative features, the ornaments are sometimes perforated and connected with a flue passing between the

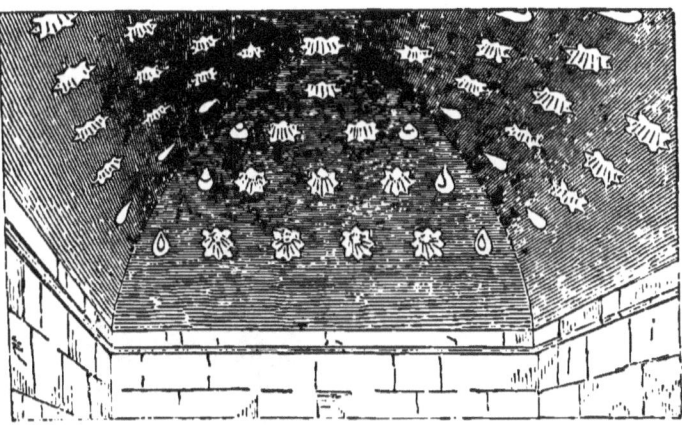

Fig. 51.—VESTIBULE OF THE BATHS, ALHAMBRA.

joists above, to the outside of the building or to a flue running alongside of the chimney flue ; by this means the room may be ventilated, fresh air being brought in from the outside, and the impure air drawn off by the draught caused by the chimney. Figs. 51 and 52 show the method adopted of ventilating the Moorish baths of the Alhambra by means of pyramidal openings in a domical ceiling.

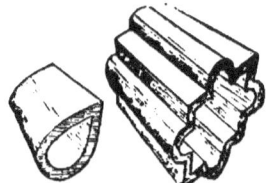

Fig. 52.—EARTHENWARE VENTILATORS OF THE ABOVE.

OUTSIDE-STUCCO to the walls of buildings is usually executed in a material called *Portland-cement*, from its having the appearance, when finished, of Portland-stone. This is gauged with fine sharp sand in various proportions, but generally about two or three parts sand to one part cement. A coarse coating of stucco is first laid on the wall, and left rough to form a key for the next coat; this is finished with fine stuff smoothed over with a trowel. When cornices or string courses have to be formed, a *core* of Yorkshire stone or plain-tiles is first built into the wall, projecting therefrom a sufficient distance to carry the required mouldings, which are run with a mould in the same manner as other plaster cornices. Where one part of the work has to be finished thicker than the rest, plain-tiles are laid in cement against the wall to give the extra thickness, and the stucco laid over; this is termed *dubbing-out*, and must be measured extra.

PORTLAND-CEMENT is formed by mixing chalk or other limestone with clay, in the proportion of three of the former to one of the latter, in a mill supplied with water, and furnished with revolving harrows to reduce the whole of the particles and secure their complete admixture. After being left for some time to evaporate its superfluous water in large tanks called *backs*, it is dried in stoves, burnt in a kiln, and reduced to fine powder by grinding in a mill.

SELENITIC-CEMENT is made by combining gypsum with common lime, which renders it hydraulic and also greatly increases its strength. This material is used for the same purpose as *Portland*, and is much cheaper, but does not set so quickly.

Cement is strongest when used *neat*, or without any admixture of sand, and the greater the quantity of sand mixed with it the weaker it becomes; sand having a good sharp grit should be used for mixing, as very fine sand has a tendency to *kill* the cement. Cement is often used for making floors upon a basis of bricks or concrete; it has the property of *setting* or becoming hard very quickly, and must therefore only be used in small quantities at a time.

In plain stuccoed work, one bushel of cement, used neat, will cover 10 square feet of work 1 inch in thickness; if gauged with an equal quantity of sand, it will cover twice as much surface; and so on, in proportion to the quantity of sand.

The external *angles* of chimney-breasts in rooms are frequently formed of cement in preference to wooden angle-staffs.

CLAIRCOLLE is washing over plaster with a solution of *size*, to prepare it for the reception of colour or paper-hanging.

WHITENING is washing over plaster, previously claircolled, with a solution of *whiting*.

LIME-WHITING is washing over walls with a solution of *lime*.

All claircolling, whitening, and colouring, is measured by the yard superficial. Old ceilings have to be *washed* or brushed over with water before they can be claircolled; and if there are any cracks, they must be *stopped* with plaster-of-paris after being washed.

External stucco is sometimes executed in BLUE-LIAS LIME gauged with sand; this being an *hydraulic* lime, or one that will set under water, is a valuable material for preventing the rain from penetrating the walls of a building; the proportion of sand to be used is about two or three parts to one part of lime. This material sets more slowly than cement.

PRICKING UP or RENDERING is the first coat of coarse

stuff, as lime and hair, laid on the walls. If intended to be floated, it is crossed, as a key for the next course or coat; if it is only intended for setting or two-coat work, then it is not crossed, as it is not necessary, and would show through the thin coat of lime and hair.

RENDER SET is two-coat work on walls; viz., one coat of rough plastering performed with lime and hair, and one coat of fine stuff, which is called setting: this is performed by laying on a very thin coat of fine mortar, denominated finishing stuff, which must be well trowelled to prevent its cracking.

RENDER, FLOAT AND SET is three-coat work: one coat of rough plastering crossed; another coat laid on ditto, and floated with a long rule to make it perfectly straight on the face; and one coat of fine stuff or setting on ditto, as 'render set.'

LATH AND PLASTER is lathing on quarter partitions, &c., and one coat of plastering only laid on the laths, as pricking up or rendering is on the walls.

LATH, PLASTER AND SET is two coats on the lathing, as 'render set' is on the walls.

LATH, PLASTER, FLOAT, AND SET, is three coat work on the laths, as 'render, float and set' is on the walls.

TROWELLED STUCCO. This work, either on walls or partitions, is performed as before described for setting; then a thin coat of stucco, which is prepared with a large portion of sand, and laid on similar to the fine coat called setting, in small portions at a time, but worked with water, and trowelled till it is perfectly hard and solid.

BASTARD STUCCO is executed in a similar way to the above, but the finishing contains a small quantity of hair behind the sand, and it is not hand floated or trowelled.

All rooms that have plaster cornices must either be floated or have a *screed* formed all round them, to obtain a straight face for running the cornice by.

ROUGH-CAST is pricked up and floated as if to be set or stuccoed, and then the rough cast (which is composed of half lime and half small stones) thrown with force against the surface; and consequently appears rough on the face when finished.

DEPETER is pricked up and floated in a similar manner, and small stones forced on dry from a board, by which the face of wall is finished rough, and the same colour as the stones used.

DEPRETOR is plastering done to represent tooled stone.

PUGGING TO FLOORS is pricking up between the joists of floors, either on laths or boards, to prevent the sound escaping from one room to another. This should be performed with coarse stuff and chopped hay, if on boards; but if on laths, with lime, sand, and hair; and not less than 1½ in. thick in either case.

Ornaments are said to be worked by hand when they are so designed that they cannot be cast; which renders the work very expensive, as every part must be performed in the plaster as if modelled in clay.

ABBREVIATIONS.

R R	Rough render.	R C B	Rough cast on brick.
R S	Render set.	R C L	Rough cast on lath.
R F S	Render float and set.	If any of these are whitened, add W.	
L O	Lath only.		
L P	Lath and plaster.	W N W	White to new work.
L P S	Lath and plaster set.	W S W	Wash stop and white.
L P F S	Lath, plaster, float, and set.	L W 1^{ce}	Lime white once done.
S B	Stucco on brick.	L W 2^{ce}	Lime white twice done.
S L	Stucco on lath.	C C	Common colouring.

ROTATION.

In measuring plasterer's work to rooms, first take the ceiling; second, the sides; third, the cornices and enrichments.

MEASURING (Plate VIII.)

Plasterer's work is taken superficially, and valued by the square yard of 9 feet.

If cornices are round the room, take the ceiling only to half the projection of the cornice, or one projection in and one out; or measure the ceilings clear of the cornices, and take the whole of their projection as lathing and pricking up.

If the cornices are bracketed (as fig. 1), measure the ceilings clear of the cornice.

The sides of the room should be taken from the ground, through the bed-mould, or half the height of the cornice.

If on brick, or bracketed (as fig. 1), take them only to the bottom of cornice.

In taking the length of cornices, measure the size of the room, taking one projection in and one out, and girt them from the mould or from the ceiling to the wall line.

Number all the angles in the room above four, as extra.

In taking cornices where there are *coves* or large hollows, take the coves as superf. of cove to cornices, and allow 1 inch extra on the girt of the cornice for the return of the mould on the cove.

All enrichments to be taken separately.

Friezes, under the cornice, must be taken as superf. of plain floated frieze. A floated ground must be taken under all enriched friezes.

If cornices are run to old ceilings, a screed must be allowed.

Enriched friezes, ceilings, or soffits must be measured first as plain work, and then the enrichments taken separately at per foot run, and a price fixed according to their description and value.

All circular mouldings and enrichments to be taken one face in and one out, fig. 3.

To explain the foregoing rules, see Section of a Cove Cornice, &c., &c., fig. 2.

Take first the ceiling through the reeds.

Second, length of cove above the cornice by 2 ft.

Third, length of moulded cornice by 1 ft. 2 in., being 1 in. extra for top on cove.

Fourth, do. of plain floated frieze by 6 in.

Fifth, do. of moulded architrave by 6 in.

Sixth, do. of moulded reeds by 9 in.

Reveals to windows taken at per foot run, price according to width.

ABSTRACT OF PLASTERING.

ABSTRACT.

YARDS.											COLOURING.					SUPERF. FEET.			FEET RUN.	
R R	R S	L O	L P	L P S	L P F S	F S B	F S L	W S W	W N W	L W	Common	Lemon	Green	Plain Frieze	Plain Cornice	Stucco groins	Quirks	B & D Quirk
											Grey	Blue	Red				Circ. do.	Circ. do.
																	Numbers	

N As some of these articles will not be whitened, as for papering, &c., place them all in the Abstract as not whitened, and the whitening under a separate head, as "W N W." — white to new work.

ROTATION.

To be attended to in bringing the quantities into Bill.

PLASTERER.

Yds. ft. in.	
	Rough render
	Render set
	Render, float and set
	Lath and plaster, one coat
	Lath, plaster and set
	Lath, plaster, float and set
	Stucco on brick
	Stucco on lath
	Pugging
	White new work
	Wash, stop, and white
	Lime white
	Colouring, *as the case may be*
	Superf. of plain cornice, &c., &c.

Then the

Run of cornices, girt, &c.
   „   reveals
   „   beads, &c.
Numbers of mitres, &c.
  Do. ornaments, centre flowers, &c.
Attending on other Trades

VALUATION OF PLASTERERS' WORK.

CALCULATION OF MATERIALS.

1 hundred of lime = 25 striked bushels (old measure)

	Materials.	Labour.
100 yards of render set require	$1\frac{1}{3}$ c. yd. of lime. $2\frac{1}{2}$ do. of sand. 4 bushels of hair.	Plasterer, labourer, and boy, 3 days each.
100 yards of lath, plaster and set require	24 bundles of laths. 12,000 nails. $1\frac{1}{2}$ c. yd. of lime. 3 do. of sand. $4\frac{1}{2}$ bushels of hair.	Plasterer, labourer, and boy, $4\frac{1}{2}$ days each.

LATHING.

1 bundle of laths and 500 nails will cover $4\frac{1}{4}$ yards.

RENDER ONLY.

100 yards require . . . $\begin{cases} \frac{3}{4} \text{ c. yd. of lime.} \\ 1\frac{1}{2} \text{ do. of sand.} \\ 2\frac{3}{4} \text{ bushels of hair.} \end{cases}$

Floating requires more labour than rendering does, but not more than half the quantity of stuff.

SETTING ONLY.

100 yards require . . $\begin{cases} \frac{3}{4} \text{ c. yd. of lime.} \\ 1\frac{1}{3} \text{ bushels of hair.} \end{cases}$

20 per cent. is always allowed on the prime cost of the materials.

CALCULATION OF LABOUR.

The decimal is to be multiplied by the rate of wages for plasterer, labourer, and boy, per hour.

	Hours.
Rough render, straight, per yard superficial. . . .	·15
Floating do.	·25
Setting, add	·05
Lathing, only	·10
If circular work, add to the lathing and also on each coat of plastering	05
If to groins, add as above	·15

External Cement Work is measured in the same way as the internal plastering, the plain surface taken by the yard super. and described as plain, jointed as masonry, rusticated, or otherwise. All moulded work exceeding 6 inches girt by the foot super. Small mouldings, arrises, angles, beads and quirks by the foot run. Columns by the foot super., as circular work, and their moulded caps and bases as circular moulded. Mitres, stops and angles to mouldings are numbered, the girt being described. Moulded balusters and ornaments are numbered and described, the sizes being given.

Cement Skirtings are taken by the foot super., the length by the girt, angles and mitres being numbered.

CHAPTER VII.

SMITHS' WORK, ENGINEERS' WORK, IRONMONGERY, BELL-HANGING.

SMITHS' WORK.

THE use of IRON in the construction of buildings of all kinds has become so extensive as to render it one of the most important materials in connection with building operations, both for structural and ornamental purposes.

CAST-IRON is obtained by pouring the molten metal into sand moulds prepared from wooden models the exact size (with a certain allowance for contraction in cooling) of the required article. It is brittle and is readily fractured by a heavy blow, presenting a crystalline appearance when broken. Cast-iron is of great value for sustaining heavy compressing loads, as in pillars or other supports; its resistance to *crushing* being from 35 to 50 tons per square inch. Its resistance to a *stretching* force is only one-fourth to one-eighth of the resistance to compression, consequently it is not so suitable for forming into beams as in using for pillars. When a large number of beams of the same size are required, great economy is obtained by using cast-iron; in which case the ⊥ form of section is employed, the lower *flange* being made to contain nearly six times as much iron as the upper, since that part has to resist a stretching force, and is made proportionately large in order to counteract the deficiency in the resisting power of the metal to an extending force.

The breaking-weight laid on the centre of a cast-iron beam of this form is found, in tons, by multiplying the area of the section of the bottom flange at the middle, in inches, by twenty-six times the depth of the beam in inches, and dividing the product by the length of the bearing or span between the supports, also in inches. The permanent load on a cast-iron beam should not be more than one-sixth of its breaking-weight. In order that a beam whose lower flange has the same width throughout, may be of uniform strength in every part, the depth should vary so as to form an ellipse, having its greatest height in the centre. If the beam is of equal depth throughout, the lower flange should be in the form of two segments of circles, having the greatest breadth in the middle. If the load on a beam is uniformly distributed over its entire length, it will bear twice as much as when all the load is in the middle.

The mode of finding the strength of cast-iron columns depends upon the relation which exists between their length and diameter. When the length is thirty times the diameter, or more, the column will break by bending only, the resistance to crushing not coming into play. If we call w the breaking-weight in tons, D the external diameter in inches, d the internal diameter for a hollow column, l the length in feet; we find the breaking-weight from the formula,

$$w = 42 \frac{D^{3.5} - d^{3.5}}{l^{1.63}}$$

The following table of values of $D^{3.5}$ and $l^{1.63}$ will facilitate the application of this formula:—

$2^{3.5} = 11.3$
$2.5^{3.5} = 24.7$
$3^{3.5} = 46.8$
$4^{3.5} = 128$
$5^{3.5} = 279$
$6^{3.5} = 529$

$5^{1.63} = 13.8$
$8^{1.63} = 29.6$
$10^{1.63} = 42.7$
$12^{1.63} = 57.4$
$15^{1.63} = 82.6$
$20^{1.63} = 132$

For a more extended table of these numbers see Tarn's "Science of Building."

The breaking-weight of a solid column can be found by the same formula, by merely putting $d = 0$

When a very inferior quality of iron is employed, the above formula will give too great a value for the breaking-weight; and the multiplier in that case should be 34 instead of 42 : for very superior qualities the multiplier will be as high as 50.

This mode of calculating does not answer for shorter columns, in which the length is between ten times and thirty times the diameter, since the resistance to crushing then comes into play. To find the breaking-weight of such a column, first apply the foregoing formula, and obtain the breaking-weight as if it were a long column; multiply that result by forty-nine times the area of the section of the column in inches, and divide the product by the first found breaking-weight added to thirty-six times the area of the section. For columns under ten times their diameter in length, we have only to take into consideration the resistance to crushing, or to multiply the area of section by 49 tons for hollow columns, and by 39 tons for solid columns. Greater strength in proportion to weight of metal is obtained in using hollow columns than solid, as the inner metal is softer and weaker in solid columns than that near the outside; and the thinner the metal the greater its comparative resistance to crushing. The permanent load upon cast-iron pillars should not exceed one-sixth of their breaking-weight.

Cast-iron should never be employed where it is liable to be subjected to sudden blows or jars, or to great variations of temperature; and if water comes in contact with it when hot, it will fly into pieces.

Another and simple method of finding the breaking-weight of a solid cast-iron column, and which will give sufficiently accurate results for any length as compared with the diameter, is as follows: take the *ratio* or proportion which the *length* bears to the *diameter* (as 5 times, 10 times, 20 times, and so on), and add 330 to the *square* of that *ratio*; then divide 270,000 by the sum, and the result is the breaking-weight in cwts. per square inch of section. The safe permanent load may be one-fifth to one-seventh of the breaking-weight obtained by this rule. The weight is supposed to act vertically down the axis of the column, whose ends are perfectly flat and parallel. If the column is hollow, the strength depends upon the thickness of the metal; when the metal is $\frac{1}{10}$th of the diameter in thickness, first multiply the *ratio* by $\frac{3}{4}$ before applying the above rule; if the thickness is $\frac{1}{7}$th of the diameter, the *ratio* must be multiplied by $\frac{4}{5}$; if $\frac{1}{6}$th of the diameter, the *ratio* is to be multiplied by $\frac{7}{8}$; if $\frac{1}{4}$th, it must be multiplied by $\frac{9}{10}$. The strength per square inch of a square column is to that of a round one as 8 to 7; so that the *ratio* being multiplied by $\frac{7}{8}$, the above rule will apply to solid square pillars.

STORY-POSTS are iron pillars, generally cast of an I section, and used to support the ends of bressummers over the openings of shop-fronts. The strength of cast-iron story-posts may be found by the rule above given for round pillars, only the *ratio of length to diameter* must be first multiplied by $1\frac{1}{8}$ for thickness of metal equal to $\frac{1}{10}$th the diameter; by $1\frac{1}{6}$ for metal of thickness $\frac{1}{7}$th the diameter; by $1\frac{1}{4}$ for thickness equal to $\frac{1}{6}$th the diameter; and by $1\frac{1}{3}$ for thickness equal to $\frac{1}{4}$th the diameter.

Cast-iron in girders, story-posts, columns, and all other heavy articles, is charged by the ton or cwt. When models have to be made expressly for iron-castings, they are charged extra according to the labour in them. The weight of a cubic foot of cast-iron is 444 lbs.; so that the weight of one foot length of a solid round column of given diameter can be found by means of the Table at page 26. The *area* expressed in parts of a foot multiplied by 444 gives the weight of 1 foot length. If the area is in square inches, multiply by 3·1 to obtain the weight in lbs. of 1 foot length. To find the weight of a *hollow* column, first find that of a solid one of the same external diameter, and also that of a solid one of the diameter of the hollow part; deduct the latter from the former, and the result is the weight of the hollow column.

The weight of a superficial foot of cast-iron 1 inch in thickness is 37 lbs.; that of a solid circular rod 1 inch in diameter and 12 inches long, is 2·44 lbs., or a little under 2½ lbs.; and since the weight increases as the square of the diameter, the weight of a rod 12 inches long of any other diameter can be found by multiplying the square of its diameter by 2·44, or, nearly, by 2½. The weight of a solid rod of cast-iron 1 inch square and 12 inches long, is 3·1 lbs.; and that of any other square rod of the same length may be found by multiplying the square of one side by 3·1.

The following Table of the weight in lbs. of one foot of cast-iron cylinders, of different diameters, will greatly facilitate the calculation of the weight and value of columns:—

Diameter of bore, in inches.	Thickness of metal, in inches.					
	3/8	1/2	5/8	3/4	7/8	1
	lbs.	lbs.	lbs.	lbs.	lbs.	lbs.	lbs.
1	2·7	4·7	7·3	9·9	12·8	16·0	19·6
1½	4·3	6·9	9·8	13·0	16·6	20·4	24·5
2	5·5	8·7	12·3	16·1	20·3	24·7	29·4
2½	6·8	10·6	14·7	19·2	23·9	29·0	34·4
3	8·0	12·4	17·2	22·2	27·6	33·3	39·3
3½	9·2	14·3	19·6	25·3	31·3	37·6	44·2
4	10·4	16·1	22·1	28·4	35·0	41·9	49·1
4½	11·7	17·9	24·5	31·4	38·7	46·2	54·0
5	12·9	19·8	27·0	34·5	42·3	50·5	58·9
5½	14·1	21·6	29·5	37·6	46·0	54·8	63·8
6	15·2	23·5	31·9	40·7	49·7	59·1	68·7
6½		25·3	34·4	43·7	53·4	63·4	73·4
7		27·2	36·8	46·8	56·8	67·7	78·5
7½		29·0	39·0	49·9	60·7	72·0	83·5
8		30·8	41·7	52·9	64·4	76·2	88·4
8½		32·9	44·4	56·2	68·3	80·8	93·5
9		34·5	46·6	59·1	71·8	84·8	98·2
9½			49·1	62·1	75·5	89·1	103·1
10			51·5	65·2	79·2	93·4	108·0
10½			54·0	68·2	82·8	97·7	112·9
11			56·4	71·3	86·5	102·0	117·8
11½			58·9	74·3	90·1	106·3	122·7
12			61·3	77·4	93·6	110·6	127·6

RAIN-WATER-PIPES, or *fall-pipes*, are made of cast-iron in 6-feet lengths, fitting one into another with a socket-joint, and used for conveying the rain-water from the gutters to the drains; when several lengths are fitted together to form one continuous pipe, it is termed a *stack*. They are sold by the yard run, according to their diameter. These pipes are fastened to the wall with large spike-nails driven through *ears* cast upon the upper end of each length of pipe; the nails being sold by number. The *Heads* of rain-water-pipes are separate castings, the upper part being much wider than the pipe, to receive the water direct from the gutter; they are numbered in the bill. *Shoes* are short lengths of pipe bent at an angle, and fitted to the lower end of the stack, for the purpose of shooting the water away from the wall, or into the drain. They are numbered in the smith's bill.

Elbows, or *bends*, are short lengths of pipe cast with a particular curve, to carry a pipe over a moulded string-course or other projection, or to divert the direction of the pipe. They are numbered and valued separately from the pipes.

EAVES-GUTTERS are cast-iron channels fixed at the eaves of a roof to receive the rain-water and convey it to the fall-pipes. They are cast in 3 ft. and 6 ft. lengths, and fitted together by sockets cast on one end of each length; they are sold by the yard, according to their size and shape. *Nozzles* are short pieces of pipe hanging down from the gutter and cast in one piece with it, to convey the water into the fall-pipe head. They are charged separately and numbered. *Elbows* are pieces of guttering cast with a bend or angle to carry the gutter round an angle of a wall. They are numbered separately. Eaves-gutters are fastened by *brackets*, every three-feet length, driven into the wall, or screwed to the feet of the rafters. Brackets are numbered: also all *stopped-ends* to gutters are numbered. Bolts, nuts, and screws for fastening the lengths of guttering together, are charged by the dozen. *Clips* are numbered at *each*.

Cast-iron railings, gates, gratings, casements, brackets, cantalivres, &c., are charged by the pound, according to the nature of the work. When lead is used for running the iron into stone-work, it is charged separately by the pound weight.

Air-bricks of cast-iron, or small gratings the size of one or more bricks, for ventilation through a wall, *Coal-plates* with hooks, and *Door-scrapers* of cast or wrought iron are all numbered.

WROUGHT-IRON is a metal which has been reduced to a fibrous condition, and rendered capable of being hammered or wrought into various shapes. It is softer than cast-iron, and its power of resisting a crushing force is one-third that of the cast metal. For very long solid columns, whose length exceeds thirty diameters, it is, however, considerably stronger than cast-iron, since its resistance to stretching is three times as great. It is not, however, often employed for columns, on account of the much greater cost of manufacture. The resistance of wrought-iron to crushing is about 16 tons per square inch of section, the resistance to stretching being from 22 to 26 tons per square inch.

The following simple rule enables us to find with considerable accuracy the strength per square inch of section of a wrought-iron pillar of circular section: take the *ratio of the length to the diameter*, square it, and add it to 2000; then divide 680,000 by the sum, and the result is the breaking-weight per *square inch of section* in cwts.; and multiplying by the number of square inches in the section gives the strength for the particular pillar. The ends are supposed to be perfectly flat and parallel, and the load acting down the axis. The safe weight may be $\frac{1}{4}$th or $\frac{1}{5}$th of the breaking-weight. If the column is square, it will bear a greater load per square inch than if round; and the above *ratio* must be multiplied by $\frac{7}{8}$ before the rule can be applied to square pillars. Pillars having L, T, or + sections are weaker in proportion than round pillars having the same area of section; and the *ratio* must be multiplied by $1\frac{1}{8}$ where the metal is $\frac{1}{16}$th of the diameter in thickness; by $1\frac{1}{6}$ if it is $\frac{1}{8}$th of the diameter; by $1\frac{1}{4}$ if it is $\frac{1}{5}$th of the diameter; and by $1\frac{1}{2}$ if it is $\frac{1}{4}$th of the diameter.

Wrought-iron is largely used in construction for forming *joists, girders, roofs, bridges*, &c.

JOISTS are rolled into a great variety of shapes and

sizes, those most commonly used having the **I** or **T** form of section. A very good FIREPROOF-FLOOR can be made with such joists (B, fig. 53), placed about two feet apart, and filled in between with 4 inches or so of cement concrete (A), laid upon temporary boarding below, and completely covering the top of the joists.

Fig. 53.

The floor may be finished in cement, (c), or light joists and floor boards laid on the concrete, and the ceiling-laths nailed to light ceiling-joists (D), laid between the lower flanges. Such a floor is thoroughly impervious to sound as well as to fire, but can only be used in buildings of a substantial character, as the weight of the floor upon the walls is very considerable.

A cheaper and lighter kind of fireproof floor may be constructed on a similar principle to the last, but with inverted tee-iron joists, as shown in fig. 54. The depth

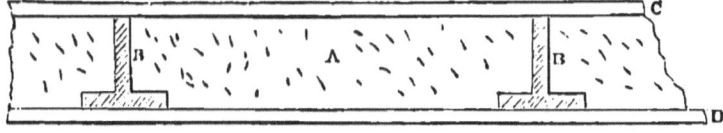

Fig. 54.

of the joists (B) will depend upon the bearing; the concrete (A), which should be made with Portland-cement, rests upon the flanges of the joists, and forms a solid mass between the floor (c) and the ceiling (D). The ceiling can be plastered against the underside of the concrete, thereby saving the cost of laths and ceiling-joists. The floor may be of cement or wood, as before mentioned. If the bearing is more than 12 feet, it will be better to use iron binders 8 or 10 feet apart, and rest the ends of the joists upon them. The load in tons that may be laid on the centre of a **T** joist, so as to produce a strain of 5 tons per square inch, is found as follows : the area of the vertical web multiplied by half the depth of the beam, multiplied by 20 and divided by the span ; all dimensions being in inches. Double this load can be borne, if equally distributed over the whole length.

GIRDERS are large beams of wrought-iron formed by riveting together by means of **L** iron thin plates of metal into the form of **I**, and called *plate-girders;* or in the shape of a long rectangular box, and called *box-girders.* The breaking-weight in tons in the middle of such beams can be approximately found by multiplying the area in inches of the bottom flange by the depth of the beam in inches, and dividing by the length of bearing in feet, the result being multiplied by 6 for plate-girders of the **I** form, by 6·5 for box-girders, and by 7 for rolled joists. Wrought-iron beams may be permanently loaded with one-third or one-fourth of their breaking-weight. When the weight is equally distributed over the entire length of a beam, it will bear twice as much load as when all is placed at the centre.

Wrought-iron is extensively employed in the construction of roofs. The parts of a roof which are subject to stretching only, as the tie-rods, king or queen rods, or other suspenders, are made of round or flat bar-iron. Those which have to sustain a longitudinal and transverse strain combined are made of angle or tee-iron. The main rafters of roofs of large span are generally plate-beams, formed in the **I** shape by riveting three plates of iron together. Cast-iron struts are sometimes introduced where resistance to compression is required. The price for iron roofing is charged by the number of squares of 100 superficial feet covered, and increases with the span.

There are several forms in which wrought-iron is employed for the main trusses of roofs, varying according to the span. Thus, for roofs of moderate span a tie-beam roof with struts and braces is employed, as shown by figures 29, 30, 31 (page 88), where the parts in tension are indicated by dotted lines. Fig. 55 also shows the

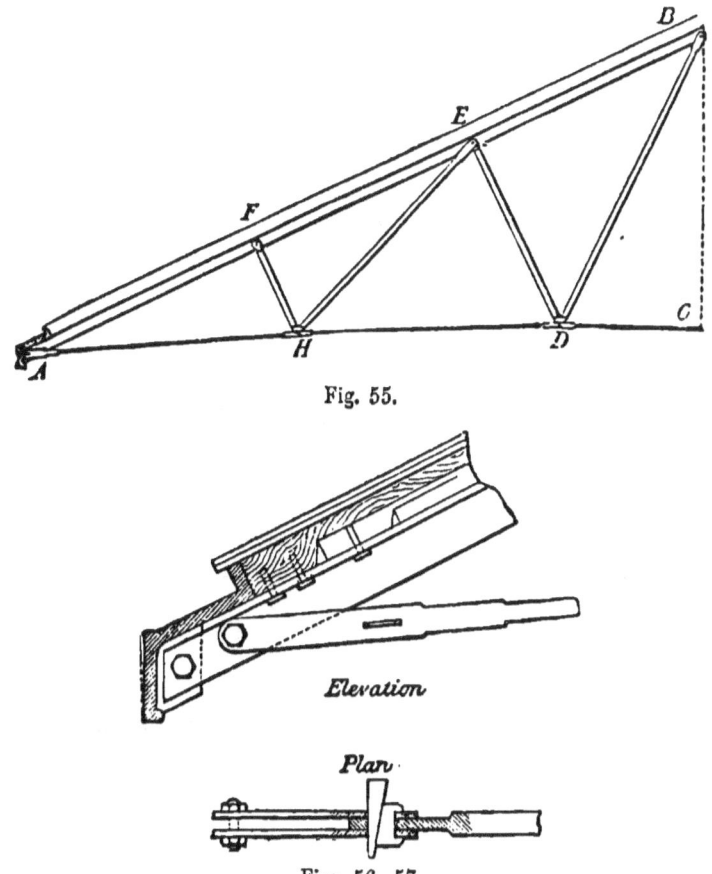

Fig. 55.

Figs. 56, 57.

half-truss of a roof of 53 ft. span, placed 12 ft. 6 in. apart, the rafters being of T iron $4'' \times 4''$ and $\frac{1}{2}$ inch thick. The struts D E, F H, are of wrought-iron piping, and the ties are of solid round iron. The tie-rod A H is $1\frac{1}{4}''$, H D $1\frac{1}{8}''$, and D C $1''$. Figs. 56, 57 show the connexion of the tie with the rafter feet.

The *bowstring-roof* is employed for long spans, and consists of a trussed principal of which the rafter or *boom* is in a curved or polygonal form, the thrust being resisted by a tie connected to the boom with struts and braces. Fig. 58 represents a truss of this kind for a span of 188

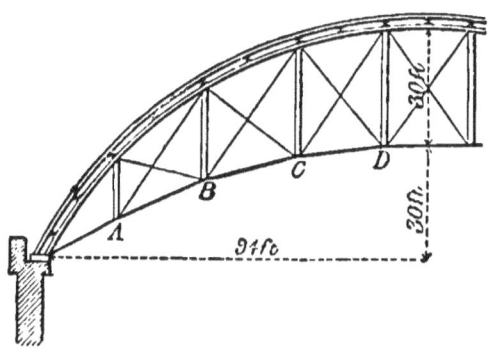

Fig. 58.

feet. The principals are 33 feet apart, and the boom is 21 in. deep of top and bottom plates connected by angle-irons to a web plate ⅜ in. thick. Purlins of similar form are placed 11 ft. apart. The tie is of 5¼ in. round-iron and the braces are flat-iron 6" × 1". The vertical ties are of two **T** irons.

Another common form of iron roof is that which consists of arched ribs without horizontal ties, sufficient abutment being provided to resist any outward thrust. These ribs are formed of considerable depth, and of **I** section, the web being either of solid plate or of open lattice-work. The depth is generally made greater at the haunches than at the centre, in order to economize material and strength.

For further information on this subject the reader is referred to an "Elementary Treatise on Roofs" (Weale's Series).

CORRUGATED-IRON is thin sheet-iron rolled into a wave form, which gives it great strength to resist bending. It is used largely for the covering of iron roofs, and is then *galvanised*, or coated with a thin layer of zinc, to preserve it from decay. Curved roofs for sheds are made very cheaply in this material, without any framework except tie-rods and suspension-rods.

Iron *tanks* or cisterns for water are formed of sheets riveted with angle-iron at the corners, or by turning in the edges at a right angle and riveting them together; and are prevented from bulging by tie-rods placed across at one-third of their depth from the bottom. The iron is galvanised to protect the tanks from decay. They are valued according to their cubical contents, and charged at a price per gallon.

The weight of one cubic foot of wrought-iron is 480 lbs.; that of a superficial foot 1 inch thick, is 40 lbs.; that of a superficial foot one-eighth of an inch thick, is 5 lbs.; that of a rod 1 inch square and 12 inches long, is $3\frac{1}{3}$ lbs.; and the weight of a square rod of the same length and any other size is found by multiplying the square of one side (in inches) by $3\frac{1}{3}$. The weight of a round rod of wrought-iron 1 inch in diameter and 12 inches long, is $2\frac{2}{3}$ lbs., and the weight of a round rod of the same length and any other diameter, is found by multiplying the square of the diameter (in inches) by $2\frac{2}{3}$ lbs.

Heavy articles in wrought-iron, as girders, joists, &c., are charged by the ton or cwt.

Galvanising is always charged extra at per foot superficial of surface covered.

Wrought-iron in straps, bolts, plates, screw-bolts, nuts, and washers, above 1 inch in diameter, are sold by the pound.

CHIMNEY-BARS are wrought-iron flat bars passed through the chimney jambs and under the arch that spans the opening for the fireplace; the two ends are turned up and down outside the wall of the chimney. They are charged by the pound.

Cores for handrails are narrow bars of iron let into the underside of the handrail of a staircase to receive the ends of the iron balusters. They are valued by the pound, and described as straight or wreathed.

Wrought-iron railings, gates, handrails, balusters, brackets, &c., are charged by the pound, unless highly ornamented, when they are charged according to the labour.

SADDLE-BARS are small wrought-iron bars let at each end into the stone jambs of a window, to which the lead lights are tied with copper. They are charged by the pound. *Casements* of wrought-iron are charged by the pound, or numbered and described.

Rivets for fastening plates of iron together are sold by the cwt.

The following is the weight of rolled ANGLE-IRON per foot run, having equal sides:—

Angle-iron with $1\frac{1}{2}$ in. sides, $\frac{1}{4}$ in. thick, weighs $2\frac{1}{2}$ lbs. per foot.
Do.	$1\frac{3}{4}$	,,	$\frac{1}{4}$	,,	,,	3	,,
Do.	2	,,	$\frac{1}{4}$	,,	,,	$3\frac{1}{4}$	,,
Do.	$2\frac{1}{4}$	,,	$\frac{5}{16}$	,,	,,	$4\frac{1}{2}$	,,
Do.	$2\frac{1}{2}$	,,	$\frac{5}{16}$	,,	,,	$5\frac{3}{4}$	,,
Do.	$2\frac{3}{4}$	,,	$\frac{3}{8}$	,,	,,	7	,,
Do.	3	,,	$\frac{3}{8}$	,,	,,	8	,,

The following is the weight of rolled TEE-IRON per foot run, of equal depth and width:—

Tee-iron with 1 in. sides, $\frac{3}{16}$ in. thick, weighs 1 lb. per foot.
Do.	$1\frac{1}{4}$	,,	$\frac{1}{4}$	,,	,,	$1\frac{3}{4}$	,,
Do.	$1\frac{1}{2}$	,,	$\frac{1}{4}$	,,	,,	$2\frac{1}{4}$	,,
Do.	$1\frac{3}{4}$	,,	$\frac{1}{4}$	,,	,,	3	,,
Do.	2	,,	$\frac{5}{16}$	,,	,,	$3\frac{3}{4}$	,,
Do.	$2\frac{1}{4}$	,,	$\frac{5}{16}$	,,	,,	$4\frac{1}{2}$	,,
Do.	$2\frac{1}{2}$	,,	$\frac{5}{16}$	,,	,,	5	,,
Do.	3	,,	$\frac{3}{8}$	,,	,,	$7\frac{1}{2}$	,,
Do.	$3\frac{1}{2}$	,,	$\frac{3}{8}$	,,	,,	$8\frac{1}{2}$	,,
Do.	4	,,	$\frac{3}{8}$	,,	,,	$9\frac{3}{4}$	,,
Do.	5	,,	$\frac{7}{16}$	,,	,,	$13\frac{3}{4}$	,,
Do.	6	,,	$\frac{1}{2}$	,,	,,	$19\frac{1}{4}$	,,

The following is the weight of rolled IRON JOISTS, of I form, like railway bars, per foot run:—

I-bars, $4\frac{1}{2}$ in. deep, $2\frac{1}{2}$ wide, $\frac{3}{4}$ thick in middle, weigh $21\frac{3}{4}$ lbs per ft.
| Do. | $4\frac{3}{4}$ | ,, | $2\frac{1}{2}$ | ,, | $\frac{3}{4}$ | ,, | ,, | $23\frac{1}{2}$ | ,, |
| Do. | 5 | ,, | $2\frac{3}{4}$ | ,, | $\frac{3}{4}$ | ,, | ,, | 25 | ,, |

WROUGHT IRON.

The following is the weight of hammered FLAT BAR-IRON, per foot run :—

Flat Bar-iron, 1 in. wide, $\frac{1}{4}$ in. thick, weighs $\frac{3}{5}$ lb. per foot.
Do. $1\frac{1}{2}$,, $\frac{3}{8}$,, ,, $\frac{3}{4}$,,
Do. 2 ,, $\frac{1}{4}$,, ,, $1\frac{2}{3}$,,
Do. $2\frac{1}{4}$,, $\frac{1}{4}$,, ,, $2\frac{1}{10}$,,
Do. $2\frac{1}{2}$,, $\frac{1}{2}$,, ,, $5\frac{1}{4}$,,
Do. 3 ,, $\frac{1}{4}$,, ,, $2\frac{1}{4}$,,
Do. 3 ,, $\frac{3}{4}$,, ,, $6\frac{1}{3}$,
Do. $3\frac{1}{4}$,, $\frac{1}{4}$,, ,, 3 ,,
Do. $3\frac{1}{2}$,, $\frac{1}{2}$,, ,, $7\frac{1}{3}$,,
Do. 4 ,, $\frac{1}{4}$,, ,, $3\frac{1}{3}$,,
Do. 4 ,, $\frac{5}{8}$,, ,, $8\frac{2}{5}$,, .

The following is the weight of ROUND BAR-IRON, per foot run :—

Round Bar-iron, $\frac{1}{2}$ in. diameter, weighs $\frac{2}{3}$ of a lb. per foot.
Do. $\frac{3}{4}$,, ,, $1\frac{1}{2}$,,
Do. 1 ,, ,, $2\frac{2}{3}$,,
Do. $1\frac{1}{4}$,, ,, $4\frac{1}{3}$,,
Do. $1\frac{1}{2}$,, ,, 6 ,,
Do. $1\frac{3}{4}$,, ,, 8 ,,
Do. 2 ,, ,, $10\frac{1}{2}$,,

The following is the weight of SHEET-IRON, per superficial foot :—

Sheet-iron ·022 in. thick, weighs 1 lb. per foot.
Do. ·032 ,, ,, $1\frac{1}{4}$,,
Do. ·042 ,, ,, $1\frac{2}{3}$,,
Do. ·056 ,, ,, $2\frac{1}{2}$,,
Do. ·065 ,, ,, $2\frac{2}{3}$,,
Do. ·083 ,, ,, $3\frac{1}{3}$,,
Do. ·095 ,, ,, $3\frac{3}{4}$,,
Do. ·125 ,, ,, 5 ,,
Do. ·137 ,, ,, $5\frac{1}{2}$,,
Do. ·158 ,, ,, $6\frac{1}{2}$,,
Do. ·187 ,, ,, $7\frac{1}{2}$,,

STABLE FITTINGS being now almost universally made of iron, should be placed, both in the Specification and in the Bill of Quantities, under the Smiths' work. Both cast and wrought iron are employed in the various fittings of stables, but the use of cast-iron must be avoided in those parts which are exposed to the kick of a horse, as any breakage will not only be injurious to the fittings themselves, but may also endanger the safety of the animal.

Stables for horses are usually divided into *stalls* about 6 feet wide, by means of parallel divisions 9 feet in length. The stall divisions are formed with a strong iron *post* at the outer end, let firmly into the ground; an iron *cill* is laid on the ground from the foot of the post to the wall forming the head of the stall; an iron *ramp-rail* is let into the head of the post at one end, the other end being firmly fixed into the wall. The post, rail, and cill have each a groove formed in them to receive the wooden matched-boarding with which the division is filled in from head to cill. This is generally 2 inches in thickness, and well tongued. The height of the division should be about 4 ft. 6 in. at the lower end, and 6 ft. 6 in. at the wall end. Instead of solid boarding the whole height of the division, a horizontal rail is sometimes fixed about 4 ft. 6 in. above the cill, and the boarding filled in between it and the cill. Above this rail is placed the ramp-rail, and the space between the two rails filled in with open iron-work, either of a plain or ornate character. By this means better ventilation is obtained than by the method of close boarding. All sharp arrises must be avoided in the open iron-work, which should be of wrought-iron, so as to avoid danger to the horses from fracture. Strong rings of iron or brass are usually attached to the top of the post, for securing the horse when occasion requires.

The food for the horse is placed in a set of fittings fixed against the wall at the head of the stall, about 3 ft. from the ground. This consists of a rack for hay, a manger for corn, and a trough for water, having a ring for securing the animal attached to the front edge.

Thorough drainage of a stall is effected by laying the paving bricks with a fall each way towards the middle of the stall, in which is placed a trough gutter of iron, covered over with a movable iron plate fitting into the rebated edges of the guttering, and laid flush with the paving. The water passes through perforations in the cover, and away into the drains. Iron gratings, with traps to prevent the smell from rising, are placed over the junction of the guttering with the drain.

A loose-box is a space cut out of the stable equal to about two ordinary stalls in area, so that a horse can move freely about therein. The height of the division is greater than that of the stalls, and uniform all round. There is an iron cill and horizontal rail, the space between which is filled in with 2-inch boarding about 5 ft. 6 in. high. Above this is open iron-work, with a top horizontal rail 2 ft. more in height. The whole is completely enclosed on all sides, having a door of the same construction as the divisions. The rack, manger, and water-trough are placed in one corner of the loose-box, and the space below them must be enclosed either with round wrought-iron bars or with wood framing, to prevent the horse from injuring itself by getting under the fittings. The drainage of a loose-box is effected by laying the paving with a slope each way towards the centre, in which a trapped *horse-pot* is fixed to receive the urine and convey it to the drain.

There are various other fittings connected with the stable, harness-room, and coach-house, which are made of iron; as the wrought-iron corn-bin, by which the food is effectually protected from vermin; iron rings for headstall chains; iron ventilating-valves for placing over the head of the stall. For the harness-room a close pedestal stove is provided, with boiler for supplying hot water for the purposes of the stable; also a saddle and harness horse, under which hot-water pipes from the boiler may be laid for drying the harness and saddles when wet; ceiling-hooks with slides, for the purpose of cleaning harness upon; open iron ventilating brackets, for hanging the harness, saddles, &c. upon, being made to the shape of the harness, and a free passage of air provided between the several parts, for the purpose of drying them underneath. These are either japanned, galvanised, or covered with leather or gutta percha.

In bringing stable-fittings into bill, each separate item must be described, the stall-posts taken at each, according to size, the rails by the foot run, the guttering at the foot run, tee-pieces, angles and traps being taken separately at each. The mangers, racks, and water-troughs are taken at each and described as cast or wrought iron, plain or enamelled. Harness fittings, stoves, brackets, &c., are charged each, or in sets, and described as japanned, galvanised, covered with gutta-percha, or covered with leather.

Cowhouses have iron extensively used as the material for their fittings, being divided into double stalls 7 or 8 ft. wide by division plates of cast-iron, the front having a round dwarf post. Each cow is fastened by a chain passing round the neck, the end of which slides up and down a vertical rod attached to the stall-division. The feeding and water-troughs are of cast-iron, and raised a little above the pavement. The fodder is placed in a wrought-iron rack, which is fixed about a foot above the trough. Behind the troughs and rack there is a passage by which the food can be brought to them without disturbing the animals.

Piggeries are also constructed with cast-iron divisions and troughs, which may be provided with swing shutters, so that the troughs may be filled on the outside, and the shutter being moved backwards enables the animal to get at the food.

The fittings for cow-houses and piggeries are charged in the Bill according to the number of animals to be accommodated.

Circular-iron-staircases are spiral stairs connected together by a central newel-post and an outer string. They are charged at per step, according to the diameter of the staircase.

Balconies and *tomb-railings*, of common pattern, are charged at per foot run, according to the height and character of the design.

Hoop-iron, used as bond for brickwork, is charged at per cwt., according to quality.

Ties and *straps* to roofs and floors are charged at per

pound; the nuts and screws to same are charged separately at per pound, but at a higher price.

Tanks of large size, made of wrought-iron plates not less than $\frac{1}{8}$ inch thick, are charged at per cwt., and described as galvanised or otherwise.

WARMING AND VENTILATING.

WARMING an apartment is effected in the simplest and most primitive mode by means of an *open stove* or *grate;* this is connected with a chimney, by which the smoke produced from the burning fuel is carried off. In order that a fire may burn properly, it is necessary that a certain amount of air should be admitted to the underside of the grate; this is commonly done by the crevices in the doors and windows through which the air finds its way, but at the same time produces a disagreeable draught to the feet of persons sitting in the room. When the crevices above named are carefully closed, and all draughts excluded from doors and windows, it is necessary to provide a supply of air to the stove by other contrivances; this can be done by gratings in the side or in the hearth, which are connected with air-pipes or flues carried to the outside of the building. As the air in a room becomes warmed it expands and rises towards the ceiling, its lightness being in proportion to its temperature, so that the air near the floor is generally colder than that above; and as the air near the ceiling becomes cool it descends towards the floor, thereby keeping up a constantly ascending and descending current. Since the air exhaled from the lungs of persons in a room is vitiated, it is desirable to remove it from the apartment; the simplest contrivance for doing which is a *valve* with a balance weight let into the chimney flue near the ceiling; the draught caused by the fire drawing the air out of the room through the valve, and taking it away up the chimney. As, however, soot is liable to return into the room through this valve when the fire is not lighted, it is sometimes let into a second flue carried up alongside of the smoke flue, the heat from which causes an ascending current in the other flue. Cold air should always be let into an apartment from the *top* of the room, since, being heavier than the warmer air therein, it gradually descends through it, becoming warmer and warmer by contact and diffusion. There are several modes of effecting this, the simplest being by opening the window at the top, but this generally causes too great a draught to be agreeable; and it is better to have a number of perforations in the ornamental parts of the ceiling, through which air is brought from the outside by pipes or flues; valves may be provided for regulating the quantity of air admitted. In order to carry off the vitiated air from rooms when the fire is not lighted, valves or gratings may be placed near the ceiling and let into flues which communicate with a furnace flue kept constantly in action.

PEDESTAL stoves of various kinds are employed for warming large rooms or halls; these, by their radiation or by currents of air carried over their heated surfaces, diffuse the heat generally through the room, the air being conveyed to them directly from the outside. The price of stoves depends on their size and the cubical contents of the rooms to be warmed by them. When these are used separate means for ventilation must be provided, and for drawing off the vitiated air, as in the manner above described. WARMING-APPARATUS for large public halls, is usually placed outside the apartment, and the warm air forced or drawn in through gratings near the floor, from which it rises in a column to the roof or ceiling, where it disperses and gradually diffuses throughout the whole building, raising the temperature of the air as it descends again; the cold air is drawn out by another grating near the floor connected by a flue with the apparatus. The *apparatus* for warming the air is placed in a chamber below the apartment to be warmed, and consists either of iron plates heated by contact with fire-brick furnaces, or of pipes filled with hot water. Apparatus of this description are charged according to the cubical contents of the building which they have to warm.

HOT-WATER PIPES are often laid under the floors of public rooms, and gratings placed over them; the heat radiated from the pipes warms the air, which ascends and diffuses through the room. These pipes vary in size according to the system employed; when the water is not raised above the boiling point, or the *low-pressure* system is used, the iron pipes are from 2 to 4 in. in bore, and in 6 ft. and 9 ft. lengths, fitted with socket joints, and having elbows, tee-pieces, syphon-bends, &c., for carrying them round corners and changing their directions. The pipes are charged at per yard length, according to diameter, and the elbows, &c., at *each*. A small wrought-iron boiler is employed for heating the water, and a small iron tank for feeding the same. When the high-pressure system is used, or the temperature of the water in the pipes raised above boiling point, iron pipes about 1 in. bore and very strong are employed, and care must be taken that they do not touch any woodwork. These pipes can be carried into all parts of a house and concealed behind gratings in the skirtings, &c., a stop-cock being provided to each room, so that the circulation may be cut off from it when required. In all these methods the ventilation must be provided for separately; and this is generally done in new buildings erected with this object in view, by air flues connected with the smoke-flue from the furnace of the apparatus, into which there are openings near the ceiling of each room. The smoke from the furnace can be taken up an iron chimney-pipe fixed inside a large air-flue, the air in which being heated thereby is kept constantly ascending with great force, and thereby drawing the vitiated air out of the several apartments from which flues are connected with it.

PRISON-CELLS are warmed and ventilated by bringing the warm or fresh air into the cell through a grating near the ceiling, and drawing it out by another grating near the floor in the opposite wall, into a large flue in which a strong ascending current is produced.

VENTILATING large rooms placed immediately under the roof of the building—as factories, meeting-halls, &c., may be effected in several ways;—one is by means of a tube of considerable size, fixed through the roof and descending into the room; this tube is divided down the middle into two compartments and termed a "syphon-ventilator," since the external air, when there is not a high wind, descends by one side of the tube, and the internal air ascends by the other side, thereby keeping up a constant change in the air of the room. The top of the tube above the roof is covered over with a cap, so as to protect it from the weather, and under this the air can pass both ways. A valve is placed in the tube so as to allow of the ingress and egress of air being regulated. When the atmosphere is in violent motion from a high wind, the syphon system often refuses to act, the external air descending *both* sides of the tube upon the heads of occupiers of the room. The diameter of the tube, which is generally made of zinc, depends on the cubical contents of the room to be ventilated. The same principle of ventilation is applied when the upper part of a sash is made to swing upon centres, a double current of ingress and egress being thereby produced.

Foul air may also be extracted from an apartment by means of a long tube carried above the roof and surmounted either by a revolving cowl or a set of *louvres*, which produces an ascending current in the tube; for a description of which method, see Buchan's "Plumbing" (Weale's Series). As, however, all these methods are dependent on the relations between the external and internal air, and also upon the force of the wind, they are uncertain in their action, and it is necessary in a regular system of ventilation, as for hospitals, concert-rooms, theatres, &c., to have an artificial mode of producing a current, which is generally by means of a revolving fan driven by machinery, so that the velocity of the current can be regulated at pleasure. The heat generated by a set of gas-burners at the bottom of a tube will also be found to produce a similar effect, and is also capable of regulation.

SUN-BURNERS are also very useful for extracting vitiated air from a room, and consists of a cluster of gas-burners arranged in circles under a reflector, above which is a tube for conveying the products of combustion to the outside; in this tube a strong upward current is produced by the great heat generated by the burners, so that the air is drawn upwards, and the room effectually ventilated and lighted at one and the same time.

STOVES for the open fireplaces of rooms vary greatly in character and cost. *Elliptic stoves* consist only of a pair of *hobs*, with bars in front of an elliptic form. They are charged according to the width of the opening, at per inch, the price varying with the quality. *Sham-fronts* are, as their name implies, only made with a front to fit into the opening, and a grate for the fire without hobs. They are sold at per inch in width. *Register-stoves* are entirely enclosed from the chimney, with which they communicate by a *valve* or *register*, which can be opened or closed at pleasure. They are generally made without hobs, and with or without fire-brick *backs* and *cheeks*. They are sold at per inch, according to width of opening, the price varying greatly, according to the thickness of metal, the ornamentation, and general finish.

Slow-combustion-stoves are made with solid fire-brick bottoms resting upon the back-hearth, air being conveyed to the fire only through the front bars. The fires in these stoves are said to burn for several hours without attention, giving out a maximum amount of heat for a minimum of fuel, and producing very little smoke. The sides and back are all of fire-brick, and an iron shutter or blower is generally provided to assist in lighting.

COOKING-RANGES are made either open or enclosed; when enclosed they are termed *kitcheners*, and are provided with an iron cover to enclose them from the chimney, with a register therein to regulate the draught. The top of a kitchener forms one continuous hot plate, and the smoke is carried away by a pipe or flue at the back. The price varies greatly, according to quality, but is generally proportioned to the width of the opening. *Open-ranges* have the fire quite open at the top, an oven being on one side, and a boiler on the other: the better kind are made to wind-up, so that the width of fire may be regulated as required. The price is charged at per inch of width, but varies greatly, according to quality. In common ranges the boiler is of cast-iron, and in the better sort it is of wrought-iron. The boiler of a range is kept constantly full by a feed-pipe from a small cistern set on the same level outside the fireplace, in which the water is kept always at the level required in the boiler by means of a ball-cock.

COPPERS are charged at per lb., when made of copper. They are frequently made of cast-iron, and called *set-pots*. These are sold by the cwt. The bearing-bars, furnace-doors, grates, &c., for coppers are charged separately; the bars by the lb., and the doors and grates at a price for each.

The SETTING of stoves, ranges, and coppers is charged in the bricklayer's account according to number.

REVOLVING-SHUTTERS.

REVOLVING-SHUTTERS are those that are made to coil upon a roller placed out of sight, and worked up and down by means of gearing. These shutters are formed of strong laths, either of iron or wood, which are attached together by hinges, so that they can be readily wound over a roller. The roller is placed either at top or bottom of the window, according to convenience, and the shutter wound upon it by means of *worm-and-wheel*, or by *chain-winding* gear. Another mode is to wind the shutter *horizontally*, and is sometimes employed in circular bay windows of first-class mansions. These shutters can be fitted either on the inside or the outside of a window, but are more commonly placed outside.

The following is the space occupied by revolving-shutters when coiled up, a small additional space being required for clearance, of one or two inches, and rather more if coiled *below* the window:—

Windows 5 ft. high require $6\frac{3}{4}$ in. for iron, and $8\frac{3}{4}$ in. for wood.
„   7 „ $7\frac{1}{2}$ „   10   „
„   10 „   9   „   12   „
„   12 „   10   „   13   „

SELF-COILING shutters are made to coil by means of a *steel-band* passing through all the laths, if of wood, and causing the shutters to coil and uncoil by merely pressing the end up or down as the case may be. *Steel self-coiling* shutters are made in one sheet of corrugated steel, and are worked up and down by the hand without balance-weights or gearing.

The following is the space required for self-coiling shutters when rolled up:—

Windows 6 ft. high require 9 in. for wood, and 8 in. for steel.
,,   8   ,,   10   ,, $9\frac{1}{2}$,,
,,   10   ,,   12   ,,   10   ,,
,,   12   ,,   13   ,,   12   ,,

Self-coiling shutters are unsuitable for windows of great width, but the revolving-shutters can be applied to any width of window in one piece.

Revolving and coiling shutters are measured by the foot superficial, and priced according to material and gearing; an additional charge per foot is made for shutters under 50 feet area. In measuring, the full width is taken, which is 2 inches more than the *sight* measure; and the height is 9 inches to 12 inches more than the sight measure, in shutters working vertically.

In measuring shutters which work horizontally, the actual dimensions are 6 inches in height and 9 inches in width greater than the sight measure.

In preparing for revolving-shutters, provision is to be made for the easy removal of all linings and casings, so as to allow of access to the shutters, shafts, and gear, which require occasionally to be oiled, cleaned, or repaired.

LIGHTNING-CONDUCTORS are formed either of copper-ropes from $\frac{3}{8}$ths in. to $\frac{3}{4}$ths in. diameter, or of flat bands of solid copper in continuous lengths without any joints, varying from $\frac{1}{16}$th to $\frac{1}{4}$th in. in thickness, and from $\frac{5}{8}$ths to 2 ins. wide, according to altitude, fixed with holdfasts of gun metal to the highest points of a building, and taken down into the ground for the purpose of carrying off the electricity from the atmosphere, and preventing it from striking the building. The rope or band is buried two or three yards below the foundation and about the same horizontal distance from it, in moist earth and surrounded with coke or cinders. The top of the rope is finished with *copper-points* gilt over. The copper-rope is charged at per foot run, according to its diameter; the points and holdfasts at *each*. Glass insulators are sometimes employed to pass the rope through and prevent its contact with the wall; these are charged *each* extra, but they are of no value in protecting the dwelling, as the electricity will always follow the copper as being the best conductor. If there are several lofty points in an extended building, such as chimney stacks, turrets, gables, &c., it is not sufficient to place a conductor on one point only, but two or more will be necessary if at a considerable distance from each other. All the external metal-work, as finials, lead-roofing, and gutters, iron rain-water piping and guttering, should be connected with the copper conductors wherever it is feasible; but the conductors should be kept clear of all gas pipes.

LIFTS are mechanical contrivances, whereby persons or goods can be raised or lowered from one story of a building to another. A common form of lift consists of a winding drum or pulley, over which is passed a rope or chain, to one end of which is attached the *cage* in which the articles to be raised or lowered are placed; and the other end is connected with a balance-weight rather heavier than the weight of the cage when empty. The winding pulley is made to revolve by means of gearing which may be driven either by manual labour exerted upon an endless rope, or by a strap moved by steam power which can be moved off and on to the revolving drum by means of a lever to which a cord is attached; a break is also applied, to prevent the cage from falling beyond the proper place. The cage moves within a framework formed of four uprights or guides which keep it in its place, and prevent it from swaying to and fro during its ascent or descent.

HYDRAULIC LIFTS are those in which the pressure of water is the raising power applied to the cage. These are frequently used in warehouses, factories, hotels, and other large establishments. The essential feature in these lifts is a good supply of water in a tank placed in the highest part of the building; when this is secured, its application to lifts is more convenient than that of any other force, since it is always available by the simple act of opening a cock or valve, as long as the supply of water remains in the tank. The construction of these lifts is as follows: a long plunger or ram of cast-iron is fitted accurately into the stuffing-box of an iron cylinder which is sunk in the ground and firmly bedded; water is brought into the cylinder from the tank by one pipe and carried off by another. On the top of the plunger is fixed a strong cast-iron plate forming the bottom of the ascending cage, which is constructed of a wrought-iron frame, and may be filled in with wood or glass if used for a passenger lift. At the top and bottom of the frame are two brass **V**-shaped guides which work against two cast-iron **A**-shaped guide bars placed perfectly upright. A counterbalance weight slides in a shaft separated for that purpose. There is a contrivance for regulating the speed of the lift, which would otherwise vary according to the weight placed in it. By opening a valve the water is let into the cylinder and causes the piston to rise, carrying the cage up with it; as the water is let out of the cylinder the piston and cage descend; and since water is incompressible, the moment the valve is closed so as to shut out the supply of water, the lift comes to a standstill.

IRONMONGERY.

Under this title are included nails, screws, bolts, pulleys, hinges, door-springs, casement and sash fastenings, locks, latches, shutter-bars, lifts, rings, knobs, buttons, brackets, hooks, hat-pins, and all other articles of iron or brass fixed by the joiner. Ironmongery is generally placed in the Bill of Quantities, immediately after the Joiner's Bill; and the articles are all numbered either at each or at per pair.

Twenty per cent. profit is allowed on the prime cost of all ironmongery.

NAILS in large quantities are sold by weight, and are

charged in the Bill at per hundred or per thousand, under the names of tacks, cut-brads, flooring-brads, cut-clasp and wrought nails, which are sold as threepenny, fourpenny, sixpenny, &c., according to the price per hundred. Brass-headed nails and screws, either of iron or brass, are charged at per dozen. *Tacks* are small nails with flat heads, and generally *tinned* over or *blued*, these are used by the Joiner for fixing small articles. *Clouts* are larger nails with rounded and pointed shank and flat head for fixing ironwork on wood. *Brads* are blunt pointed nails having the top widened so as to form a head, and are used for laying floors and fixing skirtings, mouldings, &c. *Spikes* are large nails used by the carpenter for fixing timbers to wall-plates. *Clasp*-nails have sharp point and arrow-shaped head, they are much used in joiners' work. *Holdfasts* and *Wall-hooks* are large nails used for holding woodwork to a brick wall. *Lath* nails for fixing ceiling laths are short nails with flat heads. *Brass-headed* nails have an iron shank and a brass head for fixing brass work on joinery. *Screws* are used for fixing hinges, bolts, locks, &c., upon woodwork.

BOLTS are made of iron or brass. Iron bolts for doors are of various kinds, bright-barrel, rough-rod, bright-rod, and spring-plate. They are charged at *each*, the length being described. Brass flush-bolts are iron-bolts with brass face-plates, and are let into a groove cut in the door, so as to have the face-plate *flush* or even with the surface of the door. They are charged each, the diameter of the iron bolt and the length of the face-plate being described. Espagnolette bolts are long bolts going the whole height of a French casement, and having a fastening at top and bottom, and also in the middle. They are charged according to their length at per foot.

HINGES for hanging doors, casements, flaps, &c., are of great variety, and generally charged at per *pair*. *Butt* hinges have the knuckle projecting from the face of the door, and are screwed to the edge of the door and to the rebate of the jamb; so that when the door is closed the two parts or flaps of the hinge come close together. *Back-flap* hinges are used for shutters hung in several folds or flaps. They are similar to butts, but have a longer flap, which is screwed to the back of the shutter. Butts and back-flaps are described by their height, and sold at per pair. *Cross-garnet*, or tee-hinges, are made of a **T** form, and are valued according to the length measured from the joint. *Parliament* hinges are of an **H** form, and used for outside shutters.

SPRING-HINGES are made to cause a door to fall-to of its own accord. They are of great variety of construction and cost. The better sort consist of a box containing the spring, let into the floor at the bottom of the door, and are sold either singly or by the pair.

GATE-HINGES, for gates and coach-house doors, are made with a long arm bolted to the face of the door, and turn on a pivot fixed to the post or jamb, with a cup and ball, or spherical bearing. They are valued according to the length of the arm, at per pair, and described as wrought or cast iron.

LOCKS are mechanical contrivances for securing doors when closed, and vary in quality, description, and price more than any other branch of ironmongery. They are charged according to size and description at *each*. *Cupboard-locks* are small locks with a brass plate for fixing on cupboard-doors. Iron *rim-locks* are those in which the wards are placed in an iron box or *case*, which is screwed outside the door. *Mortice-locks* are superior in quality to rim-locks, and are let into a *mortice* cut in the edge of the door, having a brass *face-plate* flush with the edge of the door. They are described by the thickness and length of their case. Mortice and rim locks may have one, two, or three bolts, and are described accordingly.

FURNITURE is the name given to the *knobs* or *handles* fitted to the lock of a door, and charged at per *set*, the quality and material being described, and also whether common or patent mode of fixing.

Escutcheon is the plate placed over or round the keyhole, made in the material of a door. A *thread* escutcheon is a narrow slip of metal let into the keyhole, and taking its exact form.

Stock-lock is a lock with a single bolt for a key only, the wards being placed in a *case* of wood or iron, and fixed on the side of the door; the price depends on the size and quality.

Draw-back lock is placed on the inside of the outer doors of houses, and furnished with a knob to *draw* the bolt back; they are charged according to length.

Master-key is a key that will fit all the locks throughout a house, although the several keys of those locks will not fit any but their own locks. An extra price is charged for locks having a master-key.

The terms *dead-lock, bushed-lock, wheel-ward, tumbler,* are applied according to the arrangement of the wards of locks. There is a great variety of patent locks, of various qualities and prices, the exact description of which must be given in the Bill.

LATCHES are lever fastenings for keeping doors closed without locking, and work on a pivot at one end which falls into a catch at the other. *Thumb-latches,* Norfolk or Suffolk, have the latch lifted by a lever pressed upon by the thumb. *Mortice-latch* is one that is let into the edge of a door, and opens by means of a knob; the price depends on the size. *Lifting-latches* are also made of large size for church and other doors, and *lifted* by an ornamental ring attached to a spindle.

Shutter-bars, made to fasten across folding shutters with a spring, are charged each, according to length.

Flush-lifts are let into the bottom rail of heavy sashes, for giving a hold to the fingers in opening them.

Shutter-lifts are rings let into the top of sliding or lifting shutters, for the purpose of raising or drawing them out.

Flush-rings are small brass rings let *flush* into a flap or door.

Sash-centres are fixed to the sides of a sash which is made to swing either vertically or horizontally.

Hat and *cloak pins* are single or double, either of brass or japanned iron, and are generally screwed to a *pin-rail* fixed to a wall.

Iron *meat-hooks,* tinned, for screwing to the joists, are

sold at each. Brass *dresser-hooks*, for screwing into the front of shelves, are charged at per dozen.

Iron *brackets* for fastening under shelves, are charged *each*, according to length.

Brass *closet-knobs* and *turnbuckles* are charged *each*; mahogany *knobs* at per dozen.

Cabin-hooks, for securing casements when open, are charged *each*, and whether iron or brass.

Sash-fastenings, for securing hung-sashes when closed, and fixed at the meeting bars, are charged *each*, and described as common, spring-roller, or patent. There are several varieties of these articles.

Thumbscrew fastenings are used to secure lifting shutters, which are hung in two heights. They are provided with brass plates let flush into each shutter at their meeting, and are charged each.

Sash-lines for connecting sashes or lifting shutters with their balance weights, are of different kinds. Those commonly used are the *flax* line and the *patent* line, which are sold in "knots" of 12 yards, and are of various thicknesses or gauge. Besides these, there are brass link chains, which are only used for very heavy sashes, and must have pulleys made to suit them. They are sold by the foot run, according to strength. Copper twisted cords and steel ribands are sometimes used for the same purpose, and are sold by the foot run.

BELL-HANGING.

Bells, for producing sounds, whether in private houses or public buildings, are made of a mixed metal, composed of copper and tin, in the proportion of four parts copper to one part tin, which is termed *bell-metal*, and is poured into moulds of a hollow cup-shape. Bells which are intended to be swung are fitted with clappers, generally made of iron, hung from the centre, and striking the lower part or *sound-bow* of the bell on the inside, this being its thickest part.

Large bells for churches, turrets, factories, &c., are sold at per lb. weight. They are attached to a horizontal

P

beam, at the top or crown, called the *stock*, at the lower ends of which are metal centres, called *gudgeons*, which are let into supports, and on which the bell swings freely. The smaller-sized turret-bells are rung by means of a rope attached to a lever projecting from the top of the stock; but in large bells this lever is replaced by a wheel having a groove round its circumference, in which the rope is fixed. The tone of a bell depends upon its size and thickness; a small bell giving a higher note when struck than a large one made in the same proportions.

HOUSE-BELLS are hung for the purpose of communicating from one room to another. They are usually tuned in peals, so that every one has a different tone, and have a curved spring fastened to the crown of the bell, and attached by a spring *carriage* and brass T-plate to a *bell-board*. Copper wires are conveyed by means of cranks from the bells to the various rooms, and are concealed from view in zinc or copper tubing, embedded in the plaster of the walls. A *pendulum* is often attached to each bell, to show which bell has rung. In estimating bell-hanger's work, it is usual to *number* the bells and describe the mode of hanging, including cranks, springs, wires, carriages, tubing, &c., in the price per bell. When the several articles are used separately, as in repairs to old work, the cranks are charged *each*, and described as purchase-cranks, mortice-cranks, either single, double, treble, or quadruple. Also the bells are charged *each*, according to size, including steel-spring, brass T-plate, and back-spring carriage. Pendulums with springs are charged *each*. Copper wire is charged at per lb.; copper or zinc tubing at per foot run. *Wheel* and *chain* apparatus, for turning corners, are charged *each*. *Lever-pulls* are the handles by which the bells are rung, and are fixed on the side of the fireplace of each room. They are generally ornamental, and with knobs to match the door furniture.

ELECTRIC BELLS are used in large houses and hotels, where the number of cranks and the great length of wire required for connecting the bells with the pulls upon the old system, renders the pulling very hard, and the liability to get out of epair very considerable. When electricity is employed there are no mechanical actions or cranks, and no motion required from the wires. The electricity is generated in a sand battery charged with dilute acid every six or twelve months. The rooms are supplied with bell-pulls, ropes, levers, or knobs, from which are brought two flexible copper wires covered with silk, and which act as conductors of the electric fluid from the battery to the bells. A single bell is sufficient for any number of rooms, and is connected with an indicating tablet, which has the name or number of each apartment marked upon it. When a lever is pulled in any room, the bell rings, and a red disc appears at the aperture on the tablet corresponding to the number or name of the room.

PNEUMATIC BELLS are generally rung by means of an india-rubber bag or bellows which being pressed upon by the hand communicates a pressure to a bell and indicator by means of small tubes. This is done either by pressing the finger on a knob, or compressing an india-rubber ball in the hand. One bell suffices as in the electric system with indicator for each pull. Another system dispenses with the india-rubber bag, and employs a metal cylinder fitted with piston and rod, the indicator being also provided with a similar apparatus which rings the bell and displays the tablets. The pneumatic system can only be employed where the distance between the pull and the bell is not very great, but with the electric bell distance is no object.

The different parts are charged separately; indicators at per number, bells each according to size, press buttons each according to quality, battery according to number of cells, wire at per 100 yds metal tubes at per yard, elastic air balls each.

CHAPTER VIII.
PLUMBING, ZINC-WORKING, GAS-FITTING.

PLUMBING.

Plumber's-work or Plumbing is all that part of the construction of an edifice which relates to the use of *lead*, and also to the supply of water to the building. *Milled-lead* or lead rolled out in sheets is used for covering roofs and flats, for lining gutters and cisterns; and for all other purposes where lead in sheets is required: it is described according to weight per superficial foot, the thinnest weighing about 4lbs. to the foot. In using lead on roofs or flats, a covering of boards is first laid by the carpenter, and on this are nailed a number of parallel *rolls* of wood, 2 or 3 ft. apart, running from top to bottom of the roof;

Fig. 59.

the lead is then cut in strips somewhat wider than the distance apart of the rolls, the edge of one strip dressed over the roll, and that of the next strip dressed over the first, so as to make a water-tight joint; the lead is kept from being raised up by the wind by *lead-headed* nails driven through the lead into the rolls, fig. 59. Lead should never be used in very large sheets, as it is liable to crack; it is therefore usual, when a large roof is to be covered, to make the sheets in short lengths of 10 or 12 ft. and to *lap* them one over the other. When laid on flats or roofs of low pitch, a *drip* or step of about 2 to 3 inches is formed at the bottom of each sheet, the upper end of the lower sheet is turned up against the drip, and the lower end of the upper one turned down over it. The same plan is adopted in forming lead-lined gutters.

Fig. 60 shows the usual mode of laying the lead over the drip B, but this has a tendency to draw the water up behind the lead, and the method shown by fig. 61 is

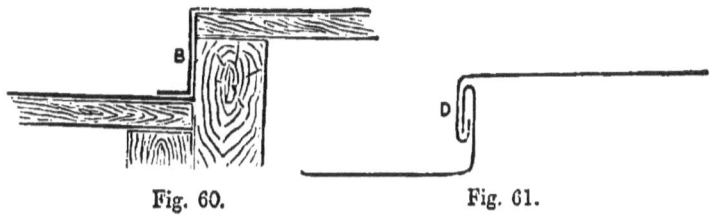

Fig. 60. Fig. 61.

preferable; the lead being turned round at D so as to form a *clinch*. (See Buchan on Plumbing.)

When lead is used to form the covering of the ridges and hips of slated-roofs, a ridge-roll of wood is first fixed;

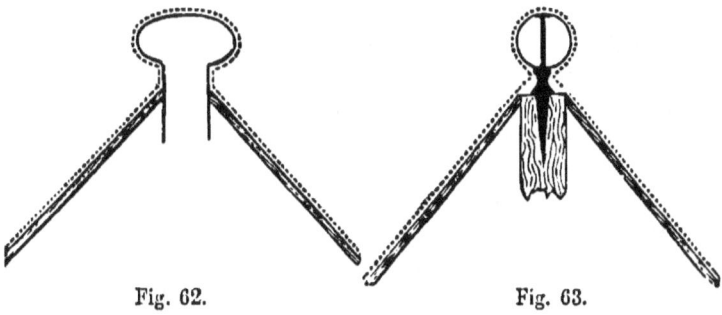

Fig. 62. Fig. 63.

then the lead is dressed over it, secured with lead-headed nails, and laid several inches over the slates on each side. Figs. 62 and 63.

FLASHING is a strip of lead laid upon a roof where it abuts against a wall; the lead is turned up against the wall, and the upper edge protected from weather by a lead *apron* let into a groove or joint in the wall, and hanging down 2 or 3 inches over the flashing. *Stepped-flashing* is where the lead is cut out in steps and turned into the joints of the wall, following the slope of the roof. Flashing is secured to the wall with *wall-hooks*. Soakers are small pieces of lead or zinc used as a flashing against a wall, half of each piece going up against the bricks, and half under the slates, and lapping one over the other.

LEAD GUTTERS are used where the eaves of two roofs meet, as in figure D, Plate III.; or where the eaves abut against a parapet wall, as in figure C, Plate III. The lead is cut of sufficient length to cover the level gutter board and to turn up from 9 to 12 inches on the slope behind the slates or tiles. It is laid in lengths of 10 to 12 feet, and secured with drips as above described for flats; a fall of about 2 inches in 10 feet is generally given to the gutter, which consequently gets wider in the upper part than in the lower; and this must be taken into consideration in measuring the lead work which is always taken nett. No soldering must be used in the lead-work of gutters.

The lowest end of a lead gutter is formed into a box called a *cesspool*, to collect the water over the fall-pipe head. The lead of a gutter must always be laid as much as half the length of a slate (at least) under the eaves of the slates; and in order to prevent melting snow from getting in between the lead and the slates, a grating of wood or metal is placed across the gutter a few inches above it, so that the snow on the top of the grating in melting is carried off by the gutter below.

LEAD CISTERNS are generally made of wood and lined with milled lead, the bottom being covered with stouter lead than the sides; the lead is *soldered* along all the joints with *solder* made of equal proportions of tin and lead; it is dressed over the top of the cistern and fastened with copper nails.

Plumber's work is valued according to the market price of lead, at per cwt., to which must be added the labour; for this, however, we have not sufficient data on which to base a set of constants for this description of work. Lead-headed nails, wall-hooks, and hold-fasts, are charged per piece; clout nails, by the hundred. Solder is sold by the pound, joints being charged separately from the lead itself; copper nails by the lb. weight.

The following is the thickness of milled lead according to its weight per foot superficial;

Milled lead weighing 4 lbs. per foot, is ·067 in. or $\frac{1}{15}$ in. thick.
,, ,, 5 ,, ·084 ,, $\frac{1}{12}$,,
,, ,, 6 ,, ·101 ,, $\frac{1}{10}$,,
,, ,, 7 ,, ·118 ,, $\frac{1}{8}$,,
,, ,, 8 ,, ·135 ,, $\frac{1}{7}$,,
,, ,, 9 ,, ·151 ,, $\frac{1}{6}$,,

TRAP-DOORS and SKYLIGHTS fixed on slated roofs require to have lead laid round them to keep them weatherproof. The frame is made to rise 2 or 3 inches above the slating, and the lead is laid 6 to 9 inches under the slating at the top of the frame and two sides, but over the slates at the bottom, being dressed over the top of the frame and nailed along the edges with copper nails, and soldered at the angles. The trap door, which is also covered with lead, is made with ledges round the edges, which lap over the sides of the frame.

DORMER-WINDOWS are those which are built out from the roof, forming valleys with it which have to be lined with lead laid under the slates as above described. The sides or *cheeks* are covered with lead, which is also laid under the slates which abut against the sides. A lead flashing is also fixed in front of the dormer and laid on top of the slates.

MEASUREMENT of sheet-lead where used in a building is by the square foot, the flat and gutters in one item, hips and ridges in another, flashings in another, and the linings of cisterns separately, as a different price is charged for each. Multiply the number of square feet by the weight per foot, as 5 lb., 6 lb., 7 lb., &c., which gives the weight in lbs., and this divided by 112 brings it into cwts., in which it is entered in the bill.

DRAWN-LEAD-PIPE, made by hydraulic pressure, is used for conveying the water supply to the different parts of a building: it is of three qualities or thicknesses, and known as common, which is the thinnest; middling, and strong. It is charged either by weight or by the foot run; all solder joints are charged extra at each, according to the size of the pipe; or they may be charged according to the weight of solder used.

The following is the weight of lead-pipe, according to strength and diameter of bore:—

				Common.	Middling.	Strong.
Lead pipe of	1/2-in.	bore,	weighs per foot,	1 lb. 0 oz.	1 lb. 2 oz.	1 lb. 7 oz.
,,	5/8	,,	,,	1 2	1 5	1 9
,,	3/4	,,	,,	1 9	1 13	2 0
,,	1	,,	,,	2 6	2 13	3 5
,,	1¼	,,	,,	3 0	3 8	4 5
,,	1½	,,	,,	4 0	4 10	5 5
,,	1¾	,,	,,	6 5	7 0	8 0
,,	2	,,	,,	7 0	8 0	9 5
,,	4	,,	,,	12 0	14 0	18 0

Iron pipes are often used for bringing water from the well or service to the house, and are made with screwed joints which are fitted with red-lead. They are charged at per foot run.

Lead should never be employed to convey or to hold soft or rain water, as it is readily attacked by the carbonic acid contained therein, and the carbonate of lead formed, which produces poisonous effects when the water is used for drinking purposes. To prevent this a lead pipe lined with *tin* is sometimes used. *Socket-pipes* are used for conveying water from sinks to the drains, and also for conveying the soil from water-closets to the drains; they are made of sheet lead, which is bent round and soldered along the edges. Seamless soil pipes are now drawn by machinery in the same way as the smaller pipes, and are much to be preferred to the old soldered pipe. They are charged at per foot run according to their diameter and their weight per foot superficial of lead. *Funnel-pipes* are made trumpet-shape, wider at top than at bottom, and act as standing-wastes in cisterns to prevent them from overflowing; they are fitted into the waste-pipe with brass *washer* and *waste*, which are ground so as to be water-tight; they can thus be easily lifted out of their place when it is required to empty the cistern. They are charged at per foot run according to diameter, and the washers and wastes at each according to size.

A TRAP is a contrivance to prevent foul air from passing into a house from a drain by means of a waste pipe; this is effected by forming a barrier of water in the middle of the pipe. A *syphon-trap* is the simplest form, being a dip made in the pipe, so that while water will pass along it to the drain, a quantity of it always remains in the dip or syphon, and prevents the foul air from returning. A *bell-trap*, used in sinks and in sink-stones to areas, consists of a raised circle let into the stone so as to form a reservoir of water; a moveable *bell* attached to a grating is dropt over it and effectually traps the waste pipe by the edge of the bell dipping into the water below; it is kept clean by occasionally lifting up the bell.

D-*traps* and **P**-*traps* are those which are made of lead of the form of the letters **D** or **P** to receive the soil pipe from the basin of a water-closet and stop the return of foul air from the drain; they are placed immediately under the closet apparatus. Cast and drawn-lead traps are now made in one piece in the syphon form, and are stronger and more effective than the other traps. Traps are numbered in the bill.

COCKS are used for turning water off and on from supply-pipes; they are of various kinds, as bib-cock, stop-cock, ball-cock, &c. *Bib-cocks* have the outlet hanging down, and are turned off and on by hand; they are used for drawing off water from pipes. *Stop-cocks* are placed in the middle of a pipe for cutting off the supply of water; they are turned by hand or by means of a spanner. *Ball-cocks* are similar to bib-cocks, and are used for supplying cisterns, having a hollow metal ball attached by means of a lever which regulates the supply; the ball being hollow floats on the surface of the water, and as it rises gradually turns the ferrule of the cock, so as to cut off the supply when the cistern is full. Cocks are generally made of gun-metal, and a cheaper sort of galvanised iron; they are numbered in the bill.

PUMPS are employed for drawing water from wells to supply a house. The common *suction-pump* is fixed in the place where the supply is required to be drawn, and is connected with the water in the well by a pipe; it will not draw water more than 30 ft. in height. When the surface of the water in the well is deeper than 30 ft. from the top, a *force-pump* is used, and is placed in the well near the surface of the water; it is worked either by an ordinary handle or by a wheel to which a handle is attached, and the water can be forced to any part of the building. The size of the pump must depend upon the supply of water required to be delivered in a given time, which is proportional to the area of section of the barrel of the pump, or to the square of the diameter. Thus, a 2-in. pump worked 25 strokes per minute with 9-in. stroke, will raise 154 gallons of water per hour; a $2\frac{1}{2}$-in. pump will raise 240 gallons; a 3-in. pump will raise 346 gallons, and so on. The pipes must also be in proportion to the size of the pump, a $2\frac{1}{2}$-in. pump requiring a $1\frac{1}{4}$-in. pipe, a 3-in. pump a $1\frac{1}{2}$-in. pipe, a $3\frac{1}{2}$-in. pump a $1\frac{3}{4}$-in. pipe. When a large supply of water is required to be raised quickly, a double or treble barrel pump is used, the quantity of water being proportional to the number of barrels. In deep wells stages are fixed at every 12 feet depth, and are fitted with *roller-guides* for the rods to pass through. A copper *air-vessel* is sometimes fitted to force-pumps, in order to give a continuous stream of water and to assist the working of the pump. Pumps are charged at each, according to their diameter and description; the rods, wheels, and framework are charged extra, according to the depth of the well.

WATER-CLOSET APPARATUS are of various kinds, the simplest being a pan or basin, made of glazed ware, having a syphon-trap attached at the bottom to connect with the soil pipe or drain. There is an opening in the upper part of the pan, to which a water pipe can be attached for the purpose of flushing out the basin.

Another kind of closet has a *valve* let into a socket in the cistern above; this is raised by means of a handle to which is also attached a *copper pan* fitting under the basin; a lead *service-box* is fixed to the bottom of the cistern under the valve, to supply a sufficient quantity of water after the valve is closed. Other kinds of apparatus are made to work without any valve in the cistern, or service-box, having a *regulator* attached to the apparatus itself; these can be placed at any distance from the cistern, and do not require to have one provided especially for them. *Self-acting* water-closets are those in which the valve which regulates the water supply is worked by means of the door or seat. When worked by the latter, the seat is hung on hinges, and a slight depression caused by the weight of the person using it opens a valve by which a regulator is filled with water, and this is discharged into the basin as soon as the pressure on the seat is removed. Water-closets are charged at *each*, according to their description and the quality of the basin and other parts.

WASTE PREVENTER is a small cistern interposed between the house cistern and the W. C. apparatus in order to regulate the quantity of water used to flush the basin and to prevent the water from being wasted. This cistern is made of cast-iron or earthenware, and capable of holding 2 gallons of water, being kept full by means of a ball-cock with supply pipe from the large cistern. When the handle of the W. C. is raised a valve is opened which discharges the whole contents of the cistern into the basin. The water then ceases to run, and the cistern is slowly filled again from the ball-cock, when another flush of water is obtained by raising the handle as before. There are several different forms of these Waste Preventers, varying considerably in price, and it is necessary to state the particular inventor's name when numbering them in the bill.

EARTH-CLOSETS are substitutes for water-closets where there is an absence of water-supply or of efficient drainage to carry off the soil. A *receptacle* or *pail* which can be readily removed is placed under the "seat," and each time the closet is used a quantity of finely-sifted dry earth or ashes mixed with earth is poured into it from a reservoir above ; by this means the *soil* is deodorised, and the contents of the receptacle can be emptied on the land and used as manure. When the closets are fixed in an upper story of a house, a straight pipe of earthenware or galvanised iron 12 inches diameter is attached to the bottom of the *pail*, to which there is a valve opened and closed by a handle; this pipe conveys the soil and earth down to a vault or receptacle placed below.

BATHS in private houses are usually supplied with hot water from a boiler in the kitchen range, for which purpose two cisterns have to be provided, one of which must be at a higher level than the bath-room ; the one for cold water is an ordinary open cistern, the other for hot water is a galvanised iron cistern, strongly riveted, and entirely closed, having a pipe passing through the top to carry off the steam to the outside of the house. A pipe descends from the bottom of the hot-water cistern to within 2 in. of the bottom of the close boiler in the range ; a pipe from the cold-water cistern joins this pipe a short distance below the hot-water cistern, and thereby keeps up the supply. Another pipe leaves the top of the boiler and conveys the hot water therefrom to the close cistern which forms a reservoir to supply the baths ; from this last pipe the hot water can be drawn off by branches to any part of the house. The ends of the pipes that enter the boiler should be of copper, and attached to the supply pipes with strong brass *unions*. The hot water cistern should be fixed as near to the kitchen range as convenient, in order to prevent the danger of the water freezing in the pipes at night in cold weather.

COPPER is sometimes used in sheets 4 ft. × 2 ft. for covering roofs, flats, or gutters ; it is charged at per foot superficial according to its thickness in ounces per foot, as 12 oz., 16 oz., and 20 oz. copper; the price includes the seams, labour, ties, and nails, and is measured on the face as finished. Copper pipes and gutters for rain water are charged at per foot run according to size. Sheet copper weighs 549 lbs. per cubic foot

ZINC-WORKING.

Zinc is a metal largely used in building operations, and especially as a light covering to roofs and flats. Its specific gravity is much lower than that of lead, and its weight per cubic foot is 439 lbs., that of lead being 713 lbs. It can also be used very much thinner than lead, and is consequently the lightest material that can be employed in roofing; but since its expansion and contraction are considerable, it must always be allowed plenty of *play* at the laps.

Zinc is supplied in sheets about 8 ft. × 3 ft., and the thicknesses which are used in covering roofs, flats, and gutters, are 16 oz. to the superficial foot, 19 oz. to the foot, 22 oz. to the foot, and 25 oz. to the foot. When used in gutters, the thickness should not be less than 22 oz. to the foot. Zinc should never be allowed to come in contact with other metals when exposed to the open air, as a galvanic action is liable to take place to the destruction of both metals. It is laid upon roofs, flats, and gutters, in the same way as sheet lead; a floor of boards being first laid to receive it, with wooden *rolls* 30 in. apart. One mode of securing it to the rolls is by a *lug* attached to the underside of the *over-lap*, which lug passes through the *under-lap* and is held by a *clip* of zinc nailed to the roll. A zinc or lead roll-cap in 8 ft. lengths is laid over the roll and soldered at the joints, being nailed to the wood-roll at the top. See Buchan's "Plumbing," where several methods of forming the roll are described. Roofs are sometimes covered with zinc without the use of any under-boarding, the sheets being secured in the above manner to the top edges of the rafters chamfered off for that purpose; in this case, the rafters must not be more than 12 in. apart.

Corrugated-zinc can be laid on roofs without any boarding, the corrugations being formed at intervals of 15 in., or two in the width of each sheet, which is called "Italian corrugation," and wooden rolls 3 in. by 2 in. fixed to the purlins are fitted into each corrugation; the purlins may be placed 10 ft. apart. If a roof has a fall of 1 in 7 no drip is required, and only a fold at the junction of the sheets. In flats and gutters there should be a drip at every 14 ft. length at the least, and a fall of 1 in 20 should be obtained where possible.

Flashings should go 3 in. into the wall, and be pointed with cement.

Eaves-gutters require stays screwed into the eaves-board or rafter-feet every 18 in. in length.

Zinc is worked into ornamental forms for ridges, hips, dormers, &c.

Zinc, when used in roofs, flats, gutters, or flashings is measured by the foot superficial, and either the *gauge* or the weight per foot described.

Eaves-gutters, rain-water-pipes, chimney pipes, &c., when made of zinc, are taken by the foot run, the diameter being described. Heads, shoes, chimney tops, cowls, &c. are charged each.

Zinc nails are sold by the lb.

Galvanising, is covering iron with a thin coating of zinc by plunging it into a bath of that metal; this process prevents for a considerable time the oxidisation of the iron.

Perforated-zinc is used for filling-in to openings where ventilation is required, and is sold by the foot superficial.

GAS-FITTING.

The work of the GAS-FITTER is to convey gas from the mains to the several parts of a building by means of pipes and tubing; and great care has to be taken that all the joints are accurately made so as to prevent escape of gas. Wrought-iron welded tubing, in lengths from 4 ft. to 12 ft. is used for the principal parts of the work, the lengths being screwed into each other with a cement of red-lead. For turning corners, and changing direction, bends, elbows, tee-pieces, cross-pieces, and connecting-pieces, are screwed on to the ends of the pipes. The branches to the different burners are made with stout tin pipe, attached to the iron tubing by union-joints.

All gas piping and tubing is measured by the foot run, the diameter and material being described. Short lengths of iron pipe under 2 ft. are charged extra. All bends, joints, elbows, &c., are charged each, according to size of pipe.

Before the gas is brought into a building it is passed through a *meter*, or apparatus for measuring the quantity of gas consumed. Meters are of two kinds, *wet* and *dry;* the former containing water which the gas displaces, thereby causing the revolution of a drum which moves an index: they are now generally discarded in favour of the *dry* meters, which indicate the quantity of gas passed through with greater accuracy. The price of meters depends upon the number of *lights* or gas-jets intended to be burnt, and are named 5-light, 10-light, 20-light meters, &c. accordingly.

Gas pipes should always be laid with a fall towards the meter, so that if any water gets into them it may be more readily passed off. The size of the pipes must be regulated by the quantity of gas required to be consumed, which is measured in cubic feet; a pipe of $\frac{1}{2}$-inch bore will deliver 90 cubic feet of gas per hour; one of $\frac{5}{8}$-in. bore, 160 ft. per hour; one of $\frac{3}{4}$-in. bore, 250 ft. per hour; one of $\frac{7}{8}$-in. bore, 380 ft. per hour; one of 1-in. bore, 500 ft. per hour; one of 2-in. bore, 2000 ft. per hour; and so on, according to the square of the internal diameter of the pipe.

GAS STOVES are employed for heating rooms, halls, conservatories, &c. A row of jets placed in an open grate and covered with asbestos has all the appearance when lighted of a clear coal fire, but without its smoke. Gas jets are also burnt in pedestal stoves, from which there should always be a flue to carry off the products of combustion. Gas is also employed for cooking purposes, as boiling, roasting, frying, baking, &c. Baths can be heated by means of gas more quickly than by any other method.

ELECTRIC LIGHTS are now superseding gas lights for many purposes, having the advantage of producing very little heat and of not vitiating the air of a room as the burning of gas or other light does. There are several different forms of this light and modes in which it is produced, but there are only two distinctive varieties, namely the *incandescent* light and the *arc* light. The former is produced by passing a current of electricity through a fine thread of carbon enclosed in a vacuum, when the resistance of the carbon causes it to become heated to whiteness. This method produces lights equal to 20 candles and upwards, so that they can be used for lighting a room of any size by employing a greater or less number of the lamps and distributing them about the room. The *Arc* light is produced by a spark of electricity passing continuously between two carbon points kept at a uniform distance apart by means of clockwork, and which become intensely heated by the electric current. This produces a much more powerful light than the incandescent method, and is suited for lighting large and lofty halls and open spaces.

The intensity of the light depends upon the power of the batteries used in generating the electricity, and where a large supply is required it is necessary to employ in addition to the batteries a magneto-electric apparatus or "dynamo" driven by a steam or gas engine, by which the power of the batteries is enormously increased. The arc light is too strong to be placed near the eye, and is therefore unsuited for ordinary rooms. The current of electricity is conveyed from the battery to the lamps by copper wires insulated in gutta percha.

CHAPTER IX.

PAINTING, GLAZING, PAPER-HANGING, DECORATING.

PAINTING.

THE work of the PAINTER is to cover with colour woodwork, ironwork, plastering, or any other material, either for the purpose of protecting it from the action of the weather and atmosphere or of producing an ornamental and decorative appearance. Work that is intended to receive paint is generally prepared beforehand, with that object in view; thus, joiners'-work which is to be painted is sand-papered *across* the grain, when finished by the joiner, so as to give a hold or *key* for the paint; if it is to be stained and varnished so as to show the grain of the wood, it must not be touched with sand-paper, or at least very slightly, and that only *with* the grain of the wood.

Iron and other metals must be carefully cleaned from all rust or oxidisation before being painted; and wrought-iron should be coated while warm with linseed oil, to protect it from the moisture of the air.

Plaster or stucco must be thoroughly dry before being painted, otherwise the work will blister and the paint be thrown off.

When deal is to be painted it is first KNOTTED, or the knots of the wood *killed* by being covered with a knotting, composed of white and red lead mixed with whiting in a solution of size.

PRIMING is the first coat of paint after the knotting is dried, and is done with red lead and linseed oil, mixed with white lead. All nail holes are at the same time stopped with putty. For the other coats white lead, linseed oil, and turpentine are the chief materials employed, with which various pigments are ground up to give the required tint to the work.

FLATTING is the finishing of painted work a *dead* colour, which is done by using a considerable proportion of turpentine in the last coat.

Each coat of paint should be thoroughly dried before the next is applied; and for good work the surface is *rubbed-down* with pumice-stone before each coat of paint is laid on, so as to impart great smoothness to the work.

When old work is to be repainted it is first washed and rubbed down with pumice-stone and water, until an even surface is obtained; and if there are great inequalities they are filled up, or *brought-forward*, with a cement or putty.

Before painting the plastering on the inside walls, it is first covered either with boiling linseed oil, or with white lead and oil mixed with litharge to the consistence of thin cream; the oil is absorbed by the plaster, and when the first coat is dry a thicker one is applied, of which the oil is only partially absorbed; each coat after is made thicker and thicker until the oil ceases to be absorbed and the surface becomes thoroughly hardened.

Carbonate of lead, called *white lead*, is the chief ingredient in all paint, but is often adulterated by the admixture of carbonates of lime and baryta, whiting, gypsum or plaster-of-paris, pipe clay, starch, flour, &c.

Zinc-white has less body than lead-white, but possesses greater whiteness and power of resisting sulphurous gases.

MASTIC is a kind of cement used to lay over other cements, or to repair defects therein when it is to be painted. It consists of a mixture of pounded brick-dust, limestone, and sand; oil is used in its mixing and using, so that it may be painted upon as soon as laid. It can only be used in thin coats, and requires frequent painting to replace the oil which is lost by evaporation.

CLEARCOLE is a cheap kind of painting done with a mixture of white lead, water, and size; a coating of this dries quickly and may be finished with a coating of white lead in a mixture of equal quantities of oil and turpentine, having some colour and a *dryer* added thereto.

DISTEMPER is whiting mixed with a solution of size, and brushed over the work.

When painted work is to be MARBLED or GRAINED, the last coat is done in equal proportions of oil and turpentine of the colour which characterises the wood or marble to be imitated. The *graining* is done by thin glazings of colour ground in water and mixed with small-beer, the grain being imitated by means of a comb. Oak-graining is done with colour in turpentine and varnish laid over the work, and combed while wet. The lights are taken out with a small brush, or a rag moistened with turpentine; it is then glazed over with colour in beer, or *over grained*, and the surface covered with copal varnish.

WOOD-WORK that is to be STAINED so as to show its natural grain, must be of material carefully selected, free from loose or dead knots, sap-wood, or other defects, and its surface finished with the tool only. It is coated with a solution of a pigment in water which is absorbed by the wood, then once or twice sized over, and varnished with copal.

FRENCH-POLISHING is a varnish rubbed over hard woods previously prepared, until they present a highly-polished surface.

To CLEAN old painted-work, soap and water is laid on with a large brush, and washed off with a sponge; it is then dried with a leather.

External painting executed in the autumn will last twice as long as that which is done earlier in the year; as the summer sun draws out the oil from that which is done early. Work done from ladders must be measured separately.

ABBREVIATIONS.

1 O 2 O 3 O 4 O 5 O 6 O	Number of times in oil common colour.
F	Flatted, as 3 O F three times in oil and flatted.
D W	Dead white.
F G	French grey, or the particular colour may be written.
C C F	Clearcole and finish.
G W	Grained wainscot.
G M	Grained mahogany.

ROTATION.

In measuring painters' work, first take the windows; secondly, the skirting, dado, or wainscoting; thirdly, the chimney-pieces, if painted; and lastly, the doors.

MEASURING.

In measuring painters' work, all work not cut-in on both edges, must be taken, including edges and projections, at per yard square of 9 superficial feet.

Work cut-in on both edges, as to skirtings, cornices, shelves, edges of stairs, handrails, arris gutters, rain water-pipes, &c., is measured at per foot run.

Ornamental work is first taken as common, and then the extra labour to ornaments at per foot superf. or run.

Sash frames, window lights, casements, bars, dormers, frontispieces, chimney-pieces, &c., numbered and valued at each. Sash squares at per doz.

Iron or wood railings, balusters to stairs, &c., are measured on both sides as solid work, to allow for the extra trouble of painting round the bars, rails, &c., at per yard.

If ornamented, add extra one face in the width of such ornamental parts.

If ornamented turned balusters, also add one extra face as far as the turned work goes.

Handrails, &c., grained mahogany, first measure them in with the balusters, and then per foot run for graining.

Soffits to windows per foot run.

Letters or figures numbered and valued at per inch in height.

Painting on iron, cement, or plaster, should be kept separate from painting on wood.

French polishing to handrails is taken by the foot run, to other work by the foot superf.

Windows and doors are measured thus :—(See Plate VIII.)

PAINTING.

Windows (fig. 4).

```
ft. in. | ft. in.
11  0  |            Window front.
 5  5  |
-------|                    ft. in.                ft. in.
       |                     7  6 height            4  6 width
       |                     0  4 edges.            1  6 boxings and edges.
 7 10  |            Shutters  ------                  ------
 6  0  |                     7 10                   6  0
-------|
       |                     ⎧ 7  6 If the backs   ⎧10  6 ⎫          ⎧      ft. in.
       |                     | 7  6 are cut away   |10  6 ⎬ linings  |  viz. 26  3
20  3  |            Linings ⎨ 0  9 the    linings ⎨ 4  9 soffit     ⎨        1  2
 1  2  |                     | 4  6 must be mea-   | 0  6 elbows     |       -----
-------|                     |  ---- sured to the  ⎩                 ⎩
       |                     ⎩20  3 floor, thus :  (26  3
       |
       |                                                      ⎧  7  6
       |                                                      |  3  9
22  6  |            Beads varnished, supposing them           | -----
       |              to be mahogany or wainscot ⎨ 11  3
       |              sashes and beads.                       | 11  3
       |                                                      | -----
       |                                                      ⎩ 22  6
       |
       |            12 squares varnished.
       |            1 locking bar.
       |
       |                                                      ⎧  1  6 boxings.
       |                                                      |  0  6 edges.
       |            Some only allow the shutters' width       | -----
       |              to be taken thus :—                    ⎨  2  0
       |                                                      |  3  9 shutters.
       |                                                      | -----
       |                                                      ⎩  5  9 instead of 6 ft.
       |
       |                                                      ⎧ N 1 frame
       |            The outside of window would be            | 1 dozen of squares.
       |              taken as  .   .   .   .   .            ⎨ 1 sill, if the stone
       |                                                      ⎩   sill is painted.
```

Doors (fig. 5).

```
2)7  0  |                                           ⎧  4  0 width.
  4  4  |                                           |  0  4 ⎧ projection of
--------|            Door fronts for both sides.  .⎨        ⎩ architraves.
        |                                           | -----
        |                                           ⎩  4  4
        |
        |                     ⎧  6  6
        |                     |  6  6                  0  6 reveal.
16  0   |            Linings ⎨  3  0                  0  2 ⎧ edge of door
 0  8   |                     | -----                       ⎩ and rebate.
--------|                     ⎩ 16  0                  -----
                                                       0  8
```

ABSTRACT.

Once in Oil.			Twice in Oil.			Three Times in Oil.			Four Times in Oil.			Five Times in Oil.		
Com.	Runs.	Nos.	Com.	Runs.	Nos.	Com.	Runs.	Nos.	Com.	Runs.	Nos.	Com.	Runs.	Nos.
And Flat.			And Flat.			And Flat.			And Flat.			And Flat.		

By this method every description of work stands in rotation in the Abstract as it should be drawn into Bill, and will likewise be found with much more facility on the Abstract.

ROTATION

To be attended to in bringing the quantities into Bill.

PAINTER.

Yds. ft. in.		
	Once in oil	
	Run of Skirting, &c.	
	No. Sashes. Doz. squares . .	
	Twice in oil	
	Runs	
	Numbers	
	Three times in oil	
	Runs	
	Numbers	
	Three times in oil and flat dead white	
	Runs	
	Numbers	

If carved work, or any other per foot superf., it must be put under the yards of painting so many times done.

Likewise party or other coloured work must be placed under the head of work according to the number of coats.

VALUATION OF PAINTERS' WORK.

CALCULATION OF MATERIALS.

45 yards of work, 1st coat, including knotting, stopping, and every preparation requisite for the second coat, will require
- 5 lbs. of white lead.
- ¼ lb. red lead.
- 5 lbs. of putty, litharge, &c.
- 1 quart of oil.
- ¼ lb. pumice stone.
- ½ quire glass paper.

Second and following coats
- 5 lbs. of white lead.
- 1 oz. litharge.
- 1 quart of oil.

20 per cent profit is always allowed on the prime cost of the materials.

CALCULATION OF LABOUR.

The decimal to be multiplied by the rate of wages for a painter per hour:—

First coat, including stopping, knotting, &c., per yard, ·27.

Second and following coats ·15.

The above data will suffice for the valuation of common work, for which alone it is possible to lay down any rules, as the value of decorative work, such as graining, imitations, &c., depends upon the ability of the artist, and the manner in which the work is executed.

GLAZIERS' WORK.

GLAZING, as the work of the glazier is called, consists in fitting glass in sashes, frames, and casements, and fixing it either in putty or in lead. The glass used in buildings is of various kinds; the commonest sort is denominated CROWN GLASS, of which there are three qualities, called best, seconds, and thirds. Being a blown glass, it is liable to air-bubbles, blisters, specks, and other defects; the *best* is that which is freest from these blemishes; *seconds* and *thirds* are commonly used for window glazing, the latter, however, only for very common windows. It is all sold at the same price per crate, but the number of *tables* is different, according to the quality. There is another quality called *fourths*, of a very coarse description, and very little used. Crown glass weighs 158 lbs. per cubic foot.

SHEET GLASS is also a blown glass, made of much larger sizes than crown, and of a purer quality or *metal*; after being blown in the form of a cylinder, it is cut lengthwise and unrolled or *flatted* in a furnace, and has rather an uneven surface; there are four qualities, named as in crown glass, and six different varieties of thickness, according to the weight per square foot—namely, 15 oz., 21 oz., 26 oz., 32 oz., 36 oz., 42 oz.; of which the thicknesses are respectively, ·071 in., ·100 in., ·124 in., ·150 in., ·171 in., ·200 in. It can be had in all sizes up to 17 sq. ft. in area, the price per foot of the squares increasing with their area. Polishing is charged extra at per foot superficial.

PATENT-PLATE GLASS is a superior quality of sheet glass polished on both sides; it is of two qualities and of four different thicknesses—namely, $\frac{1}{10}$th in., weighing 13 oz. per sq. ft.; $\frac{1}{12}$th in., weighing 17 oz.; $\frac{1}{10}$th in., weighing 21 oz.; and $\frac{1}{8}$th in., weighing 24 oz. per foot. It can be had in all sizes up to 12 square feet in area, but is not generally used of large dimensions, from being brittle, and nearly as expensive as British-plate glass.

BRITISH-PLATE GLASS is the best material for glazing purposes, and can be had in sheets of any size up to 100 superficial feet. It is of two qualities, *best* and *seconds*, the former being picked free from air-bubbles or other defects; its usual thickness is $\frac{3}{16}$ths in. to $\frac{1}{4}$th in. for the ordinary glazing sizes, but the larger sizes are thicker. The price at per foot superficial increases with the sizes of the squares and also with the thickness, so that the size must be stated in the bill. If the edges are ground or polished, they are charged at per lineal inch.

Bending glass to a particular sweep is charged extra at per foot superficial.

ROUGH-PLATE GLASS is a common kind of material used for skylights, &c.; it is made either *plain* or *fluted*, and of various thicknesses, the usual being $\frac{1}{8}$th in., $\frac{3}{16}$th in., $\frac{1}{4}$th in., $\frac{3}{8}$ths in., and $\frac{1}{2}$ in.; and in sizes up to 30 feet superficial, the price per foot increasing with the dimensions. There is also a thin rough glass made with perforations for ventilation.

ROLLED-CATHEDRAL is a rough kind of glass made in various tints, and used for forming stained-glass and other windows of churches, &c.

GLAZING or fitting glass into sashes is commonly done by bedding the glass in *putty*, formed of whiting mixed with white lead and linseed oil; when the sheets are of large size and extra thick—as plate glass—it is usual to secure them with small iron sprigs or brads in addition to the putty. When glass is fitted into iron sash-bars, a coating of two parts resin to one part tallow laid on the bars when hot, will prevent the putty from cracking by the changes of temperature.

In fixing glass into *stone-work*, a groove is cut in the stone which is coated with size, and the glass is bedded in common putty in which a little olive oil is mixed, to prevent it from getting too hard. The size prevents the oil from staining the stone. Glazing in *lead-lights* is done by fitting pieces of glass into slips of lead grooved on each side; the groove is filled with dry whiting mixed with a little white lead and painted over with oil, so as to form a cement for the glass. *Lead-lights* are let into grooves cut in the stone-work and secured to iron saddle-bars with copper ties soldered on to the leadwork. They are measured by the foot superficial, the price varying according to the quality and design of the glass used.

Hacking-out old glass from sashes is charged extra at per foot superficial.

In measuring glaziers' work, the dimensions must be taken between the rebates, and all irregular panes the extreme size each way. The price per foot of glazing with crown glass must be calculated from the prime cost per crate, allowing for carriage and 20 per cent. profit. The larger the panes are, the more difficulty, risk, and waste; consequently the price should increase in the following proportions:—

				ft. in.	ft. in.	
Panes whose superficial contents are			under		2 0 at per foot.	
Do.	do.	do.	do. from	2 0 to 2	6 add 1*d*.	Above the squares
Do.	do.	do.	do. do.	2 6 to 3	0 add 2*d*.	whose contents are
Do.	do.	do.	do. do.	3 0 to 3	6 add 3*d*.	under 2 ft.

A CRATE OF CROWN GLASS

Contains 12 tables of the best, at per crate
,, 15 ,, seconds ,,
,, 18 ,, thirds ,,
,, 18 ,, fourths ,,

Each table is from 4 ft. to 4 ft. 6 in. diameter: some

tables may be cut to within 2 in. of the centre, others not nearer than 4 inches.

	ft.	in.
Supposing a crate to be 4 ft. 6 in. diameter, and that it may be cut to 2 in. from the centre, the quantity of glass that may be cut from it, including the triangular pieces, will be	14	2
If only 4 ft. diameter, and cannot be cut nearer than 4 in. of the centre	10	10
	25	0
And deducting the triangular pieces, which are of very little value	2	6
We have as the available contents of the two tables . . .	22	6
The average contents per table	11	3

taking the sizes of squares that will cut to the most advantage: but as squares of all sizes must be cut from the tables as they are wanted, the average produce per table is not more than 10 ft. superficial.

Labour and putty per foot for glazing in new sashes with crown or sheet glass may be found by multiplying the rate of wages for a glazier per hour by ·2.

Example.—To find the value per foot of glazing, with best Newcastle crown glass, or any other kind of glass:—

	£	s.	d.	
Prime cost of crate (12 tables)	0	0	0	
Carriage, &c.	0	0	0	
	0	0	0	
20 per cent. profit	0	0	0	
Divide by No. of feet the crate will produce, for best glass . . . . . . 120)	0	0	0	
	0	0	0	per foot.
Labour and putty	0	0	0	
Total per foot . . £0	0	0		

PAPER-HANGING, DECORATING.

Before newly-plastered walls can be papered they must be thoroughly dried, then rubbed down with pumice-stone and coated with size, to prevent the paste with which the paper is attached to the walls from sinking into the plaster. Old walls that have been previously papered should be stripped of all loose paper, washed, and sized. When papers of superior quality are to be hung on a new wall, a thin lining paper is first laid on to prevent the plaster from discolouring the paper.

WALL PAPERS are sold by the *piece*; a piece of English paper is considered to be 12 yards in length and 20 inches wide: its content is equal to $6\frac{2}{3}$ square yards, or 60 feet superficial; therefore, in measuring, divide the superficial feet by 5, which will give the number of yards running, and these divided by 12 will give the number of pieces. If there are any odd yards, they are charged as one piece. It is customary to allow one piece in seven for waste. Common papers do not measure more than 11 yards nett to the piece. French papers are of various widths, but more usually about 18 inches wide, and $9\frac{1}{2}$ yards long, $4\frac{1}{2}$ sq. yards being the measurement of one French piece.

Pumicing and preparing walls for papering is charged at per piece of paper, in addition to the price of the paper. Lining paper and hanging the same, at per piece. Hanging the paper, at per piece, the price varying somewhat with the quality. Marbled papers are hung in blocks and lined round with a pencil to imitate the joints, and beds. Sizing and varnishing is charged at per yard superficial.

Borders fixed at top or bottom of the papering to a room are taken at per dozen yards run, and the hanging of them in the same way.

When walls are damp, or are liable to wet coming through from outside, they are sometimes covered with sheets of tin-foil or tar paper before being papered; this is charged according to the quantity used and the time occupied in hanging.

Where there are hollows in a wall to be papered, as the panels in old wainscoting, they are covered with canvas, on which the paper is hung.

Gold or other mouldings placed next the cornice or plinth, are charged at per foot run, and are attached by means of needle points.

DISTEMPERING is done with colour mixed with a solution of size, and is washed over walls and ceilings; it is charged at per yard superficial, the price varying with the tints used; and if the cornices and ornaments are cut-in of different tints, they are taken separately at per foot superficial. The gilding of mouldings or ornaments is taken at per foot run, according to their girt. Imitations of marbles are measured by the foot superficial.

FRESCO-PAINTING is performed by employing colours mixed and ground with water, upon a stucco, or plaster, sufficiently fresh and wet to imbibe and embody the colours with itself, so that the painting dries along with the plaster, becomes very durable, and brightens in its tones and colour as it dries. This kind of work must be done upon walls that are perfectly dry, and not liable to be affected by external damp; it is unsuitable for any but tolerably dry climates.

SCAGLIOLA is a mode of imitating in columns, pilasters, &c., the most expensive marbles, and is only used for internal decorations. The column to be decorated in this way is first formed of any material, 2 or 3 in. less in diameter than the finished column; this is covered with a coating of lime-and-hair mortar, and allowed to dry. The material used for the decoration is calcined gypsum in fine powder, mixed with various colours in a solution of glue or isinglass; this is laid on the pricking-up coat, and floated over with moulds to the required shape, the artist using the colours necessary to produce the imitation during the process of floating, so that they become incorporated with the surface. When thoroughly hard and dry, the surface is rubbed with pumice-stone, being kept damp during the process; then polished with tripoli and charcoal applied with linen rag, and after being gone over with felt dipped in a mixture of oil and tripoli, it is finished by the application of pure oil.

MOSAIC-DECORATION consists in covering flat surfaces with small squares of coloured marbles or of vitrified pastes to which every possible variety of tint is given; these squares, called *tesseræ*, are arranged in various patterns according to the design of the artist, and are firmly embedded in cement; their size will depend upon the distance which they are to be placed from the eye, and upon the character of the decoration. When mosaic-work is used for paving floors, a basis of concrete is first formed, in order to obtain an even and firm surface; and the tesseræ or tiles are cut into other shapes than squares, for convenience of arrangement and laying. Mosaic-work is measured by the superficial foot, but the price varies very greatly, according to the character and intricacy of the design.

ENAMELLED-SLATE is a material used in decoration, the surface of slabs of slate, which have been rendered perfectly smooth, being coloured and polished in imitation of various marbles. These slabs vary from 1 in. to 2 in. in thickness, and are charged at per foot superficial, the price varying somewhat according to the kind of marble imitated; there is also an extra charge for moulded work. This material is also used for boxed chimney-pieces, the price of which depends on the width of piers, jambs, and openings, and upon the amount of moulding and carving.

MARBLE is a material used for decorative purposes, but principally for *internal* work, as it loses its polish and beauty by exposure to weather, and is therefore generally unsuitable for *external* decoration. Marble in the rough is sold in blocks like any other stone, at per foot cube, the price varying according to quality and rareness; those most highly priced being the pure white *statuary*, the *sienna*, and the *verd-antique*. Marble when used for chimney-pieces and other decorations, except columns, is cut into slabs from $\frac{3}{4}$ in. to $1\frac{1}{2}$ in. in thickness, and highly polished. When used in slabs it is charged by the superficial foot, according to thickness and quality. Working or moulding the edges is charged at per foot run; rounded corners and other small matters are charged each. Copper cramps are let into the edges of slabs for fixing them to walls. Marble *columns* are turned and polished, either parallel throughout, or diminished, and with an *entasis* or curved outline. The price of marble columns is charged at per foot in length according to the diameter, and whether straight, diminished, or curved. In large columns there is a considerable difference in price, whether made in a single block or built up of several stones; also if a fillet and hollow are worked on the top and bottom of the shaft, rendering it necessary to *sink* the shaft for them; instead of being worked in with the cap and base.

POLISHED GRANITE is used for decorative purposes, both outside and inside a building. Being very durable and preserving its polish when exposed to weather, it is chiefly employed in *external* decorations. There are several kinds of granite, varying considerably in colour, the two principal ones being the *red* and the *gray* granites. It is manufactured into slabs, mouldings for doorways, cornices, columns of all sizes and forms, balusters, &c. Since granite possesses greater resistance to crushing than any other kind of stone, columns of this material are employed where a great load has to be supported on a small basis, as in piers to the nave arches of churches, and other arcades of a similar description.

LOOKING-GLASSES form an important feature in the decorator's work, being frequently let into panels, and used to fill up the spaces or piers between windows; they are made of the very best quality of plate-glass, covered at the back with an amalgam or alloy of mercury and tin, which is called "silvering," and causes the glass to reflect the image of any object placed before it. Silvered-glass is charged at per foot superficial, the price increasing with the size.

GILDING to inside work is done in oil-size on woodwork, and in water-size on plaster; it is executed with gold leaf, which is made up in books, one book containing 25 leaves $3\frac{1}{2}$ in. by 3 in., which will cover about 1 ft. superficial of plain work. Gilding in quantity is charged by the square foot, and mouldings by the foot run according to size. Iron or other metal must be painted before being gilded.

SUMMARY.

It is usual in making out a bill of quantities to add up the amounts in each trade separately, and afterwards to bring the several totals into a summary, so that the entire cost of all the works can be arrived at. In this summary are generally placed under the head of SUNDRIES, other items of expenditure which do not come under the heading of any one of the trades, such as fire insurance, District Surveyor's fees, payment for water supply during the erection of the building, erecting hoardings and scaffoldings, clearing away all rubbish and superfluous material, and leaving the building clean at the completion of the works. The following is the form of the summary, taking the several trades in the same order as in the previous pages:—

SUMMARY.	£	s.	d.
Excavating and Well Sinking			
Foundations and Brickwork			
Tiling and Slating			
Carpentry			
Joinery and Ironmongery			
Masonry			
Plastering			
Smiths' Work and Bell-hanging			
Plumbing and Gas-fitting			
Painting			
Glazing			
Paper-hanging and Decorating			
Sundries			
Total £			

NOTE.—It frequently happens that the bricklayer, mason, carpenter, joiner, and plasterer are required to cut away and make good their work for the workmen of other trades, as plumber, gasfitter, smith and bell-hanger. An item should therefore be introduced in the bill under each of the former trades as "Attending upon other trades."

CHAPTER X.

DILAPIDATIONS, REPAIRS, CONTRACTS, ANCIENT LIGHTS.

DILAPIDATIONS.

The full application of this term is to any injury which has happened to any part of a building, either from neglect, wilful or accidental damage, from fire, water, lightning or any other means, either by the action of the elements, or by the hand of man. The word DILAPIDATIONS is however more usually employed in a limited sense to the want of repair of a house held by a tenant for a term of years under the covenants of a lease, wherein it is stipulated that the tenant shall keep every part in good condition, making good all damages that may occur during his tenancy, and shall especially paint all external work, which is usually painted, once in every 3 years, and all internal work once in every 7 years, besides repapering the walls once during the same period. If a house has been newly papered and painted when the tenant enters on possession, it is sometimes stipulated that he shall do the above-named painting and papering in the *last* year of the seven, so that if he gives up possession at that time it may be in proper condition for a new tenant without any expense to the landlord. If the tenant neglects to carry out the terms of his lease and leaves the house out of repair, two surveyors are generally appointed, one by the landlord, and the other by the tenant, who shall assess the value of the dilapidations, and agree upon the sum of money to be paid by the tenant to the landlord to enable him to execute the necessary repairs without expense to himself. So much difference of opinion had been expressed as to the proper mode of valuing dilapidations, that the Council of the Institute of British Architects in the year 1842 referred the matter to a select committee of experienced surveyors to report as to the usual practice in regard to Dilapidations and Fixtures. In their report the committee define Dilapidations to be those defects only which have arisen from neglect or misuse, and not to include such as only indicate age, so long as the efficiency of the part still remains; but if any part is destroyed by use or age, this is considered as arising from neglect or misuse, the tenant being presumed to have satisfied himself that every part was strong enough to last his term before he signed the lease.

In assessing the value of Dilapidations it is generally the rule to require that the tenant should leave the premises in as good condition as he found them, as it can hardly be expected that he shall make an old house as good as a new one, or that where a dilapidation has arisen from an original defect in the mode of building, he shall not be required to pull down such portion and rebuild it in a proper manner.

If a tenant has pulled down any portion of the premises and rebuilt it in a different manner without consent of his landlord, he may be required to replace it exactly as it was when his tenancy began, even though it may have enhanced the value of the property. Any alterations that he has made, such as blocking up doors or windows, or opening out new ones, taking down partitions or erecting new ones, render him liable for the expense of reinstating the building exactly as he found it, if done without the landlord's consent. For information as to the law of Dilapidations the reader is referred to Tarbuck's "Handbook of House Property."

ECCLESIASTICAL DILAPIDATIONS apply to all such houses of residence, chancels, walls, fences, and other buildings and things as the incumbent of the benefice is by law or custom bound to maintain in repair. This is the definition given by the Act for the amendment of the law relating to ecclesiastical dilapidations, passed in the year 1871. Under this Act, official surveyors have been appointed in the several dioceses, whose business it is to inspect the buildings connected with a benefice whenever he is required to do so by the Bishop; and it is to the work of the surveyor alone that we shall confine ourselves in these pages. When the building is found to be in need of repair, the surveyor is to specify in detail the nature of the works to be done, and the probable cost of doing them, as well as the time they should take to execute. It will therefore be necessary for him not only to inspect the parts that are open to the eye, as in taking the dilapidations of civil buildings, but also to examine the state of the timbers of the floors, roofs, partitions, &c. Paper-hanging and all work that is purely decorative and not necessary for the maintenance of the building, is to be omitted from the surveyor's report.

In case of the refusal of the incumbent to execute the repairs mentioned in the report, the surveyor is to employ builders to execute them according to a specification and contract prepared by him, who shall be paid for such works under certificates from the surveyor. The amount stated in the contract must however be strictly adhered to, as no allowance is made for extra work. The incumbent may however rebuild the whole or any part of the buildings so as to render the repairs mentioned in the surveyor's report unnecessary, provided he has the consent of the Bishop and patron; in which case the surveyor shall, upon completion of such works to his satisfaction, give a special certificate, stating that they have been completed.

In case a building is damaged by fire, and the loss is not fully covered by insurance, the surveyor is to give a certificate stating the sum required, in addition to the insurance money, in order to reinstate the buildings.

REPAIRS.

Under this title may be included all such works as are necessary to the making good of whatever defects may have in any way arisen in any parts of a building, either from ordinary *wear and tear*, wilful neglect or damage, decay by age, injury by the elements, &c. In a more restricted sense it is applied only to those defects or dilapidations which a tenant or lessee is bound to make good under the covenants of a lease, as previously mentioned under "Dilapidations."

The following is an outline of the repairs generally considered to be such as the tenant is bound to do under a repairing lease:—

TILING AND SLATING.

Replace all loose, broken or defective tiles or slates, strip and retile or reslate where the laths or battens are broken, or where the timbers are broken or decayed. [If the whole roof or a considerable portion has to be stript, the work may be measured superficial by the square of 100 ft.]

Make good all defective filleting and pointing to pantiling. Secure and make good all loose or broken slate-shelves, slabs, or paving.

BRICKLAYING.

Cut out and make good all defective brickwork in walls, chimney shafts, parapets, and gables, rebuilding such portions as may be so far out of plumb, or so cracked, split, or bulged, as to be unsafe or incapable of being effectually repaired. Rake out all decayed pointing, and repoint in same manner as previously pointed, erecting all scaffolding that may be necessary for the proper execution of the work.

[The pulling down of old brickwork, cleaning and stacking the old bricks, providing scaffolding, &c., is measured by the rod as before described under BRICK-WORK. The pointing is taken by the foot superficial, and the filleting by the foot run.]

Replace split or broken chimney-pots, refixing all sound ones that are loose.

Make good all broken or defective brick pavings, taking up and relaying such parts as are sunken, and provide new paving bricks where necessary. Empty, cleanse, and repair all drains and cesspools, and clear away all accumulations of soil, earth, and rubbish.

MASONRY AND PAVING.

Make good all defective or broken portions of copings to chimneys, parapets or gables, blocking courses, curbs or copings to areas or railings, water channels, sinks, sink-stones, shelves, bearers, pavings, curbs, and other works both external and internal; take down and reset wherever displacement has occurred from settlements in brickwork. Cut out and make good such portions of the stonework as may be so damaged as not to afford a hold for ironwork.

Make good all broken or damaged portions of stone cornices, lintels, sills, balconies, string courses, plinths, and other stone dressings and ashlaring; pieces to be filled in wherever it can be done in a sound and efficient manner.

Make good all broken portions of steps and landings externally and internally, by piecing them in a sound and efficient manner; but where the nosings are broken or the treads worn so as to render them unsafe, the whole upper surface to be worked off and filled-in the depth of the nosing with a new slab to form a new nosing. Make good all broken chimney-pieces, slabs, and back hearths; take up and relay sunken slabs and hearths and remove all stains. Refix loose masonry, replace defective cramps, and the lead running where required, and point up all decayed and open joints.

Take up and relay all loose and sunken portions of pavings, and supply any deficiency of paving stones and channels.

CARPENTRY.

Secure and make good all loose, broken, or defective timbers, whether injured by weather, dry-rot, or otherwise; providing new material where necessary.

Fir-up the joists, rafters, and other timbers where

not straight, or out of level or plumb from neglect or decay.

Secure and make good all loose, broken, or decayed weather-boarding, dormer-doors, frames and windows, skylights, water trunks and gutters, boarding to dormers, roofs, gutters and flats, and other external work; taking up and relaying all sunken parts.

Secure and make good all loose, broken, decayed or defective wood fences, external doors, frames, gates, and posts; making good all imperfect hanging and defective fastenings to same. Put spurs to bottoms of posts where decayed, in case the upper parts are sound.

Take out and make good all broken, decayed, or defective camp-sheeting, timber-wharfing, piles, and land-ties.

JOINERY.

Secure and make good all loose, broken or rotten floors, fir-up and re-lay where out of level from neglect. [If the floor is out of level from the settlement of the building, the tenant cannot be called upon to put it level.] Secure and make good all loose, broken, or defective joiner's work; rehang all doors and folding shutters which do not close properly; make good with new sash-lines where broken or defective; replace broken beads to sash frames. Put new nosings to stairs where defective, and new treads where worn away. Provide new balusters to replace any that may be broken; repair defective hand-rail, clean off and repolish same if in mahogany. Where the bottom rail of a sash is loose from the stiles, it must be secured with L irons.

SMITHS' WORK, IRONMONGERY, AND BELL-HANGING.

Make good all broken parts of railings, gates and gratings, metal skylights, fanlights, sashes, casements, saddle bars, window bars, and other external iron works, refixing same where loose. Make good any imperfect hanging of gates, &c., and all defects in the fastenings. Reinstate all broken parts of internal railings to stair-cases, landings, &c., all iron beams, supports and ties.

Make good all defects in iron pipes, gutters, cistern-heads, and their appurtenances.

Make good or reinstate all broken or defective parts of iron doors, shutters, and frames; clean and repair or replace all defective door and shutter bars, locks, bolts, hinges, sash fastenings, and all other articles of ironmongery, and secure all that are loose.

Make good all defects to stoves, ranges, coppers, &c., providing new parts where broken, as new fire-brick back or iron register to stoves, or new front or bottom bars, new boiler to ranges if cracked, new fire-door and bars to copper; make good any defects in the hot-water supply to bath.

Make good all defective bell-hanging, providing new copper wires, cranks, springs, &c. where requisite; repair bell-pulls where defective, and provide new where they are too much damaged to be repaired.

PLASTERER.

Cut out all unsound, loose, damaged or defective parts of the plastering, and make good the same; stop all cracks in ceilings or walls; repair defects in ornaments or enrichments of cornices, centre flowers, &c., providing new where necessary; re-colour where partially defaced. Whiten and colour walls or ceilings wherever there has been neglect or misuse, or where the covenants of the lease have not been complied with. Cut out all loose, defective or broken external cement work or stucco, stop all cracks therein, and make good to same with new cement.

PLUMBER.

Secure and make good all loose or damaged portions of lead work, solder cracks, and replace deficiencies of lead in flats, gutters, hips, ridges, valleys, flashings, dormer tops and cheeks, aprons, cistern-heads, rain-water pipes, sinks, cisterns and pipes. Make good all damage to pumps, water-closet apparatus, soil-pipes, traps, &c., and remedy all defects therein. Repair all defective bib or ball-cocks, re-grinding same or replacing with new where necessary.

Secure and make good all loose or damaged portions of

zinc or copper flats, gutters, pipes, cistern-heads, and solder all cracks.

Clean out all gutters and rain-water pipes.

PAINTING.

Renew all defaced outside painting on wood, iron, stone, or stucco. Make good all inside painting in cases of misuse, and paint 4 oils where new wood-work has been provided to replace damages. Also repaint all external and internal work in case the covenants of the lease have not been complied with as regards painting. If any work has been originally done in party colours or ornamental graining and decoration, it must be executed in the like manner by the tenant.

PAPER-HANGING.

Secure and make good all loose and torn portions of canvas and paperings, restoring such parts as have been damaged by neglect or misuse. If the tenant has failed to comply with the covenants of his lease with regard to papering, then the whole of the old paper must be stript off and the walls be rehung with new paper of the same value as before. All gilt mouldings to be taken off, cleaned, or regilt if necessary, and refixed after the paper-hanging is completed. [Stripping and rehanging is measured by the piece, as described in the chapter on PAPER-HANGING.]

VALUATION OF REPAIRS.

In order to estimate the cost of executing repairs, it is necessary to measure the work wherever it will allow of taking a dimension, according to the rules we have laid down under the several trades; the nature of the work, however, must be fully described. Where the quantities are small it will be necessary to describe them item by item, and to estimate the time and materials that will be required in executing them. Where new material has to be provided, it can easily be measured or numbered and valued the usual manner.

CONTRACTS.

By far the greater number of buildings are now erected under a contract, or written agreement between the employer and the builder, in which a fixed sum is named, in consideration of which the builder agrees to execute the whole of the work described in the specification or shown upon the plans. When an architect is employed he is made an arbitrator between the contracting parties, and has to see that the builder strictly adheres to his part of the contract, while the employer has to pay the builder the sum stipulated upon receiving a certificate from the architect. There are certain GENERAL CONDITIONS which are always attached to a contract, the terms of which were settled in 1870 between the council of the Royal Institute of British Architects, and the committee of the London Builders' Society. These "Conditions" are given below, together with a form of contract to which they should be annexed. The specification of works is generally kept separate from the contract and conditions, but is, together with the drawings, signed by the parties to the contract. As very few buildings can be erected without alterations from the original design being made as the works proceed, it is desirable that a clause should be added to the above-named "Conditions," stating what time is to be allowed for executing any extra works according to the value of such works; that is to say, so many days for so many pounds' worth of extra work. This stipulation should be added to clause 15, in order that the contractor may not have an excuse for delaying the fulfilment of his contract in consequence of extras.

ARTICLES OF AGREEMENT made and entered into this —— day of —— one thousand eight hundred and eighty ——, BETWEEN A. B. of —— hereinafter called the EMPLOYER, of the one part, and C. D. of —— builder, hereinafter called the CONTRACTOR, of the other part. WHEREAS the said employer being desirous to rebuild his premises situate and being ——, has caused certain plans and specifications of the work to be done to be made and prepared by E. F. his architect, and the contractor has agreed to execute the said works for the sum and in manner hereinafter mentioned. Now these Presents witness that the said C. D. doth hereby agree with the said A. B. to execute the said works according to the true intent and meaning of the drawings Nos. 1, 2, &c., and the specification prepared by the said E. F., and signed by the said parties hereto, and of the CONDITIONS OF CONTRACT hereto annexed, for the sum of —— pounds, or such further or other sum as may become payable under or according to the said conditions of contract. AND the said A. B. doth hereby for himself and his heirs, administrators, and assigns agree with the contractor to pay agreeably to the said conditions of contract and by instalments as therein mentioned the said sum of —— pounds, or such further or other sum as aforesaid. As WITNESS the hands of the said parties, the day and year first above written.

Witness,   A. B.
E. F.   C. D.

GENERAL CONDITIONS FOR BUILDING CONTRACTS.

1. The contractor is to provide everything of every sort and kind which may be necessary and requisite for the due and proper execution of the several works included in the contract, according to the true intent and meaning of the drawings and specification taken together, which are to be signed by the architect and the contractor whether the same may or may not be particularly described in the specification or shown on the drawings, provided that the same are reasonably and obviously to be inferred therefrom, and in case of any discrepancy between the drawings and the specification the architect is to decide which shall be followed.

2. The contractor is to conform in all respects to the provisions and regulations of the Metropolis Local Management Act and the Metropolis Buildings Act, and to the regulations and bye-laws of the Metropolitan Board of Works and of the local authorities, and he is to give all notices required by the said Acts to be given to any local authorities, and to pay all fees payable under any of the said Acts to any such authorities or to any public officer in respect of the works.

3. The contractor is to set out the whole of the works, and during the progress of the works to amend on the requisition of the architect any errors which may arise therein, and upon request is to provide the necessary appliances or furnish the necessary vouchers to prove that the several materials are such as are described. The contractor is to provide all plant, labour, and materials which may be necessary and requisite for the works, all materials and workmanship being the best of their respective kinds; and the contractor is to leave the works in all respects clean and perfect at the completion thereof.

4. Complete copies of the drawings and specifications, signed by the architect, are to be furnished by him or by the measuring surveyor to the contractor for his own use, and the same or copies thereof are to be kept on the buildings in charge of a competent foreman, who is to be constantly kept on the ground by the contractor, and to whom instructions can be given by the architect. The contractor is not to

sublet the works or any part thereof without the consent in writing of the architect.

5. The architect is to have at all times access to the works, which are to be entirely under his control. He may require the contractor to dismiss any person in the contractor's employ upon the works, who may be incompetent or misconduct himself, and the contractor is forthwith to comply with such requirement.

6. The contractor is not to vary or deviate from the drawings or specification, or execute any extra work of any kind whatsoever, unless the same be required to comply with any of the provisions of any of the Acts of Parliament, regulations, or bye-laws hereinbefore mentioned, or unless upon the authority of the architect, to be sufficiently shown by an order in writing, or by any plan or drawing expressly given and signed or initialled by him, as an extra or variation, or by any subsequent written approval signed or initialled by him. In cases of day work, all vouchers for the same are to be delivered to the architect or clerk of the works at latest during the week following that in which the work may have been done; and only such day work is to be allowed for, as such, as may have been authorised by the architect to be so done, unless the work cannot from its character be properly measured and valued.

7. Any authority given by the architect for any alteration or addition in or to the works is not to vitiate the contract, but all additions, omissions, or variations, made in carrying out the works for which a price may not have been previously agreed upon, are to be measured and valued, and certified for by the architect, and added to or deducted from the amount of the contract as the case may be, according to the schedule of prices annexed, or where the same may not apply, at fair measure and value.

8. All work and materials brought and left upon the ground by the contractor or by his order, for the purpose of forming part of the works, are to be considered to be the property of the employer, when payment shall have been made of the amount of any certificate in which the value thereof shall be included, and in such case the same are not to be removed or taken away by the contractor or any other person without the special licence and consent of the architect, but the employer is not to be in any way answerable for any loss or damage which may happen to or in respect of any such work or materials either by the same being lost or stolen, or injured by weather or otherwis.

9. The architect is to have full power to require the removal from the premises of all materials which in his opinion are not in accordance with the specification; and in case of default the employer is to be at liberty to employ other persons to remove the same without being answerable or accountable for any loss or damage that may arise or happen to such materials; and the architect is also to have full power to require other proper materials to be substituted; and in case of default the employer may cause the

same to be supplied, and all costs which may attend such removal and substitution are to be borne by the contractor.

10. Should any of the works be, in the opinion of the architect, executed with improper materials or defective workmanship, the contractor is, when required by the architect during the progress of the work, forthwith to re-execute the same, and to substitute proper materials and workmanship, and in case of default of the contractor in so doing within a reasonable time, the architect is to have full power to employ other persons to re-execute the work, and the cost thereof is to be borne by the contractor.

11. Any defects, shrinkage, and other faults which may appear within   months from the completion of the buildings, and arising out of defective or improper materials or workmanship, are, upon the direction of the architect, to be amended and made good by the contractor at his own cost, unless the architect shall decide that he ought to be paid for the same, and, in case of default, the employer may recover from the contractor the cost of making good the works.

12. The contractor is to insure the building against loss or damage by fire, in an office to be approved, in the joint names of the employer and contractor, for half the value of the works executed until it shall be covered in, and thenceforth until completion, in three-fourths of the amount of such value, and is, upon request, to produce to the architect the policies and the receipts for the premiums for such insurance. All moneys received under any such policies are to be applied in or towards the rebuilding or reparation of the works destroyed or injured. In case of neglect the employer is to be at liberty to insure and deduct the amount of the premiums paid from any moneys payable to the contractor.

13. The building, from the commencement of the works to the completion of the same, is to be under the contractor's charge. He is to be held responsible for, and is to make good, all injuries, damages, and repairs occasioned or rendered necessary to the same by fire, or by causes over which the contractor shall have control, and he is to hold the employer harmless from any claims for injuries to persons or for structural damage to property happening from any neglect, default, want of proper care, or misconduct on the part of the contractor or of any one in his employ during the execution of the works.

14. The employer is at all times to have free access to the works, and is to have full power to send workmen upon the premises to execute fittings and other works not included in the contract, for whose operations the contractor is to afford every reasonable facility during ordinary working hours, provided that such operations shall be carried on in such a manner as not to impede the progress of the works included in the contract, but the contractor is not to be responsible for any damage which may happen to or be occasioned by any such fittings or other works.

15. The contractor is to complete the whole of the works (except painting and papering, or such other works as the architect may desire to delay) within   calendar months after the commencement of the same, unless the works are delayed by reason of any inclement weather, or causes not under the contractor's control, or in case of combination of workmen, or strikes, or lock-out affecting any of the building trades, for which due allowance shall be made by the architect, and then the contractor is to complete the works within such time as the architect shall consider to be reasonable, and shall from time to time in writing appoint, and in case of default, the contractor is to pay or allow to the employer as and by way of liquidated and agreed damages, the sum of £   per week for every week during which they shall be so in default, until the whole of the works (except as aforesaid) shall be so completed, provided the architect shall in writing certify that the works could have been reasonably completed within the time appointed.

16. If the contractor shall become bankrupt, or compound with or make any assignment for the benefit of his creditors, or shall suspend or delay the performance of his part of the contract (except on account of causes mentioned in clause 15, or on account of being restrained or hindered under any proceedings taken by parties interested in any neighbouring property, or in consequence of not having proper instructions, for which the contractor shall have duly applied), the employer, by the architect, may give to the contractor or his assignee or trustee, as the case may be, notice requiring the work to be proceeded with, and in case of default on the part of the contractor or his assignee or trustee for a period of   days, it shall be lawful for the employer, by the architect, to enter upon and take possession of the works, and to employ any other person or persons to carry on and complete the same, and to authorise him or them to use the plant, materials, and property of the contractor upon the works, and the costs and charges incurred in any way in carrying on and completing the said works are to be paid to the employer by the contractor, or may be set off by the employer against any money due, or to become due, to the contractor.

17. When the value of the works executed and not included in any former certificate shall from time to time amount to the sum of £   , or otherwise at the architect's reasonable discretion, the contractor is to be entitled to receive payment at the rate of 80 per cent. upon such value until the difference between the percentage and the value of the works executed shall amount to   per cent. upon the amount of the contract, after which time the contractor is to be entitled to receive payment of the full value of all works executed and not included in any former payment, and the architect is to give to the contractor certificates accordingly ; and when the works shall be completed, or possession of the building shall be given up to the employer, the contractor is to be entitled to receive one moiety of the amount remaining due, according to the best estimate

of the same that can then be made, and the architect is to give to the contractor certificates accordingly, and the contractor is to be entitled to receive the balance of all moneys due or payable to him under or by virtue of the contract within   months from the completion of the works, or from the date of giving up possession thereof to the employer, whichever shall first happen. The contractor is to be entitled to receive any sum reserved for painting and papering, or otherwise, on the completion thereof. Provided always that no final or other certificate is to cover or relieve the contractor from his liability under the provisions of clause No. 11, whether or not the same be notified by the architect at the time or subsequently to granting any such certificate.

18. A certificate of the architect, or an award of the referee hereinafter referred to, as the case may be, showing the final balance due or payable to the contractor, is to be conclusive evidence of the works having been duly completed, and that the contractor is entitled to receive payment of the final balance, but without prejudice to the liability of the contractor under the provisions of clause No. 11.

19. If the employer shall make default in paying any moneys to which the contractor may become entitled, for   days after the amount thereof shall have been certified, or if the works be delayed for   months by or under any proceedings taken by any other parties, the contractor is to be at liberty to suspend the works, and to require payment for all works executed, and all materials wrought up, and for any loss which he may sustain upon any goods or materials purchased for the works, and in such case the contractor is not to be bound to proceed further with the works contracted for. The contractor is to be entitled to such interest and at such rate as the architect shall certify upon all moneys payable to the contractor, payment of which may have been unduly delayed.

20. Provided always that in case any question, dispute, or difference shall arise between the employer, or the architect on his behalf, and the contractor, as to what additions, if any, ought in fairness to be made to the amount of the contract by reason of the works being delayed through no fault of the contractor, or by reason or on account of any directions or requisitions of the architect, involving increased cost to the contractor beyond the cost properly attending the carrying out the contract according to the true intent and meaning of the signed drawings and specification, or as to the works having been duly completed, or as to the construction of these presents, or as to any other matter or thing arising under or out of this contract except as to matters left during the progress of the works to the sole decision or requisition of the architect under clauses Nos. 1, 9, and 10, or in case the contractor shall be dissatisfied with any certificate of the architect under clause No. 7, or under the proviso in clause No. 15, or in case he shall withhold or not give any certificate to which he may be entitled ; then such question, dispute, or difference, or such certificate, or the value or matter which should

be certified, as the case may be, is to be from time to time referred to the arbitration and final decision of   architect, or in the event of his death or unwillingness to act, then of   architect, being a Fellow of the Royal Institute of British Architects, or in the event of his death, or unwillingness to act, then of an architect to be appointed on the request of either party, by the President for the time being of such Institute, and the award of such referee is to be equivalent to a certificate of the architect, and the contractor is to be paid accordingly.

MEASUREMENT OF THE OBSTRUCTION OF ANCIENT LIGHTS.

The term "Ancient light" is a legal phrase, applied to a window which has enjoyed a certain *amount of light* for an uninterrupted term of at least twenty years. If an adjoining owner should erect any building that would seriously diminish that *amount of light*, the owner of the window so obstructed may compel him to desist, if he can show that there will be an insufficiency of light thereto after the new building is erected. The business of the surveyor is to determine with tolerable accuracy the actual amount of light that will be abstracted by the proposed obstruction, and the proportion which it bears to the previous quantity received by the window.

In estimating the light received by a window, direct sunlight is left entirely out of consideration; the light being supposed to be derived from the quarter-sphere of sky-surface which is seen from a vertical window, if no obstacles are placed in front or on the side. In the system of measurement which gives the most exact approximation to a correct result, the sky-surface seen from the window is represented by an illuminated quarter-sphere, divided by parallel circles of latitude and vertical circles of longitude into a number of segments, each of which measures 10° vertical and 10° horizontal; and the relative value of the light from each segment, as seen from the window at the centre of the sphere, can be found by mathematical calculation, and figured upon a table divided in the same manner as the sphere. When it is required to find how much light is cut off from the window by any proposed erection, the vertical and horizontal distances thereof from the window are measured, and also the angle which its direction makes with the plane of the window; by this means the number of segments of sky-surface which are obscured can be readily seen, and a simple addition of the values of those segments enables the surveyor to compare the light lost with that obtained from the whole sky-surface.

The method by which the Table of values of the segments is calculated is as follows: when light enters horizontally through a window, a certain amount of the side light is cut off by the thickness of the wall, amounting in ordinary windows to about 10° on the extreme right and left, so that the part of the sky within 10° on each side of the plane of the window may be considered as valueless. The parts of the sky which are opposite the window give a greater amount of light for a given area of sky-surface than those which are near the sides, the quantity decreasing with the angle at which the light enters. The same holds good with respect to light coming into the window at a vertical angle, the thickness of the wall cutting off entirely the upper 5° of sky-surface; and on account of the obliquity of the rays, the value for equal areas is much less for sky-surface near the zenith than for that which is lower down and more nearly opposite the window. We must also consider the diminution of the areas of the several zones of sky-surface as they approach the zenith. Combining all these measurements, we obtain the following Table of the values of the several parts of the sky-surface measured from meridian to the extreme right or left; one-half the quarter-sphere only being represented, as the other half is exactly the same. In this Table the amount of light emitted from *every* square unit of sky-surface is assumed to be the same; this being the hypothesis upon which most surveyors ground their calculations :—

ANCIENT LIGHTS.

Table of the Relative Values of Light entering a Vertical Window from different parts of the Sky-Surface, on the hypothesis of an equal diffusion of Light.

Value of each zone.		ZENITH.								
	N. 80°	0	0	0	0	0	0	0	0	0
54·4	70°	11	10·4	9·4	8	6·6	5	3	1	0
174	60°	35·5	33·3	30	26	21	15·4	9·6	3·2	0
335·6	50°	68·4	64	58	50	41	29·5	18·5	6·2	0
544	40°	111	104	94	81	66	48	30	10	0
748·1	30°	153	143	130	111·5	91	66	39·6	14	0
944	20°	195	180	163	140	114	83	52	17	0
1088	10°	222	208	188	162	132	96	60	20	0
1177	0°	240	225	204	175	143	104	65	21	0
		H	O	R	I	Z	O	N.		
		10°	20°	30°	40°	50°	60°	70°	80°	

5065 = total value of half sky-surface.

The assumption that the illuminating power of sky-surface is everywhere the same for equal areas, whether near the horizon or near the zenith, is not in accordance either with theory or experiment, although convenient as a basis for calculation. The light obtained from the upper zones of the sky is very much more valuable than that from the lower parts for *equal areas*,

s

and its illuminating power appears to increase according to the *sine* of the angle of altitude. The following Table has been calculated upon this hypothesis, and can be applied in the same manner as the one on p. 257:—

TABLE *of the Relative Values of Light entering a Vertical Window from different parts of the Sky-surface, on the hypothesis of a variable diffusion of Light.*

Value of each zone.						ZENITH.				
	N. 80°	0	0	0	0	0	0	0	0	0
54·4	70°	11	10·4	9·4	8	6·6	5	3	1	0
157·5	60°	32·6	30·5	25	24	19·4	14	9	3	0
273·3	50°	56	52	47	41	33·3	24	15	5	0
387·9	40°	79	74	67	58	47	34·4	21·5	7	0
431	30°	88	82	75	64	52	38	24	8	0
397·3	20°	81	76	69	59	48	35	22	7·3	0
294·6	10°	60	56	51	44	36	26	16·3	5·3	0
104	0°	21	20	18	15	13	9	6	2	0
			H	O	R	I	Z	O	N.	
		10°	20°	30°	40°	50°	60°	70°	80°	

2100 = total value of half sky-surface.

The numbers in the first column on the left of each Table represent the total value of each zone from 0° up to 90°, or for *half* the quarter-sphere of sky-surface; these numbers must therefore be doubled to obtain the value of the whole zone from extreme right to extreme left. The number at the bottom of the first column represents the sum-total of all the values of the several zones of the *half* quarter-sphere, and by comparing the value of light abstracted by any obstruction with the *total*, we are enabled to see at once what proportion of the whole of the light from the sky-surface is destroyed thereby. For an account of the methods by which the above tables are calculated, the reader is referred to a paper by the present Editor in the Transactions of the Institute of British Architects for the year 1870.

In order to illustrate the practical application of these Tables, we will employ them to find the loss of light in a case which is of common occurrence in large towns, and which frequently leads to expensive litigation. Suppose there is a row of houses in a street having open gardens at the back, and with nothing to impede the light to the back windows. The owner, A, of one of these houses desires to enlarge his premises, and builds out a projection of 10 feet the whole width and height of his house. His next door neighbour, B, has windows at the back, the middle of which is 10 feet from the wall of A's projection, consequently the whole of the side light from the angle of 45° has been cut off by this projection, at least from the lower windows, if the building is several stories in height. Take the window on the lowest story of B's house, and suppose that it is impossible to see the top of the wall of A's projection from the inside of the window; then it is clear that no direct light comes to that window beyond 45° from the centre or "meridian" line. In order to find the proportion of light thus cut off, we add up the columns in the Tables between 70° and 80°, between 60° and 70°, between 50° and 60°, and between 40° and 50°, taking only half the sum of the last-named column, as the angle is 45°. The sum of all these quantities from the first table is 1124·3, while the total value of the whole sky-surface is 10,130; so that the proportion of light lost in this case is ·111, or one-ninth of the whole. A similar result is obtained by using the second Table, the sum of the above-named quantities being 468·4, while the total value of the whole sky-surface is 4200, and the proportion of light lost is ·1115, or a very little more than one-ninth of the whole. For the windows of the upper stories the loss will be less, as they will obtain light sideways from over the top of the projection.

Where the obstruction is erected in front of the window, it is necessary to take the angular height of the wall at different points, as well as its horizontal angle on each side of the meridian of the centre of the window; and then by reference to the Tables it is easily seen how many segments of the sphere are obscured thereby.

This subject is one that is constantly engaging the attention of architects who build much in towns, where scarcely any enlargement can be made without some encroachment on a neighbour's rights to light and air. The subject has frequently been discussed at the Royal Institute of British Architects, and the law will be found laid down in the Transactions of the Institute for 1866, as well as in those for 1870 alluded to above.

APPENDIX.

HOPPUS'S TABLES.

TABLE I.

SQUARE OF UNEQUAL-SIDED TIMBER, ETC.

By this Table the side of a square piece equivalent to any piece of timber, stone, &c., broader one way than the other, may be found, from 2 inches to 18 inches, the broadest side ; and therefore, by addition only, may serve to any greater breadth, if there should ever be occasion.

The figures at the left hand of the top of each Table are the thickness or lesser side, and the figures in the first column are the breadth or larger side, of the end of a piece of timber of which the square is required. The second column of figures shows the side of a square piece, in inches and quarters of an inch, answering to the contents of the unequal-sided piece.

Example I.—To find the side of the square equivalent to a piece of timber or stone, whose scantling or size is 3 inches thick and 7 inches broad. Look in the column of scantling for 3 inches, the thickness (or lesser side), keeping the eye down the column till you come to 7 inches the breadth (or greater side), and over against that will be found $4\frac{1}{2}$ inches, and that is the side (nearly) of the square of a piece equal to one 3 inches thick, and 7 inches broad.

Example II.—To find the square side of a piece of stone too large for the extent of this Table, say, 28 inches by 34 inches. Take half the thickness of 28, which is 14, and half the breadth of 34, which is 17 ; look in the Table for 14 by 17, and over against that is $15\frac{1}{2}$, which $15\frac{1}{2}$ is one-half of the square side of the piece ; and that doubled makes 31, which is the side of the square piece required.

In.		In.	Sqr.
2	by	2½	2¼
		3	2½
		3½	2¾
		4	2¾
		4½	3
		5	3¼
		5½	3½
		6	3½
		6½	3¾
		7	3¾
		7½	3¾
		8	4
		8½	4
		9	4¼
		9½	4¼
		10	4½
		10½	4½
		11	4¾
		11½	4¾
		12	5
		12½	5
		13	5¼
		13½	5¼
		14	5¼
		14½	5½
		15	5½
		15½	5½
		16	5¾
		16½	5¾
		17	5¾
		17½	6
		18	6
2½	by	3	2¾
		3½	3
		4	3¼
		4½	3½
		5	3½
		5½	3¾
		6	3¾
		6½	4
		7	4¼
		7½	4¼
		8	4½
		8½	4½
		9	4¾
		9½	4¾
		10	5
		10½	5
		11	5¼
		11½	5¼
		12	5½
		12½	5½
		13	5¾
		13½	5¾
		14	6
		14½	6
		15	6¼
		15½	6¼
		16	6¼
		16½	6½
		17	6½
		17½	6½
		18	6¾
3	by	3½	3¼
		4	3½
		4½	3¾
		5	3¾
		5½	4
		6	4¼
		6½	4¼
		7	4½
		7½	4¾
		8	4¾
		8½	5
		9	5¼
		9½	5¼
		10	5½
		10½	[illegible]
		11	5¾
		11½	6
		12	6
		12½	6¼
		13	6¼
		13½	6¼
		14	6½
		14½	6½
		15	6¾
		15½	6¾
		16	7
		16½	7
		17	7¼
		17½	7¼
		18	7½
3½	by	4	3¾
		4½	3¾
		5	4
		5½	4¼
		6	4½
		6½	4¾
		7	5
		7½	5
		8	5¼
		8½	5½
		9	5½
		9½	5¾
		10	6
		10½	6
		11	6¼
		11½	6¼
		12	6½
		12½	6½
		13	6¾
		13½	7
		14	7
		14½	7
		15	7¼
		15½	7¼
		16	7½
		16½	7½
		17	7¾
		17½	7¾
		18	8
4	by	4½	4¼
		5	4½
		5½	4¾
		6	5
		6½	5
		7	5¼
		7½	5½
		8	5½
		8½	5¾
		9	6
		9½	6¼
		10	6¼
		10½	6½
		11	6½
		11½	6¾
		12	7
		12½	7
		13	7¼
		13½	7¼
		14	7½
		14½	7½
		15	7¾
		15½	7¾
		16	8
		16½	8¼
		17	8¼
		17½	8½
		18	8½
4½	by	5	4¾
		5½	5
		6	5¼
		6½	5½
		7	5½
		7½	5¾
		8	6
		8½	6¼
		9	6¼
		9½	6½
		10	6¾
		10½	6¾
		11	7
		11½	7¼
		12	7¼
		12½	7½
		13	7¾
		13½	7¾
		14	8
		14½	8
		15	8¼
		15½	8¼
		16	8½
		16½	8½
		17	8¾
		17½	8¾
		18	9
5	by	5½	5¼
		6	5½
		6½	5¾
		7	6
		7½	6
		8	6¼
		8½	6½
		9	6¾
		9½	6¾
		10	7
		10½	7¼
		11	7½
		11½	7½
		12	7¾
		12½	8
		13	8
		13½	8¼
		14	8¼
		14½	8½
		15	8¾
		15½	8¾
		16	9
		16½	9¼
		17	9¼
		17½	9½
		18	9½
5½	by	6	5¾
		6½	6

In.		*In.*	*Sqr.*
5½	by	7	6¼
		7½	6½
		8	6¾
		8½	6¾
		9	7
		9½	7¼
		10	7½
		10½	7½
		11	7¾
		11½	8
		12	8
		12½	8¼
		13	8½
		13½	8½
		14	8¾
		14½	9
		15	9
		15½	9¼
		16	9½
		16½	9½
		17	9¾
		17½	9¾
		18	10
6	by	6½	6¼
		7	6½
		7½	6¾
		8	7
		8½	7¼
		9	7¼
		9½	7½
		10	7¾
		10½	8
		11	8
		11½	8¼
		12	8½
		12½	8¾
		13	8¾
		13½	9
		14	9¼
		14½	9¼
		15	9½
		15½	9¾
		16	9¾
		16½	10
		17	10
		17½	10¼
		18	10½
6½	by	7	6¾
		7½	7
		8	7¼
		8½	7½
		9	7¾
		9½	7¾
		10	8
		10½	8¼
		11	8½
		11½	8¾
		12	8¾
		12½	9
		13	9¼
		13½	9¼
		14	9½
		14½	9¾
		15	10
		15½	10
		16	10¼
		16½	10¼
		17	10½
		17½	10¾
		18	11
7	by	7½	7¼
		8	7½
		8½	7¾
		9	8
		9½	8¼
		10	8¼
		10½	8½
		11	8¾
		11½	9
		12	9¼
		12½	9¼
		13	9½
		13½	9¾
		14	10
		14½	10
		15	10¼
		15½	10½
		16	10½
		16½	10¾
		17	11
		17½	11
		18	11¼
7½	by	8	7¾
		8½	8
		9	8¼
		9½	8½
		10	8¾
		10½	8¾
		11	9
		11½	9¼
		12	9½
		12½	9¾
		13	9¾
		13½	10
		14	10¼
		14½	10½
		15	10½
		15½	10¾
		16	11
		16½	11
		17	11¼
		17½	11½
		18	11¾
8	by	8½	8¼
		9	8½
		9½	8¾
		10	9
		10½	9¼
		11	9½
		11½	9½
		12	9¾
		12½	10
		13	10¼
		13½	10½
		14	10½
		14½	10¾
		15	11
		15½	11¼
		16	11¼
		16½	11½
		17	11¾
		17½	11¾
		18	12
8½	by	9	8¾
		9½	9
		10	9¼
		10½	9½
		11	9¾
		11½	10
		12	10
		12½	10¼
		13	10½
		13½	10¾
		14	11
		14½	11
		15	11¼
		15½	11½
		16	11¾
		16½	11¾
		17	12
		17½	12¼
		18	12¼
9	by	9½	9¼
		10	9½
		10½	9¾
		11	10
		11½	10¼
		12	10½
		12½	10½
		13	10¾
		13½	11
		14	11¼
		14½	11½
		15	11½
		15½	11¾
		16	12
		16½	12¼
		17	12½
		17½	12½
		18	12¾
9½	by	10	9¾
		10½	10
		11	10¼
		11½	10½
		12	10¾
		12½	11
		13	11
		13½	11¼
		14	11½
		14½	11¾
		15	12
		15½	12¼
		16	12¼
		16½	12½
		17	12¾
		17½	13
		18	13
10	by	10½	10¼
		11	10½
		11½	10¾
		12	11
		12½	11¼
		13	11½
		13½	11¾
		14	12
		14½	12¼
		15	12¼
		15½	12½
		16	12¾
		16½	13
		17	13¼
		17½	13¼
		18	13½

In.		In.	Sqr.	In.		In.	Sqr.	In.		In.	Sqr.	In.		In.	Sqr.
10½	by	11	10¾	11½	by	13½	12¼	12½	by	17½	14¾	14½	by	15	14¾
		11½	11			14	12½			18	15			15½	15
		12	11¼			14½	12¾							16	15¼
		12½	11½			15	13	13	by	13½	13⅛			16½	15½
		13	11¾			15½	13¼			14	13½			17	15¾
		13½	12			16	13½			14½	13¾			17½	16
		14	12¼			16½	13¾			15	14			18	16¼
		14½	12½			17	14			15½	14¼				
		15	12¾			17½	14¼			16	14½	15	by	15½	15¼
		15½	13			18	14½			16½	14¾			16	15½
		16	13							17	14¾			16½	15¾
		16½	13¼	12	by	12½	12¼			17½	15			17	16
		17	13½			13	12½			18	15¼			17½	16¼
		17½	13¾			13½	12¾							18	16½
		18	14			14	13	13½	by	14	13¾				
						14½	13¼			14½	14	15½	by	16	15¾
11	by	11½	11¼			15	13½			15	14¼			16½	16
		12	11½			15½	13¾			15½	14½			17	16¼
		12½	11¾			16	13¾			16	14¾			17½	16½
		13	12			16½	14			16½	15			18	16¾
		13½	12¼			17	14¼			17	15¼				
		14	12½			17½	14½			17½	15¼	16	by	16½	16¼
		14½	12¾			18	14¾			18	15½			17	16½
		15	13											17½	16¾
		15½	13	12½	by	13	12¾	14	by	14½	14¼			18	17
		16	13¼			13½	13			15	14½				
		16½	13½			14	13¼			15½	14¾	16½	by	17	16¾
		17	13¾			14½	13½			16	15			17½	17
		17½	14			15	13¾			16½	15¼			18	17¼
		18	14¼			15½	14			17	15½				
						16	14¼			17½	15¾	17	by	17½	17¼
11½	by	12	11¾			16½	14¼			18	16			18	17½
		12½	12			17	14½								
		13	12¼									17½	by	18	17¾

TABLE II.

SOLID OR CUBICAL MEASURE OF TIMBER, ETC.

By this Table the solid content, and consequently the value, of any piece or quantity of timber, stone, &c. may be found, AT SIGHT, from 2 inches to 18 inches, the side of the square (or one-fourth of the girt), and from one-quarter of a foot to 45 feet, the length; and therefore, by addition only, may serve to any greater breadth, if there should ever be occasion.

The Table begins with two inches, the side of the square, and by the addition of a quarter of an inch enlarges itself to the extent of 18 inches, the side of the square. Every page consists of four distinct parts, divided from each other by a thick line; and at the top of each of the said parts is set down the side of the square piece that has been found by Table I. to be equal to the piece intended to be measured.

The first column to the left hand shows the several lengths, in feet, from one quarter of a foot to 45 feet, and of such a square piece of timber, whose side is set down at the top. The three figures in the second column, marked at the top with Ft. In. Pa., give the solid content in cubic feet, 12ths of a cubic foot, and 144ths of a cubic foot, answering to the length in the left-hand column.

To measure square timber or stone, measure the length of the piece in feet, and set it down in the memorandum book. Then, if the timber be equal-sided, take the side of the square in inches, and quarters, and set that down likewise. (If the timber is unequal-sided, first reduce it to a square by Table I.) Look at the top of the Table for the side of the square; and having found that, keep the eye down the left-hand column until the length of the piece is found in feet; and the three rows of figures which stand over against the length in feet, represent the solid content.

Example I.—To find the solid content of a piece 19 feet long, of which the side of the square is 14¾ inches. Having found 14¾, the side of the square, look for 19 feet in the left-hand column, and over against it stands 28 ft. 8 in. 5 pa.

Example II.—To find the solid content of a piece of stone whose side is greater than the largest in the Table, say 26 inches, the length being 37 feet. Take the half of 26, or 13; find 13 the side of the square, and over against 37 feet is 43.5.1, which multiply by 4, and the product is the content of the piece 26 inches square and 37 feet long—namely, 173 ft. 8 in. 4 pa.

Example III.—To find the solid content of a piece 14 inches square and 22½ feet long. Seek in the Table for 14 inches, the side of the square, and 22 feet the length, which set down; then find the ½ foot at the bottom of the Table, and set that down under the former quantity; add these together, and the sum is the content required—namely—

	ft.	in.	pa.
22 ft. long is	29	11	4
½ ft.	0	8	2
The content required	30	7	6

[TABLE II.]

Ft. long	Side 2 in.			Ft. long	Side 2¼ in.			Ft. long	Side 2½ in.			Ft. long	Side 2¾ in.		
	Ft.	In.	Pa.		Ft.	In.	Pa.		Ft.	In.	Pa.		Ft.	In.	Pa.
1	0	0	4	1	0	0	5	1	0	0	6	1	0	0	7
2	0	0	8	2	0	0	10	2	0	1	0	2	0	1	3
3	0	1	0	3	0	1	3	3	0	1	6	3	0	1	10
4	0	1	4	4	0	1	8	4	0	2	1	4	0	2	6
5	0	1	8	5	0	2	1	5	0	2	7	5	0	3	1
6	0	2	0	6	0	2	6	6	0	3	1	6	0	3	9
7	0	2	4	7	0	2	11	7	0	3	7	7	0	4	4
8	0	2	8	8	0	3	4	8	0	4	2	8	0	5	0
9	0	3	0	9	0	3	9	9	0	4	8	9	0	5	8
10	0	3	4	10	0	4	2	10	0	5	2	10	0	6	3
11	0	3	8	11	0	4	7	11	0	5	8	11	0	6	11
12	0	4	0	12	0	5	0	12	0	6	3	12	0	7	6
13	0	4	4	13	0	5	5	13	0	6	9	13	0	8	2
14	0	4	8	14	0	5	10	14	0	7	3	14	0	8	9
15	0	5	0	15	0	6	3	15	0	7	9	15	0	9	5
16	0	5	4	16	0	6	8	16	0	8	4	16	0	10	1
17	0	5	8	17	0	7	1	17	0	8	10	17	0	10	8
18	0	6	0	18	0	7	6	18	0	9	4	18	0	11	4
19	0	6	4	19	0	7	11	19	0	9	10	19	0	11	11
20	0	6	8	20	0	8	4	20	0	10	5	20	1	0	7
21	0	7	0	21	0	8	9	21	0	10	11	21	1	1	2
22	0	7	4	22	0	9	2	22	0	11	5	22	1	1	10
23	0	7	8	23	0	9	7	23	0	11	11	23	1	2	5
24	0	8	0	24	0	10	0	24	1	0	6	24	1	3	1
25	0	8	4	25	0	10	5	25	1	1	0	25	1	3	9
26	0	8	8	26	0	10	10	26	1	1	6	26	1	4	4
27	0	9	0	27	0	11	3	27	1	2	0	27	1	5	0
28	0	9	4	28	0	11	8	28	1	2	7	28	1	5	7
29	0	9	8	29	1	0	1	29	1	3	1	29	1	6	3
30	0	10	0	30	1	0	6	30	1	3	7	30	1	6	10
31	0	10	4	31	1	0	11	31	1	4	1	31	1	7	6
32	0	10	8	32	1	1	4	32	1	4	8	32	1	8	2
33	0	11	0	33	1	1	9	33	1	5	2	33	1	8	9
34	0	11	4	34	1	2	2	34	1	5	8	34	1	9	5
35	0	11	8	35	1	2	7	35	1	6	2	35	1	10	0
36	1	0	0	36	1	3	0	36	1	6	9	36	1	10	8
37	1	0	4	37	1	3	5	37	1	7	3	37	1	11	3
38	1	0	8	38	1	3	10	38	1	7	9	38	1	11	10
39	1	1	0	39	1	4	3	39	1	8	3	39	2	0	6
40	1	1	4	40	1	4	8	40	1	8	10	40	2	1	2
41	1	1	8	41	1	5	1	41	1	9	4	41	2	1	10
42	1	2	0	42	1	5	6	42	1	9	10	42	2	2	5
43	1	2	4	43	1	5	11	43	1	10	4	43	2	3	1
44	1	2	8	44	1	6	4	44	1	10	11	44	2	3	8
45	1	3	0	45	1	6	9	45	1	11	5	45	2	4	4

	Quarters of a Foot.				Quarters of a Foot.				Quarters of a Foot.				Quarters of a Foot.		
	Ft.	In.	Pa.	S.	Ft.	In.	Pa.	S.	Ft.	In.	Pa.	S.	Ft.	In.	Pa.	S.
¼	0	0	1	0	0	0	1	3	0	0	1	6	0	0	1	9
½	0	0	2	0	0	0	2	6	0	0	2	0	0	0	3	6
¾	0	0	3	0	0	0	3	9	0	0	3	6	0	0	5	3

[TABLE II.

Ft. long	Side 3 in.			Ft. long	Side 3¼ in.			Ft. long	Side 3½ in.			Ft. long	Side 3¾ in.		
	Ft.	In.	Pa.		Ft.	In.	Pa.		Ft.	In.	Pa.		Ft.	In.	Pa.
1	0	0	9	1	0	0	10	1	0	1	0	1	0	1	2
2	0	1	6	2	0	1	9	2	0	2	0	2	0	2	4
3	0	2	3	3	0	2	7	3	0	3	0	3	0	3	6
4	0	3	0	4	0	3	6	4	0	4	1	4	0	4	8
5	0	3	9	5	0	4	4	5	0	5	1	5	0	5	10
6	0	4	6	6	0	5	3	6	0	6	1	6	0	7	0
7	0	5	3	7	0	6	1	7	0	7	1	7	0	8	2
8	0	6	0	8	0	7	0	8	0	8	2	8	0	9	4
9	0	6	9	9	0	7	11	9	0	9	2	9	0	10	6
10	0	7	6	10	0	8	9	10	0	10	2	10	0	11	8
11	0	8	3	11	0	9	8	11	0	11	2	11	1	0	10
12	0	9	0	12	0	10	6	12	1	0	3	12	1	2	0
13	0	9	9	13	0	11	5	13	1	1	3	13	1	3	2
14	0	10	6	14	1	0	3	14	1	2	3	14	1	4	4
15	0	11	3	15	1	1	2	15	1	3	3	15	1	5	6
16	1	0	0	16	1	2	1	16	1	4	4	16	1	6	8
17	1	0	9	17	1	2	11	17	1	5	4	17	1	7	11
18	1	1	6	18	1	3	10	18	1	6	4	18	1	9	1
19	1	2	3	19	1	4	8	19	1	7	4	19	1	10	3
20	1	3	0	20	1	5	7	20	1	8	5	20	1	11	5
21	1	3	9	21	1	6	5	21	1	9	5	21	2	0	7
22	1	4	6	22	1	7	4	22	1	10	5	22	2	1	9
23	1	5	3	23	1	8	2	23	1	11	5	23	2	2	11
24	1	6	0	24	1	9	1	24	2	0	6	24	2	4	1
25	1	6	9	25	1	10	0	25	2	1	6	25	2	5	3
26	1	7	6	26	1	10	10	26	2	2	6	26	2	6	5
27	1	8	3	27	1	11	9	27	2	3	6	27	2	7	7
28	1	9	0	28	2	0	7	28	2	4	7	28	2	8	9
29	1	9	9	29	2	1	6	29	2	5	7	29	2	9	11
30	1	10	6	30	2	2	4	30	2	6	7	30	2	11	1
31	1	11	3	31	2	3	3	31	2	7	7	31	3	0	3
32	2	0	0	32	2	4	2	32	2	8	8	32	3	1	6
33	2	0	9	33	2	5	0	33	2	9	8	33	3	2	8
34	2	1	6	34	2	5	11	34	2	10	8	34	3	3	10
35	2	2	3	35	2	6	9	35	2	11	8	35	3	5	0
36	2	3	0	36	2	7	8	36	3	0	9	36	3	6	2
37	2	3	9	37	2	8	6	37	3	1	9	37	3	7	4
38	2	4	6	38	2	9	5	38	3	2	9	38	3	8	6
39	2	5	3	39	2	10	3	39	3	3	9	39	3	9	8
40	2	6	0	40	2	11	2	40	3	4	10	40	3	10	10
41	2	6	9	41	3	0	1	41	3	5	10	41	4	0	0
42	2	7	6	42	3	0	11	42	3	6	10	42	4	1	2
43	2	8	3	43	3	1	10	43	3	7	10	43	4	2	4
44	2	9	0	44	3	2	8	44	3	8	11	44	4	3	6
45	2	9	9	45	3	3	7	45	3	9	11	45	4	4	8

Quarters of a Foot.				Quarters of a Foot.				Quarters of a Foot.				Quarters of a Foot.			
	Ft.	In.	Pa.	S.	Ft.	In.	Pa.	S.	Ft.	In.	Pa.	S.	Ft.	In.	Pa.	S.
¼	0	0	2	3	0	0	2	6	0	0	3	0	0	0	3	6
½	0	0	4	6	0	0	5	0	0	0	6	0	0	0	7	0
¾	0	0	6	9	0	0	7	6	0	0	9	0	0	0	10	6

[TABLE II.]

Ft. long	Side 4 in.			Ft. long	Side 4¼ in.			Ft. long	Side 4½ in.			Ft. long	Side 4¾ in.		
	Ft.	In.	Pa.		Ft.	In.	Pa.		Ft.	In.	Pa.		Ft.	In.	Pa.
1	0	1	4	1	0	1	6	1	0	1	8	1	0	1	10
2	0	2	8	2	0	3	0	2	0	3	4	2	0	3	9
3	0	4	0	3	0	4	6	3	0	5	0	3	0	5	7
4	0	5	4	4	0	6	0	4	0	6	9	4	0	7	6
5	0	6	8	5	0	7	6	5	0	8	5	5	0	9	4
6	0	8	0	6	0	9	0	6	0	10	1	6	0	11	3
7	0	9	4	7	0	10	6	7	0	11	9	7	1	1	1
8	0	10	8	8	1	0	0	8	1	1	6	8	1	3	0
9	1	0	0	9	1	1	6	9	1	3	2	9	1	4	11
10	1	1	4	10	1	3	0	10	1	4	10	10	1	6	9
11	1	2	8	11	1	4	6	11	1	6	6	11	1	8	8
12	1	4	0	12	1	6	0	12	1	8	3	12	1	10	6
13	1	5	4	13	1	7	6	13	1	9	11	13	2	0	5
14	1	6	8	14	1	9	0	14	1	11	7	14	2	2	3
15	1	8	0	15	1	10	6	15	2	1	3	15	2	4	2
16	1	9	4	16	2	0	1	16	2	3	0	16	2	6	1
17	1	10	8	17	2	1	7	17	2	4	8	17	2	7	11
18	2	0	0	18	2	3	1	18	2	6	4	18	2	9	10
19	2	1	4	19	2	4	7	19	2	8	0	19	2	11	8
20	2	2	8	20	2	6	1	20	2	9	9	20	3	1	7
21	2	4	0	21	2	7	7	21	2	11	5	21	3	3	5
22	2	5	4	22	2	9	1	22	3	1	1	22	3	5	4
23	2	6	8	23	2	10	7	23	3	2	9	23	3	7	2
24	2	8	0	24	3	0	1	24	3	4	6	24	3	9	1
25	2	9	4	25	3	1	7	25	3	6	2	25	3	11	0
26	2	10	8	26	3	3	1	26	3	7	10	26	4	0	10
27	3	0	0	27	3	4	7	27	3	9	6	27	4	2	9
28	3	1	4	28	3	6	1	28	3	11	3	28	4	4	7
29	3	2	8	29	3	7	7	29	4	0	11	29	4	6	6
30	3	4	0	30	3	9	1	30	4	2	7	30	4	8	4
31	3	5	4	31	3	10	7	31	4	4	3	31	4	10	3
32	3	6	8	32	4	0	2	32	4	6	0	32	5	0	2
33	3	8	0	33	4	1	8	33	4	7	8	33	5	2	0
34	3	9	4	34	4	3	2	34	4	9	4	34	5	3	11
35	3	10	8	35	4	4	8	35	4	11	0	35	5	5	9
36	4	0	0	36	4	6	2	36	5	0	9	36	5	7	8
37	4	1	4	37	4	7	8	37	5	2	5	37	5	9	6
38	4	2	8	38	4	9	2	38	5	4	1	38	5	11	5
39	4	4	0	39	4	10	8	39	5	5	9	39	6	1	3
40	4	5	4	40	5	0	2	40	5	7	6	40	6	3	2
41	4	6	8	41	5	1	8	41	5	9	2	41	6	5	1
42	4	8	0	42	5	3	2	42	5	10	10	42	6	6	11
43	4	9	4	43	5	4	8	43	6	0	6	43	6	8	10
44	4	10	8	44	5	6	2	44	6	2	3	44	6	10	8
45	5	0	0	45	5	7	8	45	6	3	11	45	7	0	7
Quarters of a Foot.				Quarters of a Foot.				Quarters of a Foot.				Quarters of a Foot.			
	Ft.	In.	Pa. S.		Ft.	In.	Pa. S.		Ft.	In.	Pa. S.		Ft.	In.	Pa. S.
¼	0	0	4 0	¼	0	0	4 6	¼	0	0	5 0	¼	0	0	5 6
½	0	0	8 0	½	0	0	9 0	½	0	0	10 0	½	0	0	11 0
¾	0	1	0 0	¾	0	1	1 6	¾	0	1	3 0	¾	0	1	4 6

Ft. long	Side 5 in.			Ft. long	Side 5¼ in.			Ft. long	Side 5½ in.			Ft. long	Side 5¾ in.		
	Ft.	In.	Pa.		Ft.	In.	Pa.		Ft.	In.	Pa.		Ft.	In.	Pa.
1	0	2	1	1	0	2	3	1	0	2	6	1	0	2	9
2	0	4	2	2	0	4	7	2	0	5	0	2	0	5	6
3	0	6	3	3	0	6	10	3	0	7	6	3	0	8	3
4	0	8	4	4	0	9	2	4	0	10	1	4	0	11	0
5	0	10	5	5	0	11	5	5	1	0	7	5	1	1	9
6	1	0	6	6	1	1	9	6	1	3	1	6	1	4	6
7	1	2	7	7	1	4	0	7	1	5	7	7	1	7	3
8	1	4	8	8	1	6	4	8	1	8	2	8	1	10	0
9	1	6	9	9	1	8	8	9	1	10	8	9	2	0	9
10	1	8	10	10	1	10	11	10	2	1	2	10	2	3	6
11	1	10	11	11	2	1	3	11	2	3	8	11	2	6	3
12	2	1	0	12	2	3	6	12	2	6	3	12	2	9	0
13	2	3	1	13	2	5	10	13	2	8	9	13	2	11	9
14	2	5	2	14	2	8	1	14	2	11	3	14	3	2	6
15	2	7	3	15	2	10	5	15	3	1	9	15	3	5	3
16	2	9	4	16	3	0	9	16	3	4	4	16	3	8	1
17	2	11	5	17	3	3	0	17	3	6	10	17	3	10	10
18	3	1	6	18	3	5	4	18	3	9	4	18	4	1	7
19	3	3	7	19	3	7	7	19	3	11	10	19	4	4	4
20	3	5	8	20	3	9	11	20	4	2	5	20	4	7	1
21	3	7	9	21	4	0	2	21	4	4	11	21	4	9	10
22	3	9	10	22	4	2	6	22	4	7	5	22	5	0	7
23	3	11	11	23	4	4	9	23	4	9	11	23	5	3	4
24	4	2	0	24	4	7	1	24	5	0	6	24	5	6	1
25	4	4	1	25	4	9	5	25	5	3	0	25	5	8	10
26	4	6	2	26	4	11	8	26	5	5	6	26	5	11	7
27	4	8	3	27	5	2	0	27	5	8	0	27	6	2	4
28	4	10	4	28	5	4	3	28	5	10	6	28	6	5	1
29	5	0	5	29	5	6	7	29	6	1	1	29	6	7	10
30	5	2	6	30	5	8	10	30	6	3	7	30	6	10	7
31	5	4	7	31	5	11	2	31	6	6	1	31	7	1	4
32	5	6	8	32	6	1	6	32	6	8	8	32	7	4	2
33	5	8	9	33	6	3	9	33	6	11	2	33	7	6	11
34	5	10	10	34	6	6	1	34	7	1	8	34	7	9	8
35	6	0	11	35	6	8	4	35	7	4	2	35	8	0	5
36	6	3	0	36	6	10	8	36	7	6	9	36	8	3	2
37	6	5	1	37	7	0	11	37	7	9	3	37	8	6	11
38	6	7	2	38	7	3	3	38	7	11	9	38	8	8	8
39	6	9	3	39	7	5	6	39	8	2	3	39	8	11	5
40	6	11	4	40	7	7	10	40	8	4	10	40	9	2	2
41	7	1	5	41	7	10	2	41	8	7	4	41	9	4	11
42	7	3	6	42	8	0	5	42	8	9	10	42	9	7	8
43	7	5	7	43	8	2	9	43	9	0	4	43	9	10	5
44	7	7	8	44	8	5	0	44	9	2	11	44	10	1	2
45	7	9	9	45	8	7	4	45	9	5	5	45	10	3	11

Quarters of a Foot.				Quarters of a Foot.				Quarters of a Foot.				Quarters of a Foot.			
	Ft.	In.	Pa.	S.	Ft.	In.	Pa.	S.	Ft.	In.	Pa.	S.	Ft.	In.	Pa.	S.
¼	0	0	6	3	0	0	6	9	0	0	7	6	0	0	8	3
½	0	1	0	6	0	1	1	6	0	1	3	0	0	1	4	6
¾	0	1	6	9	0	1	8	3	0	1	10	6	0	2	0	9

Ft. long	Side 6 in.			Ft. long	Side 6¼ in.			Ft. long	Side 6½ in.			Ft. long	Side 6¾ in.		
	Ft.	In.	Pa.		Ft.	In.	Pa.		Ft.	In.	Pa.		Ft.	In.	Pa.
1	0	3	0	1	0	3	3	1	0	3	6	1	0	3	9
2	0	6	0	2	0	6	6	2	0	7	0	2	0	7	7
3	0	9	0	3	0	9	9	3	0	10	6	3	0	11	4
4	1	0	0	4	1	1	0	4	1	2	1	4	1	3	2
5	1	3	0	5	1	4	3	5	1	5	7	5	1	6	11
6	1	6	0	6	1	7	6	6	1	9	1	6	1	10	9
7	1	9	0	7	1	10	9	7	2	0	7	7	2	2	6
8	2	0	0	8	2	2	0	8	2	4	2	8	2	6	4
9	2	3	0	9	2	5	3	9	2	7	8	9	2	10	2
10	2	6	0	10	2	8	6	10	2	11	2	10	3	1	11
11	2	9	0	11	2	11	9	11	3	2	8	11	3	5	9
12	3	0	0	12	3	3	0	12	3	6	3	12	3	9	6
13	3	3	0	13	3	6	3	13	3	9	9	13	4	1	4
14	3	6	0	14	3	9	6	14	4	1	3	14	4	5	1
15	3	9	0	15	4	0	9	15	4	4	9	15	4	8	11
16	4	0	0	16	4	4	1	16	4	8	4	16	5	0	9
17	4	3	0	17	4	7	4	17	4	11	10	17	5	4	6
18	4	6	0	18	4	10	7	18	5	3	4	18	5	8	4
19	4	9	0	19	5	1	10	19	5	6	10	19	6	0	1
20	5	0	0	20	5	5	1	20	5	10	5	20	6	3	11
21	5	3	0	21	5	8	4	21	6	1	11	21	6	7	8
22	5	6	0	22	5	11	7	22	6	5	5	22	6	11	6
23	5	9	0	23	6	2	10	23	6	8	11	23	7	3	3
24	6	0	0	24	6	6	1	24	7	0	6	24	7	7	1
25	6	3	0	25	6	9	4	25	7	4	0	25	7	10	11
26	6	6	0	26	7	0	7	26	7	7	6	26	8	2	4
27	6	9	0	27	7	3	10	27	7	11	0	27	8	6	8
28	7	0	0	28	7	7	1	28	8	2	7	28	8	10	3
29	7	3	0	29	7	10	4	29	8	6	1	29	9	2	1
30	7	6	0	30	8	1	7	30	8	9	7	30	9	5	10
31	7	9	0	31	8	4	10	31	9	1	1	31	9	9	8
32	8	0	0	32	8	8	2	32	9	4	8	32	10	1	6
33	8	3	0	33	8	11	5	33	9	8	2	33	10	5	5
34	8	6	0	34	9	2	8	34	9	11	8	34	10	9	1
35	8	9	0	35	9	5	11	35	10	3	2	35	11	0	10
36	9	0	0	36	9	9	2	36	10	6	9	36	11	4	8
37	9	3	0	37	10	0	5	37	10	10	3	37	11	8	5
38	9	6	0	38	10	3	8	38	11	1	9	38	12	0	3
39	9	9	0	39	10	6	11	39	11	5	3	39	12	4	0
40	10	0	0	40	10	10	2	40	11	8	10	40	12	7	10
41	10	3	0	41	11	1	5	41	12	0	4	41	12	11	8
42	10	6	0	42	11	4	8	42	12	3	10	42	13	3	5
43	10	9	0	43	11	7	11	43	12	7	4	43	13	7	3
44	11	0	0	44	11	11	2	44	12	10	11	44	13	11	0
45	11	3	0	45	12	2	5	45	13	2	5	45	14	2	10

Quarters of a Foot.				Quarters of a Foot.				Quarters of a Foot.				Quarters of a Foot.			
	Ft.	In.	Pa.	S.	Ft.	In.	Pa.	S.	Ft.	In.	Pa.	S.	Ft.	In.	Pa.	S.
¼	0	0	9	0	0	0	9	9	0	0	10	6	0	0	11	3
½	0	1	6	0	0	1	7	6	0	1	9	0	0	1	10	6
¾	0	2	3	0	0	2	5	3	0	2	7	6	0	2	9	9

[TABLE II.

Ft. long	Side 7 in.			Ft. long	Side 7¼ in.			Ft. long	Side 7½ in.			Ft. long	Side 7¾ in.		
	Ft.	In.	Pa.		Ft.	In.	Pa.		Ft.	In.	Pa.		Ft.	In.	Pa.
1	0	4	1	1	0	4	4	1	0	4	8	1	0	5	0
2	0	8	2	2	0	8	9	2	0	9	4	2	0	10	0
3	1	0	3	3	1	1	1	3	1	2	0	3	1	3	0
4	1	4	4	4	1	5	6	4	1	6	9	4	1	8	0
5	1	8	5	5	1	9	10	5	1	11	5	5	2	1	0
6	2	0	6	6	2	2	3	6	2	4	1	6	2	6	0
7	2	4	7	7	2	6	7	7	2	8	9	7	2	11	0
8	2	8	8	8	2	11	0	8	3	1	6	8	3	4	0
9	3	0	9	9	3	3	5	9	3	6	2	9	3	9	0
10	3	4	10	10	3	7	9	10	3	10	10	10	4	2	0
11	3	8	11	11	4	0	2	11	4	3	6	11	4	7	0
12	4	1	0	12	4	4	6	12	4	8	3	12	5	0	0
13	4	5	1	13	4	8	11	13	5	0	11	13	5	5	0
14	4	9	2	14	5	1	3	14	5	5	7	14	5	10	0
15	5	1	3	15	5	5	8	15	5	10	3	15	6	3	0
16	5	5	4	16	5	10	1	16	6	3	0	16	6	8	1
17	5	9	5	17	6	2	5	17	6	7	8	17	7	1	1
18	6	1	6	18	6	6	10	18	7	0	4	18	7	6	1
19	6	5	7	19	6	11	2	19	7	5	0	19	7	11	1
20	6	9	8	20	7	3	7	20	7	9	9	20	8	4	1
21	7	1	9	21	7	7	11	21	8	2	5	21	8	9	1
22	7	5	10	22	8	0	4	22	8	7	1	22	9	2	1
23	7	9	11	23	8	4	8	23	8	11	9	23	9	7	1
24	8	2	0	24	8	9	1	24	9	4	6	24	10	0	1
25	8	6	1	25	9	1	6	25	9	9	2	25	10	5	1
26	8	10	2	26	9	5	10	26	10	1	10	26	10	10	1
27	9	2	3	27	9	10	3	27	10	6	6	27	11	3	1
28	9	6	4	28	10	2	7	28	10	11	3	28	11	8	1
29	9	10	5	29	10	7	0	29	11	3	11	29	12	1	1
30	10	2	6	30	10	11	4	30	11	8	7	30	12	6	1
31	10	6	7	31	11	3	9	31	12	1	3	31	12	11	1
32	10	10	8	32	11	8	2	32	12	6	0	32	13	4	2
33	11	2	9	33	12	0	6	33	12	10	8	33	13	9	2
34	11	6	10	34	12	4	11	34	13	3	4	34	14	2	2
35	11	10	11	35	12	9	3	35	13	8	0	35	14	7	2
36	12	3	0	36	13	1	8	36	14	0	9	36	15	0	2
37	12	7	1	37	13	6	0	37	14	5	5	37	15	5	2
38	12	11	2	38	13	10	5	38	14	10	1	38	15	10	2
39	13	3	3	39	14	2	9	39	15	2	9	39	16	3	2
40	13	7	4	40	14	7	2	40	15	7	6	40	16	8	2
41	13	11	5	41	14	11	7	41	16	0	2	41	17	1	2
42	14	3	6	42	15	3	11	42	16	4	10	42	17	6	2
43	14	7	7	43	15	8	4	43	16	9	6	43	17	11	2
44	14	11	8	44	16	0	8	44	17	2	3	44	18	4	2
45	15	3	9	45	16	5	1	45	17	6	11	45	18	9	2

Quarters of a Foot.				Quarters of a Foot.				Quarters of a Foot.				Quarters of a Foot.			
	Ft.	In.	Pa.	S.	Ft.	In.	Pa.	S.	Ft.	In.	Pa.	S.	Ft.	In.	Pa.	S.
¼	0	1	0	3	0	1	1	0	0	1	2	0	0	1	3	0
½	0	2	0	6	0	2	2	0	0	2	4	0	0	2	6	0
¾	0	3	0	9	0	3	3	0	0	3	6	0	0	3	9	0

[Table II.]

Ft. long	Side 8 in.			Ft. long	Side 8¼ in.			Ft. long	Side 8½ in.			Ft. long	Side 8¾ in.		
	Ft.	In.	Pa.		Ft.	In.	Pa.		Ft.	In.	Pa.		Ft.	In.	Pa.
1	0	5	4	1	0	5	8	1	0	6	0	1	0	6	0
2	0	10	8	2	0	11	4	2	1	0	0	2	1	0	9
3	1	4	0	3	1	5	0	3	1	6	0	3	1	7	1
4	1	9	4	4	1	10	8	4	2	0	1	4	2	1	6
5	2	2	8	5	2	4	4	5	2	6	1	5	2	7	10
6	2	8	0	6	2	10	0	6	3	0	1	6	3	2	3
7	3	1	4	7	3	3	8	7	3	6	1	7	3	8	7
8	3	6	8	8	3	9	4	8	4	0	2	8	4	3	0
9	4	0	0	9	4	3	0	9	4	6	2	9	4	9	5
10	4	5	4	10	4	8	8	10	5	0	2	10	5	3	9
11	4	10	8	11	5	2	4	11	5	6	2	11	5	10	2
12	5	4	0	12	5	8	0	12	6	0	3	12	6	4	6
13	5	9	4	13	6	1	8	13	6	6	3	13	6	10	11
14	6	2	8	14	6	7	4	14	7	0	3	14	7	5	3
15	6	8	0	15	7	1	0	15	7	6	3	15	7	11	8
16	7	1	4	16	7	6	9	16	8	0	4	16	8	6	1
17	7	6	8	17	8	0	5	17	8	6	4	17	9	0	5
18	8	0	0	18	8	6	1	18	9	0	4	18	9	6	10
19	8	5	4	19	8	11	9	19	9	6	4	19	10	1	2
20	8	10	8	20	9	5	5	20	10	0	5	20	10	7	7
21	9	4	0	21	9	11	1	21	10	6	5	21	11	2	11
22	9	9	4	22	10	4	9	22	11	0	5	22	11	8	4
23	10	2	8	23	10	10	5	23	11	6	5	23	12	2	8
24	10	8	0	24	11	4	1	24	12	0	6	24	12	9	1
25	11	1	4	25	11	9	9	25	12	6	6	25	13	3	6
26	11	6	8	26	12	3	5	26	13	0	6	26	13	9	10
27	12	0	0	27	12	9	1	27	13	6	6	27	14	4	3
28	12	5	4	28	13	2	9	28	14	0	7	28	14	10	7
29	12	10	8	29	13	8	5	29	14	6	7	29	15	5	0
30	13	4	0	30	14	2	1	30	15	0	7	30	15	11	4
31	13	9	4	31	14	7	9	31	15	6	7	31	16	5	9
32	14	2	8	32	15	1	6	32	16	0	8	32	17	0	2
33	14	8	0	33	15	7	2	33	16	6	8	33	17	6	6
34	15	1	4	34	16	0	10	34	17	0	8	34	18	0	11
35	15	6	8	35	16	6	6	35	17	6	8	35	18	7	3
36	16	0	0	36	17	0	2	36	18	0	9	36	19	1	8
37	16	5	4	37	17	5	10	37	18	6	9	37	19	8	0
38	16	10	8	38	17	11	6	38	19	0	9	38	20	2	5
39	17	4	0	39	18	5	2	39	19	6	9	39	20	8	9
40	17	9	4	40	18	10	10	40	20	0	10	40	21	3	2
41	18	2	8	41	19	4	6	41	20	6	10	41	21	9	7
42	18	8	0	42	19	10	2	42	21	0	10	42	22	3	11
43	19	1	4	43	20	3	10	43	21	6	10	43	22	10	4
44	19	6	8	44	20	9	6	44	22	0	11	44	23	4	8
45	20	0	0	45	21	3	2	45	22	6	11	45	23	11	0

Quarters of a Foot.				Quarters of a Foot.				Quarters of a Foot.				Quarters of a Foot.			
	Ft.	In.	Pa.	S.	Ft.	In.	Pa.	S.	Ft.	In.	Pa.	S.	Ft.	In.	Pa.	S.
¼	0	1	4	0	0	1	5	0	0	1	6	0	0	1	7	0
½	0	2	8	0	0	2	10	0	0	3	0	0	0	3	2	0
¾	0	4	0	0	0	4	3	0	0	4	6	0	0	4	9	0

[TABLE II

Ft. long	Side 9 in.			Ft. long	Side 9¼ in.			Ft. long	Side 9½ in.			Ft. long	Side 9¾ in.		
	Ft.	In.	Pa.		Ft.	In.	Pa.		Ft.	In.	Pa.		Ft.	In.	Pa.
1	0	6	9	1	0	7	1	1	0	7	6	1	0	7	11
2	1	1	6	2	1	2	3	2	1	3	0	2	1	3	10
3	1	8	3	3	1	9	4	3	1	10	6	3	1	11	9
4	2	3	0	4	2	4	6	4	2	6	1	4	2	7	8
5	2	9	9	5	2	11	7	5	3	1	7	5	3	3	7
6	3	4	6	6	3	6	9	6	3	9	1	6	3	11	6
7	3	11	3	7	4	1	10	7	4	4	7	7	4	7	5
8	4	6	0	8	4	9	0	8	5	0	2	8	5	3	4
9	5	0	9	9	5	4	2	9	5	7	8	9	5	11	3
10	5	7	6	10	5	11	3	10	6	3	2	10	6	7	2
11	6	2	3	11	6	6	5	11	6	10	8	11	7	3	1
12	6	9	0	12	7	1	6	12	7	6	3	12	7	11	0
13	7	3	9	13	7	8	8	13	8	1	9	13	8	6	11
14	7	10	6	14	8	3	9	14	8	9	3	14	9	2	10
15	8	5	3	15	8	10	11	15	9	4	9	15	9	10	9
16	9	0	0	16	9	6	1	16	10	0	4	16	10	6	9
17	9	6	9	17	10	1	2	17	10	7	10	17	11	2	8
18	10	1	6	18	10	8	4	18	11	3	4	18	11	10	7
19	10	8	3	19	11	3	5	19	11	10	10	19	12	6	6
20	11	3	0	20	11	10	7	20	12	6	5	20	13	2	5
21	11	9	9	21	12	5	8	21	13	1	11	21	13	10	4
22	12	4	6	22	13	0	10	22	13	9	5	22	14	6	3
23	12	11	3	23	13	7	11	23	14	4	11	23	15	2	2
24	13	6	0	24	14	3	1	24	15	0	6	24	15	10	1
25	14	0	9	25	14	10	3	25	15	8	0	25	16	6	0
26	14	7	6	26	15	5	0	26	16	3	6	26	17	1	11
27	15	2	3	27	16	0	6	27	16	11	0	27	17	9	10
28	15	9	0	28	16	7	7	28	17	6	7	28	18	5	9
29	16	3	9	29	17	2	9	29	18	2	1	29	19	1	8
30	16	10	6	30	17	9	10	30	18	9	7	30	19	9	7
31	17	5	3	31	18	5	0	31	19	5	1	31	20	5	6
32	18	0	0	32	19	0	2	32	20	0	8	32	21	1	6
33	18	6	9	33	19	7	3	33	20	8	2	33	21	9	5
34	19	1	6	34	20	2	5	34	21	3	8	34	22	5	4
35	19	8	3	35	20	9	6	35	21	11	2	35	23	1	3
36	20	3	0	36	21	4	8	36	22	6	9	36	23	9	2
37	20	9	9	37	21	11	9	37	23	2	3	37	24	5	1
38	21	4	6	38	22	6	11	38	23	9	9	38	25	1	0
39	21	11	3	39	23	2	0	39	24	5	3	39	25	8	11
40	22	6	0	40	23	9	2	40	25	0	10	40	26	4	10
41	23	0	9	41	24	4	4	41	25	8	4	41	27	0	9
42	23	7	6	42	24	11	5	42	26	3	10	42	27	8	8
43	24	2	3	43	25	6	7	43	26	11	4	43	28	4	7
44	24	9	0	44	26	1	8	44	27	6	11	44	29	0	6
45	25	3	9	45	26	8	10	45	28	2	5	45	29	8	5

Quarters of a Foot.				Quarters of a Foot.				Quarters of a Foot.				Quarters of a Foot.			
	Ft.	In.	Pa.	S.	Ft.	In.	Pa.	S.	Ft.	In.	Pa.	S.	Ft.	In.	Pa.	S.
¼	0	1	8	3	0	1	9	3	0	1	10	6	0	1	11	9
½	0	3	4	6	0	3	6	6	0	3	9	0	0	3	11	6
¾	0	5	0	9	0	5	3	9	0	5	7	6	0	5	11	3

TABLE II.]

Ft. long	Side 10 in.			Ft. long	Side 10¼ in.			Ft. long	Side 10½ in.			Ft. long	Side 10¾ in.		
	Ft.	In.	Pa.		Ft.	In.	Pa.		Ft.	In.	Pa.		Ft.	In.	Pa.
1	0	8	4	1	0	8	9	1	0	9	2	1	0	9	7
2	1	4	8	2	1	5	6	2	1	6	4	2	1	7	3
3	2	1	0	3	2	2	3	3	2	3	6	3	2	4	10
4	2	9	4	4	2	11	0	4	3	0	9	4	3	2	6
5	3	5	8	5	3	7	9	5	3	9	11	5	4	0	1
6	4	2	0	6	4	4	6	6	4	7	1	6	4	9	9
7	4	10	4	7	5	1	3	7	5	4	3	7	5	7	4
8	5	6	8	8	5	10	0	8	6	1	6	8	6	5	0
9	6	3	0	9	6	6	9	9	6	10	8	9	7	2	7
10	6	11	4	10	7	3	6	10	7	7	10	10	8	0	3
11	7	7	8	11	8	0	3	11	8	5	0	11	8	9	11
12	8	4	0	12	8	9	0	12	9	2	3	12	9	7	6
13	9	0	4	13	9	5	9	13	9	11	5	13	10	5	2
14	9	8	8	14	10	2	5	14	10	8	7	14	11	2	9
15	10	5	0	15	10	11	3	15	11	5	9	15	12	0	5
16	11	1	4	16	11	8	1	16	12	3	0	16	12	10	0
17	11	9	8	17	12	4	10	17	13	0	2	17	13	7	8
18	12	6	0	18	13	1	7	18	13	9	4	18	14	5	4
19	13	2	4	19	13	10	4	19	14	6	6	19	15	2	11
20	13	10	8	20	14	7	1	20	15	3	9	20	16	0	7
21	14	7	0	21	15	3	10	21	16	0	11	21	16	10	2
22	15	3	4	22	16	0	7	22	16	10	1	22	17	7	10
23	15	11	8	23	16	9	4	23	17	7	3	23	18	5	5
24	16	8	0	24	17	6	1	24	18	4	6	24	19	3	1
25	17	4	4	25	18	2	10	25	19	1	8	25	20	0	9
26	18	0	8	26	18	11	7	26	19	10	10	26	20	10	4
27	18	9	0	27	19	8	4	27	20	8	0	27	21	8	0
28	19	5	4	28	20	5	1	28	21	5	3	28	22	5	7
29	20	1	8	29	21	1	10	29	22	2	5	29	23	3	3
30	20	10	0	30	21	10	7	30	22	11	7	30	24	0	10
31	21	6	4	31	22	7	4	31	23	8	9	31	24	10	6
32	22	2	8	32	23	4	2	32	24	6	0	32	25	8	2
33	22	11	0	33	24	0	11	33	25	3	2	33	26	5	9
34	23	7	4	34	24	9	8	34	26	0	4	34	27	3	5
35	24	3	8	35	25	6	5	35	26	9	6	35	28	1	0
36	25	0	0	36	26	3	2	36	27	6	9	36	28	10	8
37	25	8	4	37	26	11	11	37	28	3	11	37	29	8	3
38	26	4	8	38	27	8	8	38	29	1	1	38	30	5	11
39	27	1	0	39	28	5	5	39	29	10	3	39	31	3	6
40	27	9	4	40	29	2	2	40	30	7	6	40	32	1	2
41	28	5	8	41	29	10	11	41	31	4	8	41	32	10	10
42	29	2	0	42	30	7	8	42	32	1	10	42	33	8	5
43	29	10	4	43	31	4	5	43	32	11	0	43	34	6	1
44	30	6	8	44	32	1	2	44	33	8	3	44	35	3	8
45	31	4	0	45	32	9	11	45	34	5	5	45	36	1	4
Quarters of a Foot.				Quarters of a Foot.				Quarters of a Foot.				Quarters of a Foot.			
	Ft.	In.	Pa. S.		Ft.	In.	Pa. S.		Ft.	In.	Pa. S.		Ft.	In.	Pa. S.
¼	0	2	1 0	¼	0	2	2 3	¼	0	2	3 6	¼	0	2	4 9
½	0	4	2 0	½	0	4	4 6	½	0	4	7 0	½	0	4	9 6
¾	0	6	3 0	¾	0	6	6 9	¾	0	6	10 6	¾	0	7	2 3

Ft. long	Side 11 in.			Ft. long	Side 11¼ in.			Ft. long	Side 11½ in.			Ft. long	Side 11¾ in.		
	Ft.	In.	Pa.		Ft.	In.	Pa.		Ft.	In.	Pa.		Ft.	In.	Pa.
1	0	10	1	1	0	10	6	1	0	11	0	1	0	11	6
2	1	8	2	2	1	9	1	2	1	10	0	2	1	11	0
3	2	6	3	3	2	7	7	3	2	9	0	3	2	10	6
4	3	4	4	4	3	6	2	4	3	8	1	4	3	10	0
5	4	2	5	5	4	4	8	5	4	7	1	5	4	9	6
6	5	0	6	6	5	3	3	6	5	6	1	6	5	9	0
7	5	10	7	7	6	1	9	7	6	5	1	7	6	8	6
8	6	8	8	8	7	0	4	8	7	4	2	8	7	8	0
9	7	6	9	9	7	10	11	9	8	3	2	9	8	7	6
10	8	4	10	10	8	9	5	10	9	2	2	10	9	7	0
11	9	2	11	11	9	8	0	11	10	1	2	11	10	6	6
12	10	1	0	12	10	6	6	12	11	0	3	12	11	6	0
13	10	11	1	13	11	5	1	13	11	11	3	13	12	5	6
14	11	9	2	14	12	3	7	14	12	10	3	14	13	5	0
15	12	7	3	15	13	2	2	15	13	9	3	15	14	4	6
16	13	5	4	16	14	0	9	16	14	8	4	16	15	4	1
17	14	3	5	17	14	11	3	17	15	7	4	17	16	3	7
18	15	1	6	18	15	9	10	18	16	6	4	18	17	3	1
19	15	11	7	19	16	8	4	19	17	5	4	19	18	2	7
20	16	9	8	20	17	6	11	20	18	4	5	20	19	2	1
21	17	7	9	21	18	5	5	21	19	3	5	21	20	1	7
22	18	5	10	22	19	4	0	22	20	2	5	22	21	1	1
23	19	3	11	23	20	2	6	23	21	1	5	23	22	0	7
24	20	2	0	24	21	1	1	24	22	0	6	24	23	0	1
25	21	0	1	25	21	11	8	25	22	11	6	25	23	11	7
26	21	10	2	26	22	10	2	26	23	10	6	26	24	11	1
27	22	8	3	27	23	8	9	27	24	9	6	27	25	10	7
28	23	6	4	28	24	7	3	28	25	8	7	28	26	10	1
29	24	4	5	29	25	5	10	29	26	7	7	29	27	9	7
30	25	2	6	30	26	4	4	30	27	6	7	30	28	9	1
31	26	0	7	31	27	2	11	31	28	5	7	31	29	8	7
32	26	10	8	32	28	1	6	32	29	4	8	32	30	8	2
33	27	8	9	33	29	0	0	33	30	3	8	33	31	7	8
34	28	6	10	34	29	10	7	34	31	2	8	34	32	7	2
35	29	4	11	35	30	9	1	35	32	1	8	35	33	6	8
36	30	3	0	36	31	7	8	36	33	0	9	36	34	6	2
37	31	1	1	37	32	6	2	37	33	11	9	37	35	5	8
38	31	11	2	38	33	4	9	38	34	10	9	38	36	5	2
39	32	9	3	39	34	3	3	39	35	9	9	39	37	4	8
40	33	7	4	40	35	1	10	40	36	8	10	40	38	4	2
41	34	5	5	41	36	0	5	41	37	7	10	41	39	3	8
42	35	3	6	42	36	10	11	42	38	6	10	42	40	3	2
43	36	1	7	43	37	9	6	43	39	5	10	43	41	2	8
44	36	11	8	44	38	8	0	44	40	4	11	44	42	2	2
45	37	9	9	45	39	6	7	45	41	3	11	45	43	1	8
Quarters of a Foot.				Quarters of a Foot.				Quarters of a Foot.				Quarters of a Foot.			
	Ft.	In.	Pa. S.		Ft.	In.	Pa. S.		Ft.	In.	Pa. S.		Ft.	In.	Pa. S.
¼	0	2	6 3	¼	0	2	7 6	¼	0	2	9 0	¼	0	2	10 6
½	0	5	0 6	½	0	5	3 0	½	0	5	6 0	½	0	5	9 0
¾	0	7	6 9	¾	0	7	10 6	¾	0	8	3 0	¾	0	8	8 6

Ft. long	Side 12 in.			Ft. long	Side 12¼ in.			Ft. long	Side 12½ in.			Ft. long	Side 12¾ in.		
	Ft.	In.	Pa.		Ft.	In.	Pa.		Ft.	In.	Pa.		Ft.	In.	Pa.
1	1	0	0	1	1	0	6	1	1	1	0	1	1	1	6
2	2	0	0	2	2	1	0	2	2	2	0	2	2	3	1
3	3	0	0	3	3	1	6	3	3	3	0	3	3	4	7
4	4	0	0	4	4	2	0	4	4	4	1	4	4	6	2
5	5	0	0	5	5	2	6	5	5	5	1	5	5	7	8
6	6	0	0	6	6	3	0	6	6	6	1	6	6	9	3
7	7	0	0	7	7	3	6	7	7	7	1	7	7	10	9
8	8	0	0	8	8	4	0	8	8	8	2	8	9	0	4
9	9	0	0	9	9	4	6	9	9	9	2	9	10	1	11
10	10	0	0	10	10	5	0	10	10	10	2	10	11	3	5
11	11	0	0	11	11	5	6	11	11	11	2	11	12	5	0
12	12	0	0	12	12	6	0	12	13	0	3	12	13	6	6
13	13	0	0	13	13	6	6	13	14	1	3	13	14	8	1
14	14	0	0	14	14	7	0	14	15	2	3	14	15	9	7
15	15	0	0	15	15	7	6	15	16	3	3	15	16	11	2
16	16	0	0	16	16	8	1	16	17	4	4	16	18	0	9
17	17	0	0	17	17	8	7	17	18	5	4	17	19	2	3
18	18	0	0	18	18	9	1	18	19	6	4	18	20	3	10
19	19	0	0	19	19	9	7	19	20	7	4	19	21	5	4
20	20	0	0	20	20	10	1	20	21	8	5	20	22	6	11
21	21	0	0	21	21	10	7	21	22	9	5	21	23	8	5
22	22	0	0	22	22	11	1	22	23	10	5	22	24	10	0
23	23	0	0	23	23	11	7	23	24	11	5	23	25	11	6
24	24	0	0	24	25	0	1	24	26	0	6	24	27	1	1
25	25	0	0	25	26	0	7	25	27	1	6	25	28	2	8
26	26	0	0	26	27	1	1	26	28	2	6	26	29	4	2
27	27	0	0	27	28	1	7	27	29	3	6	27	30	5	9
28	28	0	0	28	29	2	1	28	30	4	7	28	31	7	3
29	29	0	0	29	30	2	7	29	31	5	7	29	32	8	10
30	30	0	0	30	31	3	1	30	32	6	7	30	33	10	4
31	31	0	0	31	32	3	8	31	33	7	7	31	34	11	11
32	32	0	0	32	33	4	2	32	34	8	8	32	36	1	6
33	33	0	0	33	34	4	8	33	35	9	8	33	37	3	0
34	34	0	0	34	35	5	2	34	36	10	8	34	38	4	7
35	35	0	0	35	36	5	8	35	37	11	8	35	39	6	1
36	36	0	0	36	37	6	2	36	39	0	9	36	40	7	8
37	37	0	0	37	38	6	8	37	40	1	9	37	41	9	2
38	38	0	0	38	39	7	2	38	41	2	9	38	42	10	9
39	39	0	0	39	40	7	8	39	42	3	9	39	44	0	3
40	40	0	0	40	41	8	2	40	43	4	10	40	45	1	10
41	41	0	0	41	42	8	8	41	44	5	10	41	46	3	5
42	42	0	0	42	43	9	2	42	45	6	10	42	47	4	11
43	43	0	0	43	44	9	8	43	46	7	10	43	48	6	6
44	44	0	0	44	45	10	2	44	47	8	11	44	49	8	0
45	45	0	0	45	46	10	8	45	48	9	11	45	50	9	7

Quarters of a Foot.				Quarters of a Foot.				Quarters of a Foot.				Quarters of a Foot.			
	Ft.	In.	Pa.	S.	Ft.	In.	Pa.	S.	Ft.	In.	Pa.	S.	Ft.	In.	Pa.	S.
¼	0	3	0	0	0	3	1	6	0	3	3	0	0	3	4	6
½	0	6	0	0	0	6	3	0	0	6	6	0	0	6	9	0
¾	0	9	0	0	0	9	4	6	0	9	9	0	0	10	1	6

[TABLE II.

Ft. long	Side 13 in.			Ft. long	Side 13¼ in.			Ft. long	Side 13½ in.			Ft. long	Side 13¾ in.		
	Ft.	In.	Pa.		Ft.	In.	Pa.		Ft.	In.	Pa.		Ft.	In.	Pa.
1	1	2	1	1	1	2	7	1	1	3	2	1	1	3	9
2	2	4	2	2	2	5	8	2	2	6	4	2	2	7	6
3	3	6	3	3	3	7	10	3	3	9	6	3	3	11	3
4	4	8	4	4	4	10	6	4	5	0	9	4	5	3	0
5	5	10	5	5	6	1	1	5	6	3	11	5	6	6	9
6	7	0	6	6	7	3	9	6	7	7	1	6	7	10	6
7	8	2	7	7	8	6	4	7	8	10	3	7	9	2	3
8	9	4	8	8	9	9	0	8	10	1	6	8	10	6	0
9	10	6	9	9	10	11	8	9	11	4	8	9	11	9	9
10	11	8	10	10	12	2	3	10	12	7	10	10	13	1	6
11	12	10	11	11	13	4	11	11	13	11	0	11	14	5	3
12	14	1	0	12	14	7	6	12	15	2	3	12	15	9	0
13	15	3	1	13	15	10	2	13	16	5	5	13	17	0	9
14	16	5	2	14	17	0	9	14	17	8	7	14	18	4	6
15	17	7	3	15	18	3	5	15	18	11	9	15	19	8	3
16	18	9	4	16	19	6	1	16	20	3	0	16	21	0	1
17	19	11	5	17	20	8	8	17	21	6	2	17	22	3	10
18	21	1	6	18	21	11	4	18	22	9	4	18	23	7	7
19	22	3	7	19	23	1	11	19	24	0	6	19	24	11	4
20	23	5	8	20	24	4	7	20	25	3	9	20	26	3	1
21	24	7	9	21	25	7	2	21	26	6	11	21	27	6	10
22	25	9	10	22	26	9	10	22	27	10	1	22	28	10	7
23	26	11	11	23	28	0	5	23	29	1	3	23	30	2	4
24	28	2	0	24	29	3	1	24	30	4	6	24	31	6	1
25	29	4	1	25	30	5	9	25	31	7	8	25	32	9	10
26	30	6	2	26	31	8	4	26	32	10	10	26	34	1	7
27	31	8	3	27	32	11	0	27	34	2	0	27	35	5	4
28	32	10	4	28	34	1	7	28	35	5	3	28	36	9	1
29	34	0	5	29	35	4	3	29	36	8	5	29	38	0	10
30	35	2	6	30	36	6	10	30	37	11	7	30	39	4	7
31	36	4	7	31	37	9	6	31	39	2	9	31	40	8	4
32	37	6	8	32	39	0	2	32	40	6	0	32	42	0	2
33	38	8	9	33	40	2	9	33	41	9	2	33	43	3	11
34	39	10	10	34	41	5	5	34	43	0	4	34	44	7	8
35	41	0	11	35	42	8	0	35	44	3	6	35	45	11	5
36	42	3	0	36	43	10	8	36	45	6	9	36	47	3	2
37	43	5	1	37	45	1	3	37	46	9	11	37	48	6	11
38	44	7	2	38	46	3	11	38	48	1	1	38	49	10	8
39	45	9	3	39	47	6	6	39	49	4	3	39	51	2	5
40	46	11	4	40	48	9	2	40	50	7	6	40	52	6	2
41	48	1	5	41	49	11	10	41	51	10	8	41	53	9	11
42	49	3	6	42	51	2	5	42	53	1	10	42	55	1	8
43	50	5	7	43	52	5	1	43	54	5	0	43	56	5	5
44	51	7	8	44	53	7	8	44	55	8	3	44	57	9	2
45	52	9	9	45	54	10	4	45	56	11	5	45	59	0	11

Quarters of a Foot.				Quarters of a Foot.				Quarters of a Foot.				Quarters of a Foot.			
	Ft.	In.	Pa.	S.	Ft.	In.	Pa.	S.	Ft.	In.	Pa.	S.	Ft.	In.	Pa.	S.
¼	0	3	6	3	0	3	7	3	0	3	9	6	0	3	11	3
½	0	7	0	6	0	7	3	6	0	7	7	0	0	7	10	6
¾	0	10	6	9	0	10	11	9	0	11	4	6	0	11	9	9

Ft. long	Side 14 in.			Ft. long	Side 14¼ in.			Ft. long	Side 14½ in.			Ft. long	Side 14¾ in.		
	Ft.	In.	Pa.		Ft.	In.	Pa.		Ft.	In.	Pa.		Ft.	In.	Pa.
1	1	4	4	1	1	4	11	1	1	5	6	1	1	6	1
2	2	8	8	2	2	9	10	2	2	11	0	2	3	0	3
3	4	1	0	3	4	2	9	3	4	4	6	3	4	6	4
4	5	5	4	4	5	7	8	4	5	10	1	4	6	0	6
5	6	9	8	5	7	0	7	5	7	3	7	5	7	6	7
6	8	2	0	6	8	5	6	6	8	9	1	6	9	0	9
7	9	6	4	7	9	10	5	7	10	2	7	7	10	6	10
8	10	10	8	8	11	3	4	8	11	8	2	8	12	1	0
9	12	3	0	9	12	8	3	9	13	1	8	9	13	7	2
10	13	7	4	10	14	1	2	10	14	7	2	10	15	1	3
11	14	11	8	11	15	6	1	11	16	0	8	11	16	7	5
12	16	4	0	12	16	11	0	12	17	6	3	12	18	1	6
13	17	8	4	13	18	3	11	13	18	11	9	13	19	7	8
14	19	0	8	14	19	8	10	14	20	5	3	14	21	1	9
15	20	5	0	15	21	1	9	15	21	10	9	15	22	7	11
16	21	9	4	16	22	6	9	16	23	4	4	16	24	2	1
17	23	1	8	17	23	11	8	17	24	9	10	17	25	8	2
18	24	6	0	18	25	4	7	18	26	3	4	18	27	2	4
19	25	10	4	19	26	9	6	19	27	8	10	19	28	8	5
20	27	2	8	20	28	2	5	20	29	2	5	20	30	2	7
21	28	7	0	21	29	7	4	21	30	7	11	21	31	8	1
22	29	11	4	22	31	0	3	22	32	1	5	22	33	2	10
23	31	3	8	23	32	5	2	23	33	6	11	23	34	8	11
24	32	8	0	24	33	10	1	24	35	0	6	24	36	3	1
25	34	0	4	25	35	3	0	25	36	6	0	25	37	9	3
26	35	4	8	26	36	7	11	26	37	11	6	26	39	3	4
27	36	9	0	27	38	0	10	27	39	5	0	27	40	9	6
28	38	1	4	28	39	5	9	28	40	10	7	28	42	3	7
29	39	5	8	29	40	10	8	29	42	4	1	29	43	9	9
30	40	10	0	30	42	3	7	30	43	9	7	30	45	3	10
31	42	2	4	31	43	8	6	31	45	3	1	31	46	10	0
32	43	6	8	32	45	1	6	32	46	8	8	32	48	4	2
33	44	11	0	33	46	6	5	33	48	2	2	33	49	10	3
34	46	3	4	34	47	11	4	34	49	7	8	34	51	4	5
35	47	7	8	35	49	4	3	35	51	1	2	35	52	10	6
36	49	0	0	36	50	9	2	36	52	6	9	36	54	4	8
37	50	4	4	37	52	2	1	37	54	0	3	37	55	10	9
38	51	8	8	38	53	7	0	38	55	5	9	38	57	4	11
39	53	1	0	39	54	11	11	39	56	11	3	39	58	11	0
40	54	5	4	40	56	4	10	40	58	4	10	40	60	5	2
41	55	9	8	41	57	9	9	41	59	10	4	41	61	11	4
42	57	2	0	42	59	2	8	42	61	3	10	42	63	5	5
43	58	6	4	43	60	7	7	43	62	9	4	43	64	11	7
44	59	10	8	44	62	0	6	44	64	2	11	44	66	5	8
45	61	3	0	45	63	5	5	45	65	8	5	45	67	11	10
Quarters of a Foot.				Quarters of a Foot.				Quarters of a Foot.				Quarters of a Foot.			
	Ft.	In.	Pa.	S.	Ft.	In.	Pa.	S.	Ft.	In.	Pa.	S.	Ft.	In.	Pa.	S.
¼	0	4	1	0	0	4	2	9	0	4	4	6	0	4	6	3
½	0	8	2	0	0	8	5	6	0	8	9	0	0	9	0	6
¾	1	0	3	0	1	0	8	3	1	1	1	6	1	1	6	9

Ft. long	Side 15 in.			Ft. long	Side 15¼ in.			Ft. long	Side 15½ in.			Ft. long	Side 15¾ in.		
	Ft.	In.	Pa.		Ft.	In.	Pa.		Ft.	In.	Pa.		Ft.	In.	Pa.
1	1	6	9	1	1	7	4	1	1	8	0	1	1	8	8
2	3	1	6	2	3	2	0	2	3	4	0	2	3	5	4
3	4	8	3	3	4	10	1	3	5	0	0	3	5	2	0
4	6	3	0	4	6	5	6	4	6	8	1	4	6	10	8
5	7	9	9	5	8	0	10	5	8	4	1	5	8	7	4
6	9	4	6	6	9	8	3	6	10	0	1	6	10	4	0
7	10	11	3	7	11	3	7	7	11	8	1	7	12	0	8
8	12	6	0	8	12	11	0	8	13	4	2	8	13	9	4
9	14	0	9	9	14	6	5	9	15	0	2	9	15	6	0
10	15	7	6	10	16	1	9	10	16	8	2	10	17	2	8
11	17	2	3	11	17	9	2	11	18	4	2	11	18	11	4
12	18	9	0	12	19	4	6	12	20	0	3	12	20	8	0
13	20	3	9	13	20	11	11	13	21	8	3	13	22	4	8
14	21	10	6	14	22	7	3	14	23	4	3	14	24	1	4
15	23	5	3	15	24	2	8	15	25	0	3	15	25	10	0
16	25	0	0	16	25	10	1	16	26	8	4	16	27	6	9
17	26	6	9	17	27	5	5	17	28	4	4	17	29	3	5
18	28	1	6	18	29	0	10	18	30	0	4	18	31	0	1
19	29	8	3	19	30	8	2	19	31	8	4	19	32	8	9
20	31	3	0	20	32	3	7	20	33	4	5	20	34	5	5
21	32	9	9	21	33	10	11	21	35	0	5	21	36	2	1
22	34	4	6	22	35	6	4	22	36	8	5	22	37	10	9
23	35	11	3	23	37	1	8	23	38	4	5	23	39	7	5
24	37	6	0	24	38	9	1	24	40	0	6	24	41	4	1
25	39	0	9	25	40	4	6	25	41	8	6	25	43	0	9
26	40	7	6	26	41	11	10	26	43	4	6	26	44	9	5
27	42	2	3	27	43	7	3	27	45	0	6	27	46	6	1
28	43	9	0	28	45	2	7	28	46	8	7	28	48	2	9
29	45	3	9	29	46	10	0	29	48	4	7	29	49	11	5
30	46	10	6	30	48	5	4	30	50	0	7	30	51	8	1
31	48	5	3	31	50	0	9	31	51	8	7	31	53	4	9
32	50	0	0	32	51	8	2	32	53	4	8	32	55	1	6
33	51	6	9	33	53	3	6	33	55	0	8	33	56	10	2
34	53	1	6	34	54	10	11	34	56	8	8	34	58	6	10
35	54	8	3	35	56	6	3	35	58	4	8	35	60	3	6
36	56	3	0	36	58	1	8	36	60	0	9	36	62	0	2
37	57	9	9	37	59	9	0	37	61	8	9	37	63	8	10
38	59	4	6	38	61	4	5	38	63	4	9	38	65	5	6
39	60	11	3	39	62	11	9	39	65	0	9	39	67	2	2
40	62	6	0	40	64	7	2	40	66	8	10	40	68	10	10
41	64	0	9	41	66	2	7	41	68	4	10	41	70	7	6
42	65	7	6	42	67	9	11	42	70	0	10	42	72	4	2
43	67	2	3	43	69	5	4	43	71	8	10	43	74	0	10
44	68	9	0	44	71	0	8	44	73	4	11	44	75	9	6
45	70	3	9	45	72	8	1	45	75	0	11	45	77	6	2

Quarters of a Foot.				Quarters of a Foot.				Quarters of a Foot.				Quarters of a Foot.			
	Ft.	In.	Pa.	S.	Ft.	In.	Pa.	S.	Ft.	In.	Pa.	S.	Ft.	In.	Pa.	S.
¼	0	4	8	3	0	4	10	0	0	5	0	0	0	5	2	0
½	0	9	4	6	0	9	8	0	0	10	0	0	0	10	4	0
¾	1	2	0	9	1	2	6	0	1	3	0	0	1	3	6	0

Ft. long	Side 16 in.			Ft. long	Side 16¼ in.			Ft. long	Side 16½ in.			Ft. long	Side 16¾ in.		
	Ft.	In.	Pa.		Ft.	In.	Pa.		Ft.	In.	Pa.		Ft.	In.	Pa.
1	1	9	4	1	1	10	0	1	1	10	8	1	1	11	4
2	3	6	8	2	3	8	0	2	3	9	4	2	3	10	9
3	5	4	0	3	5	6	0	3	5	8	0	3	5	10	1
4	7	1	4	4	7	4	0	4	7	6	9	4	7	9	6
5	8	10	8	5	9	2	0	5	9	5	5	5	9	8	10
6	10	8	0	6	11	0	0	6	11	4	1	6	11	9	3
7	12	5	4	7	12	10	0	7	13	2	9	7	13	7	7
8	14	2	8	8	14	8	0	8	15	1	6	8	15	7	0
9	16	0	0	9	16	6	0	9	17	0	2	9	17	6	5
10	17	9	4	10	18	4	0	10	18	10	10	10	19	5	9
11	19	6	8	11	20	2	0	11	20	9	6	11	21	5	2
12	21	4	0	12	22	0	0	12	22	8	3	12	23	4	6
13	23	1	4	13	23	10	0	13	24	6	11	13	25	3	11
14	24	10	8	14	25	8	0	14	26	5	7	14	27	3	3
15	26	8	0	15	27	6	0	15	28	4	3	15	29	2	8
16	28	5	4	16	29	4	1	16	30	3	0	16	31	2	1
17	30	2	8	17	31	2	1	17	32	1	8	17	33	1	5
18	32	0	0	18	33	0	1	18	34	0	3	18	35	0	10
19	33	9	4	19	34	10	1	19	35	11	0	19	37	0	2
20	35	6	8	20	36	8	1	20	37	9	9	20	38	11	7
21	37	4	0	21	38	6	1	21	39	8	5	21	40	10	11
22	39	1	4	22	40	4	1	22	41	7	1	22	42	10	4
23	40	10	8	23	42	2	1	23	43	5	9	23	44	9	8
24	42	8	0	24	44	0	1	24	45	4	6	24	46	9	1
25	44	5	4	25	45	10	1	25	47	3	2	25	48	8	6
26	46	2	8	26	47	8	1	26	49	1	10	26	50	7	10
27	48	0	0	27	49	6	1	27	51	0	6	27	52	7	3
28	49	9	4	28	51	4	1	28	52	11	3	28	54	6	7
29	51	6	8	29	53	2	1	29	54	9	11	29	56	6	0
30	53	4	0	30	55	0	1	30	56	8	7	30	58	5	4
31	55	1	4	31	56	10	1	31	58	7	3	31	60	4	9
32	56	10	8	32	58	8	2	32	60	6	0	32	62	4	2
33	58	8	0	33	60	6	2	33	62	4	8	33	64	3	6
34	60	5	4	34	62	4	2	34	64	3	4	34	66	2	11
35	62	2	8	35	64	2	2	35	66	2	0	35	68	2	3
36	64	0	0	36	66	0	2	36	68	0	9	36	70	1	8
37	65	9	4	37	67	10	2	37	69	11	5	37	72	1	0
38	67	6	8	38	69	8	2	38	71	10	1	38	74	0	5
39	69	4	0	39	71	6	2	39	73	8	9	39	75	11	9
40	71	1	4	40	73	4	2	40	75	7	6	40	77	11	2
41	72	10	8	41	75	2	2	41	77	6	2	41	79	10	7
42	74	8	0	42	77	0	2	42	79	4	10	42	81	9	11
43	76	4	4	43	78	10	2	43	81	3	6	43	83	9	4
44	78	2	8	44	80	8	2	44	83	2	3	44	85	8	8
45	80	0	0	45	82	6	2	45	85	0	11	45	87	8	1
Quarters of a Foot.				Quarters of a Foot.				Quarters of a Foot.				Quarters of a Foot.			
	Ft.	In.	Pa.	S.	Ft.	In.	Pa.	S.	Ft.	In.	Pa.	S.	Ft.	In.	Pa.	S.
¼	0	5	4	0	0	5	6	0	0	5	8	0	0	5	10	0
½	0	10	8	0	0	11	0	0	0	11	4	0	0	11	8	0
¾	1	4	0	0	1	4	6	0	1	5	0	0	1	5	6	0

282   [TABLE II.

Ft. long	Side 17 in.			Ft. long	Side 17¼ in.			Ft. long	Side 17½ in.			Ft. long	Side 17¾ in.		
	Ft.	In.	Pa.		Ft.	In.	Pa.		Ft.	In.	Pa.		Ft.	In.	Pa.
1	2	0	1	1	2	0	9	1	2	1	6	1	2	2	3
2	4	0	2	2	4	1	7	2	4	3	0	2	4	4	6
3	6	0	3	3	6	2	4	3	6	4	6	3	6	6	9
4	8	0	4	4	8	3	2	4	8	6	1	4	8	9	0
5	10	0	5	5	10	3	11	5	10	7	7	5	10	11	3
6	12	0	6	6	12	4	9	6	12	9	1	6	13	1	6
7	14	0	7	7	14	5	6	7	14	10	7	7	15	3	9
8	16	0	8	8	16	6	4	8	17	0	2	8	17	6	0
9	18	0	9	9	18	7	2	9	19	1	8	9	19	8	3
10	20	0	10	10	20	7	11	10	21	3	2	10	21	10	6
11	22	0	11	11	22	8	9	11	23	4	8	11	24	0	9
12	24	1	0	12	24	9	6	12	25	6	3	12	26	3	0
13	26	1	1	13	26	10	4	13	27	7	9	13	28	5	3
14	28	1	2	14	28	11	1	14	29	9	3	14	30	7	6
15	30	1	3	15	30	11	11	15	31	10	9	15	32	9	9
16	32	1	4	16	33	0	9	16	34	0	4	16	35	0	1
17	34	1	5	17	35	1	6	17	36	1	10	17	37	2	4
18	36	1	6	18	37	2	4	18	38	3	4	18	39	4	7
19	38	1	7	19	39	3	1	19	40	4	10	19	41	6	10
20	40	1	8	20	41	3	11	20	42	6	5	20	43	9	1
21	42	1	9	21	43	4	8	21	44	7	11	21	45	11	4
22	44	1	10	22	45	5	6	22	46	9	5	22	48	1	7
23	46	1	11	23	47	6	3	23	48	10	11	23	50	3	10
24	48	2	0	24	49	7	1	24	51	0	6	24	52	6	1
25	50	2	1	25	51	7	11	25	53	2	0	25	54	8	4
26	52	2	2	26	53	8	8	26	55	3	6	26	56	10	7
27	54	2	3	27	55	9	6	27	57	5	0	27	59	0	10
28	56	2	4	28	57	10	3	28	59	6	7	28	61	3	1
29	58	2	5	29	59	11	1	29	61	8	1	29	63	5	4
30	60	2	6	30	61	11	10	30	63	9	7	30	65	7	7
31	62	2	7	31	64	0	8	31	65	11	1	31	67	9	10
32	64	2	8	32	66	1	6	32	68	0	8	32	70	0	2
33	66	2	9	33	68	2	3	33	70	2	2	33	72	2	5
34	68	2	10	34	70	3	1	34	72	3	8	34	74	4	8
35	70	2	11	35	72	3	10	35	74	5	2	35	76	6	11
36	72	3	0	36	74	4	8	36	76	6	9	36	78	9	2
37	74	3	1	37	76	5	5	37	78	8	3	37	80	11	5
38	76	3	2	38	78	6	3	38	80	9	9	38	83	1	8
39	78	3	3	39	80	7	0	39	82	11	3	39	85	3	11
40	80	3	4	40	82	7	10	40	85	0	10	40	87	6	2
41	82	3	5	41	84	8	8	41	87	2	4	41	89	8	5
42	84	3	6	42	86	9	5	42	89	3	10	42	91	10	8
43	86	3	7	43	88	10	3	43	91	5	4	43	94	0	11
44	88	3	8	44	90	11	0	44	93	6	11	44	96	3	2
45	90	3	9	45	92	11	10	45	95	8	5	45	98	5	5
Quarters of a Foot.				Quarters of a Foot.				Quarters of a Foot.				Quarters of a Foot.			
	Ft.	In.	Pa.	S.	Ft.	In.	Pa.	S.	Ft.	In.	Pa.	S.	Ft.	In.	Pa.	S.
¼	0	6	0	3	0	6	2	3	0	6	4	6	0	6	6	9
½	1	0	0	6	1	0	4	6	1	0	9	0	1	1	1	6
¾	1	6	0	9	1	6	6	9	1	7	1	6	1	7	8	3

Ft. long	Side 18 in.			Ft. long	Side 18 in.			Ft. long	Side 18 in.			Ft. long	Side 18 in.		
	Ft.	In.	Pa.		Ft.	In.	Pa.		Ft.	In.	Pa.		Ft.	In.	Pa.
1	2	3	0	14	31	6	0	27	60	9	0	40	90	0	0
2	4	6	0	15	33	9	0	28	63	0	0	41	92	3	0
3	6	9	0	16	36	0	0	29	65	3	0	42	94	6	0
4	9	0	0	17	38	3	0	30	67	6	0	43	96	9	0
5	11	3	0	18	40	6	0	31	69	9	0	44	99	0	0
6	13	6	0	19	42	9	0	32	72	0	0	45	101	3	0
7	15	9	0	20	45	0	0	33	74	3	0				
8	18	0	0	21	47	3	0	34	76	6	0	Quarters of a Foot.			
9	20	3	0	22	49	6	0	35	78	9	0		Ft.	In.	Pa. S.
10	22	6	0	23	51	9	0	36	81	0	0				
11	24	9	0	24	54	0	0	37	83	3	0	$\frac{1}{4}$	0	6	9 0
12	27	0	0	25	56	3	0	38	85	6	0	$\frac{1}{2}$	1	1	6 0
13	29	3	0	26	58	6	0	39	87	9	0	$\frac{3}{4}$	1	8	3 0

TABLE III.

SUPERFICIAL MEASURE.

By this Table the superficial contents, and consequently the value, of any quantity of board, glass, &c., may be found, AT SIGHT, from 1 inch to 24 inches, the breadth, and from ¼ of an inch to 24 feet, the length; and therefore, by addition only, may serve to any greater breadth or length, if there should ever be occasion.

Each page of this Table is divided into three distinct parts; and each part consists of four columns. The first part of the narrow column to the left-hand shows the several lengths in feet, from 1 to 24; the second part the odd inches, from 1 to 11; and the third part, the odd quarters of an inch. The three or four figures in the larger column on the right-hand, marked at the top Ft. In. Pa., and lower Ft. In. Pa. S. T., are the several contents in square feet, twelfths of a square foot, square inches, seconds, and thirds answering to the length in the left-hand column.

Example I.—A board is five inches broad and 9 feet long, to find how many square feet it contains; look at the top of the Table for 5 in., and then keep the eye down the left-hand column until you come to 9 ft., over against which stands 3ft. 9 in. 0 pa.

Example II.—To find the square feet in a plank 17 ft. 9 in. long and 21½ in. broad. At the top of the Table find 21½ in., and then look down the left-hand column for 17 ft., against which stands 30 ft. 5in. 6 pa. Then look for 9 in., against which stands 1 ft. 4 in. 1 pa. 6 s.; add the two together, and the content is found to be 31 ft.

9 in. 7 pa. The quarters of an inch (when you have occasion for them) must be found, and added to the feet and inches. If a plank is broader at one end than the other, either take the breadth at the middle, or at each end, and add them together, and take the half sum for the true breadth (provided both edges are straight).

Example III.—To find the area of a plank broader than 24 in.—say 27 in.—and 19 ft. long, proceed thus:—

		ft.	in.	pa.
19 ft. long, and 24 in. broad, is	. . . .	28	0	0
Ditto 3 ditto	. . . . .	4	9	0
Content required is	. . .	42	9	0

Example IV.—To find how many feet of glass are contained in a sash-window having 18 squares, each square being 20 in. high, and 11¾ in. broad. First add together the height of the 18 squares, which is 360 in., or 30 ft.; then look in the Table for 11¾ in.; but, since it does not extend to 30 ft., take the answer out at twice; thus:—

		ft.	in.	pa.
20 ft. long, and 11¾ in. in breadth is	. .	19	7	0
10 ft. ditto ditto ditto	. . .	9	9	6
Content required is	. . .	29	4	6

[TABLE III.

Ft. long	1 in. broad.			Ft. long	1¼ in. broad.			Ft. long	1½ in. broad.			Ft. long	1¾ in. broad.		
	Ft.	In.	Pa.		Ft.	In.	Pa.		Ft.	In.	Pa.		Ft.	In.	Pa.
1	0	1	0	1	0	1	3	1	0	1	6	1	0	1	9
2	0	2	0	2	0	2	6	2	0	3	0	2	0	3	6
3	0	3	0	3	0	3	9	3	0	4	6	3	0	5	3
4	0	4	0	4	0	5	0	4	0	6	0	4	0	7	0
5	0	5	0	5	0	6	3	5	0	7	6	5	0	8	9
6	0	6	0	6	0	7	6	6	0	9	0	6	0	10	6
7	0	7	0	7	0	8	9	7	0	10	6	7	1	0	3
8	0	8	0	8	0	10	0	8	1	0	0	8	1	2	0
9	0	9	0	9	0	11	3	9	1	1	6	9	1	3	9
10	0	10	0	10	1	0	6	10	1	3	0	10	1	5	6
11	0	11	0	11	1	1	9	11	1	4	6	11	1	7	3
12	1	0	0	12	1	3	0	12	1	6	0	12	1	9	0
13	1	1	0	13	1	4	3	13	1	7	6	13	1	10	9
14	1	2	0	14	1	5	6	14	1	9	0	14	2	0	6
15	1	3	0	15	1	6	9	15	1	10	6	15	2	2	3
16	1	4	0	16	1	8	0	16	2	0	0	16	2	4	0
17	1	5	0	17	1	9	3	17	2	1	6	17	2	5	9
18	1	6	0	18	1	10	6	18	2	3	0	18	2	7	6
19	1	7	0	19	1	11	9	19	2	4	6	19	2	9	3
20	1	8	0	20	2	1	0	20	2	6	0	20	2	11	0
21	1	9	0	21	2	2	3	21	2	7	6	21	3	0	9
22	1	10	0	22	2	3	6	22	2	9	0	22	3	2	6
23	1	11	0	23	2	4	9	23	2	10	6	23	3	4	3
24	2	0	0	24	2	6	0	24	3	0	0	24	3	6	0

In. long	Ft.	In.	Pa.	S.	In. long	Ft.	In.	Pa.	S.	In. long	Ft.	In.	Pa.	S.	In. long	Ft.	In.	Pa.	S.
1	0	0	1	0	1	0	0	1	3	1	0	0	1	6	1	0	0	1	9
2	0	0	2	0	2	0	0	2	6	2	0	0	3	0	2	0	0	3	6
3	0	0	3	0	3	0	0	3	9	3	0	0	4	6	3	0	0	5	3
4	0	0	4	0	4	0	0	5	0	4	0	0	6	0	4	0	0	7	0
5	0	0	5	0	5	0	0	6	3	5	0	0	7	6	5	0	0	8	9
6	0	0	6	0	6	0	0	7	6	6	0	0	9	0	6	0	0	10	6
7	0	0	7	0	7	0	0	8	9	7	0	0	10	6	7	0	1	0	3
8	0	0	8	0	8	0	0	10	0	8	0	1	0	0	8	0	1	2	0
9	0	0	9	0	9	0	0	11	3	9	0	1	1	6	9	0	1	3	9
10	0	0	10	0	10	0	1	0	6	10	0	1	3	0	10	0	1	5	6
11	0	0	11	0	11	0	1	1	9	11	0	1	4	6	11	0	1	7	3

Qrs. in. long	In.	Pa.	S.	T.	Qrs. in. long	In.	Pa.	S.	T.	Qrs. in. long	In.	Pa.	S.	T.	Qrs. in. long	In.	Pa.	S.	T.
¼	0	0	3	0	¼	0	0	3	9	¼	0	0	4	6	¼	0	0	5	3
½	0	0	6	0	½	0	0	7	6	½	0	0	9	0	½	0	0	10	6
¾	0	0	9	0	¾	0	0	11	3	¾	0	1	1	6	¾	0	1	3	9

Ft. long	2 in. broad.			Ft. long	2¼ in. broad.			Ft. long	2½ in. broad.			Ft. long	2¾ in. broad.		
	Ft.	In.	Pa.		Ft.	In.	Pa.		Ft.	In.	Pa.		Ft.	In.	Pa.
1	0	2	0	1	0	2	3	1	0	2	6	1	0	2	9
2	0	4	0	2	0	4	6	2	0	5	0	2	0	5	6
3	0	6	0	3	0	6	9	3	0	7	6	3	0	8	3
4	0	8	0	4	0	9	0	4	0	10	0	4	0	11	0
5	0	10	0	5	0	11	3	5	1	0	6	5	1	1	9
6	1	0	0	6	1	1	6	6	1	3	0	6	1	4	6
7	1	2	0	7	1	3	9	7	1	5	6	7	1	7	3
8	1	4	0	8	1	6	0	8	1	8	0	8	1	10	0
9	1	6	0	9	1	8	3	9	1	10	6	9	2	0	9
10	1	8	0	10	1	10	6	10	2	1	0	10	2	3	6
11	1	10	0	11	2	0	9	11	2	3	6	11	2	6	3
12	2	0	0	12	2	3	0	12	2	6	0	12	2	9	0
13	2	2	0	13	2	5	3	13	2	8	6	13	2	11	9
14	2	4	0	14	2	7	6	14	2	11	0	14	3	2	6
15	2	6	0	15	2	9	9	15	3	1	6	15	3	5	3
16	2	8	0	16	3	0	0	16	3	4	0	16	3	8	0
17	2	10	0	17	3	2	3	17	3	6	6	17	3	10	9
18	3	0	0	18	3	4	6	18	3	9	0	18	4	1	6
19	3	2	0	19	3	6	9	19	3	11	6	19	4	4	3
20	3	4	0	20	3	9	0	20	4	2	0	20	4	7	0
21	3	6	0	21	3	11	3	21	4	4	6	21	4	9	9
22	3	8	0	22	4	1	6	22	4	7	0	22	5	0	6
23	3	10	0	23	4	3	9	23	4	9	6	23	5	3	3
24	4	0	0	24	4	6	0	24	5	0	0	24	5	6	0

In. long	Ft.	In.	Pa.	S.	In. long	Ft.	In.	Pa.	S.	In. long	Ft.	In.	Pa.	S.	In. long	Ft.	In.	Pa.	S.
1	0	0	2	0	1	0	0	2	3	1	0	0	2	6	1	0	0	2	9
2	0	0	4	0	2	0	0	4	6	2	0	0	5	0	2	0	0	5	6
3	0	0	6	0	3	0	0	6	9	3	0	0	7	6	3	0	0	8	3
4	0	0	8	0	4	0	0	9	0	4	0	0	10	0	4	0	0	11	0
5	0	0	10	0	5	0	0	11	3	5	0	1	0	6	5	0	1	1	9
6	0	1	0	0	6	0	1	1	6	6	0	1	3	0	6	0	1	4	6
7	0	1	2	0	7	0	1	3	9	7	0	1	5	6	7	0	1	7	3
8	0	1	4	0	8	0	1	6	0	8	0	1	8	0	8	0	1	10	0
9	0	1	6	0	9	0	1	8	3	9	0	1	10	6	9	0	2	0	9
10	0	1	8	0	10	0	1	10	6	10	0	2	1	0	10	0	2	3	6
11	0	1	10	0	11	0	2	0	9	11	0	2	3	6	11	0	2	6	3

Qrs. in. long	In.	Pa.	S.	T.	Qrs. in. long	In.	Pa.	S.	T.	Qrs. in. long	In.	Pa.	S.	T.	Qrs. in. long	In.	Pa.	S.	T.
¼	0	0	6	0	¼	0	0	6	9	¼	0	0	7	6	¼	0	0	8	3
½	0	1	0	0	½	0	1	1	6	½	0	1	3	0	½	0	1	4	6
¾	0	1	6	0	¾	0	1	8	3	¾	0	1	10	6	¾	0	2	0	9

Ft. long	3 in. broad.			Ft. long	3¼ in. broad.			Ft. long	3½ in. broad.			Ft. long	3¾ in. broad.		
	Ft.	In.	Pa.		Ft.	In.	Pa.		Ft.	In.	Pa.		Ft.	In.	Pa.
1	0	3	0	1	0	3	3	1	0	3	6	1	0	3	9
2	0	6	0	2	0	6	6	2	0	7	0	2	0	7	6
3	0	9	0	3	0	9	9	3	0	10	6	3	0	11	3
4	1	0	0	4	1	1	0	4	1	2	0	4	1	3	0
5	1	3	0	5	1	4	3	5	1	5	6	5	1	6	9
6	1	6	0	6	1	7	6	6	1	9	0	6	1	10	6
7	1	9	0	7	1	10	9	7	2	0	6	7	2	2	3
8	2	0	0	8	2	2	0	8	2	4	0	8	2	6	0
9	2	3	0	9	2	5	3	9	2	7	6	9	2	9	9
10	2	6	0	10	2	8	6	10	2	11	0	10	3	1	6
11	2	9	0	11	2	11	9	11	3	2	6	11	3	5	3
12	3	0	0	12	3	3	0	12	3	6	0	12	3	9	0
13	3	3	0	13	3	6	3	13	3	9	6	13	4	0	9
14	3	6	0	14	3	9	6	14	4	1	0	14	4	4	6
15	3	9	0	15	4	0	9	15	4	4	6	15	4	8	3
16	4	0	0	16	4	4	0	16	4	8	0	16	5	0	0
17	4	3	0	17	4	7	3	17	4	11	6	17	5	3	9
18	4	6	0	18	4	10	6	18	5	3	0	18	5	7	6
19	4	9	0	19	5	1	9	19	5	6	6	19	5	11	3
20	5	0	0	20	5	5	0	20	5	10	0	20	6	3	0
21	5	3	0	21	5	8	3	21	6	1	6	21	6	6	9
22	5	6	0	22	5	11	6	22	6	5	0	22	6	10	6
23	5	9	0	23	6	2	9	23	6	8	6	23	7	2	3
24	6	0	0	24	6	6	0	24	7	0	0	24	7	6	0

In. long					In. long					In. long					In. long				
	Ft.	In.	Pa.	S.		Ft.	In.	Pa.	S.		Ft.	In.	Pa.	S.		Ft.	In.	Pa.	S.
1	0	0	3	0	1	0	0	3	3	1	0	0	3	6	1	0	0	3	9
2	0	0	6	0	2	0	0	6	6	2	0	0	7	0	2	0	0	7	6
3	0	0	9	0	3	0	0	9	9	3	0	0	10	6	3	0	0	11	3
4	0	1	0	0	4	0	1	1	0	4	0	1	2	0	4	0	1	3	0
5	0	1	3	0	5	0	1	4	3	5	0	1	5	6	5	0	1	6	9
6	0	1	6	0	6	0	1	7	6	6	0	1	9	0	6	0	1	10	6
7	0	1	9	0	7	0	1	10	9	7	0	2	0	6	7	0	2	2	3
8	0	2	0	0	8	0	2	2	0	8	0	2	4	0	8	0	2	6	0
9	0	2	3	0	9	0	2	5	3	9	0	2	7	6	9	0	2	9	9
10	0	2	6	0	10	0	2	8	6	10	0	2	11	0	10	0	3	1	6
11	0	2	9	0	11	0	2	11	9	11	0	3	2	6	11	0	3	5	3

Qrs. in. long					Qrs. in. long					Qrs. in. long					Qrs. in. long				
	In.	Pa.	S.	T.		In.	Pa.	S.	T.		In.	Pa.	S.	T.		In.	Pa.	S.	T.
¼	0	0	9	0	¼	0	0	9	9	¼	0	0	10	6	¼	0	0	11	3
½	0	1	6	0	½	0	1	7	6	½	0	1	9	0	½	0	1	10	6
¾	0	2	3	0	¾	0	2	5	3	¾	0	2	7	6	¾	0	2	9	9

Ft. long	4 in. broad.			Ft. long	4¼ in. broad.			Ft. long	4½ in. broad.			Ft. long	4¾ in. broad.		
	Ft.	In.	Pa.		Ft.	In.	Pa.		Ft.	In.	Pa.		Ft.	In.	Pa.
1	0	4	0	1	0	4	3	1	0	4	6	1	0	4	9
2	0	8	0	2	0	8	6	2	0	9	0	2	0	9	6
3	1	0	0	3	1	0	9	3	1	1	6	3	1	2	3
4	1	4	0	4	1	5	0	4	1	6	0	4	1	7	0
5	1	8	0	5	1	9	3	5	1	10	6	5	1	11	9
6	2	0	0	6	2	1	6	6	2	3	0	6	2	4	6
7	2	4	0	7	2	5	9	7	2	7	6	7	2	9	3
8	2	8	0	8	2	10	0	8	3	0	0	8	3	2	0
9	3	0	0	9	3	2	3	9	3	4	6	9	3	6	9
10	3	4	0	10	3	6	6	10	3	9	0	10	3	11	6
11	3	8	0	11	3	10	9	11	4	1	6	11	4	4	3
12	4	0	0	12	4	3	0	12	4	6	0	12	4	9	0
13	4	4	0	13	4	7	3	13	4	10	6	13	5	1	9
14	4	8	0	14	4	11	6	14	5	3	0	14	5	6	6
15	5	0	0	15	5	3	9	15	5	7	6	15	5	11	3
16	5	4	0	16	5	8	0	16	6	0	0	16	6	4	0
17	5	8	0	17	6	0	3	17	6	4	6	17	6	8	9
18	6	0	0	18	6	4	6	18	6	9	0	18	7	1	6
19	6	4	0	19	6	8	9	19	7	1	6	19	7	6	3
20	6	8	0	20	7	1	0	20	7	6	0	20	7	11	0
21	7	0	0	21	7	5	3	21	7	10	6	21	8	3	9
22	7	4	0	22	7	9	6	22	8	3	0	22	8	8	6
23	7	8	0	23	8	1	9	23	8	7	6	23	9	1	3
24	8	0	0	24	8	6	0	24	9	0	0	24	9	6	0

In. long	Ft.	In.	Pa.	S.	In. long	Ft.	In.	Pa.	S.	In. long	Ft.	In.	Pa.	S.	In. long	Ft.	In.	Pa.	S.
1	0	0	4	0	1	0	0	4	3	1	0	0	4	6	1	0	0	4	9
2	0	0	8	0	2	0	0	8	6	2	0	0	9	0	2	0	0	9	6
3	0	1	0	0	3	0	1	0	9	3	0	1	1	6	3	0	1	2	3
4	0	1	4	0	4	0	1	5	0	4	0	1	6	0	4	0	1	7	0
5	0	1	8	0	5	0	1	9	3	5	0	1	10	6	5	0	1	11	9
6	0	2	0	0	6	0	2	1	6	6	0	2	3	0	6	0	2	4	6
7	0	2	4	0	7	0	2	5	9	7	0	2	7	6	7	0	2	9	3
8	0	2	8	0	8	0	2	10	0	8	0	3	0	0	8	0	3	2	0
9	0	3	0	0	9	0	3	2	3	9	0	3	4	6	9	0	3	6	9
10	0	3	4	0	10	0	3	6	6	10	0	3	9	0	10	0	3	11	6
11	0	3	8	0	11	0	3	10	9	11	0	4	1	6	11	0	4	4	3

Qrs. in. long	In.	Pa.	S.	T.	Qrs. in. long	In.	Pa.	S.	T.	Qrs. in. long	In.	Pa.	S.	T.	Qrs. in. long	In.	Pa.	S.	T.
¼	0	1	0	0	¼	0	1	0	9	¼	0	1	1	6	¼	0	1	2	3
½	0	2	0	0	½	0	2	1	6	½	0	2	3	0	½	0	2	4	6
¾	0	3	0	0	¾	0	3	2	3	¾	0	3	4	6	¾	0	3	6	9

[TABLE III.

Ft. long	5 in. broad.			Ft. long	5¼ in. broad.			Ft. long	5½ in. broad.			Ft. long	5¾ in. broad.		
	Ft.	In.	Pa.		Ft.	In.	Pa.		Ft.	In.	Pa.		Ft.	In.	Pa.
1	0	5	0	1	0	5	3	1	0	5	6	1	0	5	9
2	0	10	0	2	0	10	6	2	0	11	0	2	0	11	6
3	1	3	0	3	1	3	9	3	1	4	6	3	1	5	3
4	1	8	0	4	1	9	0	4	1	10	0	4	1	11	0
5	2	1	0	5	2	2	3	5	2	3	6	5	2	4	9
6	2	6	0	6	2	7	6	6	2	9	0	6	2	10	6
7	2	11	0	7	3	0	9	7	3	2	6	7	3	4	3
8	3	4	0	8	3	6	0	8	3	8	0	8	3	10	0
9	3	9	0	9	3	11	3	9	4	1	6	9	4	3	9
10	4	2	0	10	4	4	6	10	4	7	0	10	4	9	6
11	4	7	0	11	4	9	9	11	5	0	6	11	5	3	3
12	5	0	0	12	5	3	0	12	5	6	0	12	5	9	0
13	5	5	0	13	5	8	3	13	5	11	6	13	6	2	9
14	5	10	0	14	6	1	6	14	6	5	0	14	6	8	6
15	6	3	0	15	6	6	9	15	6	10	6	15	7	2	3
16	6	8	0	16	7	0	0	16	7	4	0	16	7	8	0
17	7	1	0	17	7	5	3	17	7	9	6	17	8	1	9
18	7	6	0	18	7	10	6	18	8	3	0	18	8	7	6
19	7	11	0	19	8	3	9	19	8	8	6	19	9	1	3
20	8	4	0	20	8	9	0	20	9	2	0	20	9	7	0
21	8	9	0	21	9	2	3	21	9	7	6	21	10	0	9
22	9	2	0	22	9	7	6	22	10	1	0	22	10	6	6
23	9	7	0	23	10	0	9	23	10	6	6	23	11	2	3
24	10	0	0	24	10	6	0	24	11	0	0	24	11	6	0

In. long	Ft.	In.	Pa.	S.	In. long	Ft.	In.	Pa.	S.	In. long	Ft.	In.	Pa.	S.	In. long	Ft.	In.	Pa.	S.
1	0	0	5	0	1	0	0	5	3	1	0	0	5	6	1	0	0	5	9
2	0	0	10	0	2	0	0	10	6	2	0	0	11	0	2	0	0	11	6
3	0	1	3	0	3	0	1	3	9	3	0	1	4	6	3	0	1	5	3
4	0	1	8	0	4	0	1	9	0	4	0	1	10	0	4	0	1	11	0
5	0	2	1	0	5	0	2	2	3	5	0	2	3	6	5	0	2	4	9
6	0	2	6	0	6	0	2	7	6	6	0	2	9	0	6	0	2	10	6
7	0	2	11	0	7	0	3	0	9	7	0	3	2	6	7	0	3	4	3
8	0	3	4	0	8	0	3	6	0	8	0	3	8	0	8	0	3	10	0
9	0	3	9	0	9	0	3	11	3	9	0	4	1	6	9	0	4	3	9
10	0	4	2	0	10	0	4	4	6	10	0	4	7	0	10	0	4	9	6
11	0	4	7	0	11	0	4	9	9	11	0	5	0	6	11	0	5	3	3

Qrs. in. long	In.	Pa.	S.	T.	Qrs. in. long	In.	Pa.	S.	T.	Qrs. in. long	In.	Pa.	S.	T.	Qrs. in. long	In.	Pa.	S.	T.
¼	0	1	3	0	¼	0	1	3	9	¼	0	1	4	6	¼	0	1	5	3
½	0	2	6	0	½	0	2	7	6	½	0	2	9	0	½	0	2	10	6
¾	0	3	9	0	¾	0	3	11	3	¾	0	4	1	6	¾	0	4	3	9

[TABLE III.] 291

Ft. long	6 in. broad.			Ft. long	6¼ in. broad.			Ft. long	6½ in. broad.			Ft. long	6¾ in. broad.		
	Ft.	In.	Pa.		Ft.	In.	Pa.		Ft.	In.	Pa.		Ft.	In.	Pa.
1	0	6	0	1	0	6	3	1	0	6	6	1	0	6	9
2	1	0	0	2	1	0	6	2	1	1	0	2	1	1	6
3	1	6	0	3	1	6	9	3	1	7	6	3	1	8	3
4	2	0	0	4	2	1	0	4	2	2	0	4	2	3	0
5	2	6	0	5	2	7	3	5	2	8	6	5	2	9	9
6	3	0	0	6	3	1	6	6	3	3	0	6	3	4	6
7	3	6	0	7	3	7	9	7	3	9	6	7	3	11	3
8	4	0	0	8	4	2	0	8	4	4	0	8	4	6	0
9	4	6	0	9	4	8	3	9	4	10	6	9	5	0	9
10	5	0	0	10	5	2	6	10	5	5	0	10	5	7	6
11	5	6	0	11	5	8	9	11	5	11	6	11	6	2	3
12	6	0	0	12	6	3	0	12	6	6	0	12	6	9	0
13	6	6	0	13	6	9	3	13	7	0	6	13	7	3	9
14	7	0	0	14	7	3	6	14	7	7	0	14	7	10	6
15	7	6	0	15	7	9	9	15	8	1	6	15	8	5	3
16	8	0	0	16	8	4	0	16	8	8	0	16	9	0	0
17	8	6	0	17	8	10	3	17	9	2	6	17	9	6	9
18	9	0	0	18	9	4	6	18	9	9	0	18	10	1	6
19	9	6	0	19	9	10	9	19	10	3	6	19	10	8	3
20	10	0	0	20	10	5	0	20	10	10	0	20	11	3	0
21	10	6	0	21	10	11	3	21	11	4	6	21	11	9	9
22	11	0	0	22	11	5	6	22	11	11	0	22	12	4	6
23	11	6	0	23	11	11	9	23	12	5	6	23	12	11	3
24	12	0	0	24	12	6	0	24	13	0	0	24	13	6	0

In. long	Ft.	In.	Pa.	S.	In. long	Ft.	In.	Pa.	S.	In. long	Ft.	In.	Pa.	S.	In. long	Ft.	In.	Pa.	S.
1	0	0	6	0	1	0	0	6	3	1	0	0	6	6	1	0	0	6	9
2	0	1	0	0	2	0	1	0	6	2	0	1	1	0	2	0	1	1	6
3	0	1	6	0	3	0	1	6	9	3	0	1	7	6	3	0	1	8	3
4	0	2	0	0	4	0	2	1	0	4	0	2	2	0	4	0	2	3	0
5	0	2	6	0	5	0	2	7	3	5	0	2	8	6	5	0	2	9	9
6	0	3	0	0	6	0	3	1	6	6	0	3	3	0	6	0	3	4	6
7	0	3	6	0	7	0	3	7	9	7	0	3	9	6	7	0	3	11	3
8	0	4	0	0	8	0	4	2	0	8	0	4	4	0	8	0	4	6	0
9	0	4	6	0	9	0	4	8	3	9	0	4	10	6	9	0	5	0	9
10	0	5	0	0	10	0	5	2	6	10	0	5	5	0	10	0	5	7	6
11	0	5	6	0	11	0	5	8	9	11	0	5	11	6	11	0	6	2	3

Qrs. in. long	In.	Pa.	S.	T.	Qrs. in. long	In.	Pa.	S.	T.	Qrs. in. long	In.	Pa.	S.	T.	Qrs. in. long	In.	Pa.	S.	T.
¼	0	1	6	0	¼	0	1	6	9	¼	0	1	7	6	¼	0	1	8	3
½	0	3	0	0	½	0	3	1	6	½	0	3	3	0	½	0	3	4	6
¾	0	4	6	0	¾	0	4	8	3	¾	0	4	10	6	¾	0	5	0	9

Ft. long	7 in. broad.			Ft. long	7¼ in. broad.			Ft. long	7½ in. broad.			Ft. long	7¾ in. broad.		
	Ft.	In.	Pa.		Ft.	In.	Pa.		Ft.	In.	Pa.		Ft.	In.	Pa.
1	0	7	0	1	0	7	3	1	0	7	6	1	0	7	9
2	1	2	0	2	1	2	6	2	1	3	0	2	1	3	6
3	1	9	0	3	1	9	9	3	1	10	6	3	1	11	3
4	2	4	0	4	2	5	0	4	2	6	0	4	2	7	0
5	2	11	0	5	3	0	3	5	3	1	6	5	3	2	9
6	3	6	0	6	3	7	6	6	3	9	0	6	3	10	6
7	4	1	0	7	4	2	9	7	4	4	6	7	4	6	3
8	4	8	0	8	4	10	0	8	5	0	0	8	5	2	0
9	5	3	0	9	5	5	3	9	5	7	6	9	5	9	9
10	5	10	0	10	6	0	6	10	6	3	0	10	6	5	6
11	6	5	0	11	6	7	9	11	6	10	6	11	7	1	3
12	7	0	0	12	7	3	0	12	7	6	0	12	7	9	0
13	7	7	0	13	7	10	3	13	8	1	6	13	8	4	9
14	8	2	0	14	8	5	6	14	8	9	0	14	9	0	6
15	8	9	0	15	9	0	9	15	9	4	6	15	9	8	3
16	9	4	0	16	9	8	0	16	10	0	0	16	10	4	0
17	9	11	0	17	10	3	3	17	10	7	6	17	10	11	9
18	10	6	0	18	10	10	6	18	11	3	0	18	11	7	6
19	11	1	0	19	11	5	9	19	11	10	6	19	12	3	3
20	11	8	0	20	12	1	0	20	12	6	0	20	12	11	0
21	12	3	0	21	12	8	3	21	13	1	6	21	13	6	9
22	12	10	0	22	13	3	6	22	13	9	0	22	14	2	6
23	13	5	0	23	13	10	9	23	14	4	6	23	14	10	3
24	14	0	0	24	14	6	0	24	15	0	0	24	15	6	0

In. long	Ft.	In.	Pa.	S.	In. long	Ft.	In.	Pa.	S.	In. long	Ft.	In.	Pa.	S.	In. long	Ft.	In.	Pa.	S.
1	0	0	7	0	1	0	0	7	3	1	0	0	7	6	1	0	0	7	9
2	0	1	2	0	2	0	1	2	6	2	0	1	3	0	2	0	1	3	6
3	0	1	9	0	3	0	1	9	9	3	0	1	10	6	3	0	1	11	3
4	0	2	4	0	4	0	2	5	0	4	0	2	6	0	4	0	2	7	0
5	0	2	11	0	5	0	3	0	3	5	0	3	1	6	5	0	3	2	9
6	0	3	6	0	6	0	3	7	6	6	0	3	9	0	6	0	3	10	6
7	0	4	1	0	7	0	4	2	9	7	0	4	4	6	7	0	4	6	3
8	0	4	8	0	8	0	4	10	0	8	0	5	0	0	8	0	5	2	0
9	0	5	3	0	9	0	5	5	3	9	0	5	7	6	9	0	5	9	9
10	0	5	10	0	10	0	6	0	6	10	0	6	3	0	10	0	6	5	6
11	0	6	5	0	11	0	6	7	9	11	0	6	10	6	11	0	7	1	3

Qrs. in. long	In.	Pa.	S.	T.	Qrs. in. long	In.	Pa.	S.	T.	Qrs. in. long	In.	Pa.	S.	T.	Qrs. in. long	In.	Pa.	S.	T.
¼	0	1	9	0	¼	0	1	9	9	¼	0	1	10	6	¼	0	1	11	3
½	0	3	6	0	½	0	3	7	6	½	0	3	9	0	½	0	3	10	6
¾	0	5	3	0	¾	0	5	5	3	¾	0	5	7	6	¾	0	5	9	9

[Table III.] 293

Ft. long	8 in. broad.			Ft. long	8¼ in. broad.			Ft. long	8½ in. broad.			Ft. long	8¾ in. broad.		
	Ft.	In.	Pa.		Ft.	In.	Pa.		Ft.	In.	Pa.		Ft.	In.	Pa.
1	0	8	0	1	0	8	3	1	0	8	6	1	0	8	9
2	1	4	0	2	1	4	6	2	1	5	0	2	1	5	6
3	2	0	0	3	2	0	9	3	2	1	6	3	2	2	3
4	2	8	0	4	2	9	0	4	2	10	0	4	2	11	0
5	3	4	0	5	3	5	3	5	3	6	6	5	3	7	9
6	4	0	0	6	4	1	6	6	4	3	0	6	4	4	6
7	4	8	0	7	4	9	9	7	4	11	6	7	5	1	3
8	5	4	0	8	5	6	0	8	5	8	0	8	5	10	0
9	6	0	0	9	6	2	3	9	6	4	6	9	6	6	9
10	6	8	0	10	6	10	6	10	7	1	0	10	7	3	6
11	7	4	0	11	7	6	9	11	7	9	6	11	8	0	3
12	8	0	0	12	8	3	0	12	8	6	0	12	8	9	0
13	8	8	0	13	8	11	3	13	9	2	6	13	9	5	9
14	9	4	0	14	9	7	6	14	9	11	0	14	10	2	6
15	10	0	0	15	10	3	9	15	10	7	6	15	10	11	3
16	10	8	0	16	11	0	0	16	11	4	0	16	11	8	0
17	11	4	0	17	11	8	3	17	12	0	6	17	12	4	9
18	12	0	0	18	12	4	6	18	12	9	0	18	13	1	6
19	12	8	0	19	13	0	9	19	13	5	6	19	13	10	3
20	13	4	0	20	13	9	0	20	14	2	0	20	14	7	0
21	14	0	0	21	14	5	3	21	14	10	6	21	15	3	9
22	14	8	0	22	15	1	6	22	15	7	0	22	16	0	6
23	15	4	0	23	15	9	9	23	16	3	6	23	16	9	3
24	16	0	0	24	16	6	0	24	17	0	0	24	17	6	0

In. long	Ft.	In.	Pa.	S.	In. long	Ft.	In.	Pa.	S.	In. long	Ft.	In.	Pa.	S.	In. long	Ft.	In.	Pa.	S.
1	0	0	8	0	1	0	0	8	3	1	0	0	8	6	1	0	0	8	9
2	0	1	4	0	2	0	1	4	6	2	0	1	5	0	2	0	1	5	6
3	0	2	0	0	3	0	2	0	9	3	0	2	1	6	3	0	2	2	3
4	0	2	8	0	4	0	2	9	0	4	0	2	10	0	4	0	2	11	0
5	0	3	4	0	5	0	3	5	3	5	0	3	6	6	5	0	3	7	9
6	0	4	0	0	6	0	4	1	6	6	0	4	3	0	6	0	4	4	6
7	0	4	8	0	7	0	4	9	9	7	0	4	11	6	7	0	5	1	3
8	0	5	4	0	8	0	5	6	0	8	0	5	8	0	8	0	5	10	0
9	0	6	0	0	9	0	6	2	3	9	0	6	4	6	9	0	6	6	9
10	0	6	8	0	10	0	6	10	6	10	0	7	1	0	10	0	7	3	6
11	0	7	4	0	11	0	7	6	9	11	0	7	9	6	11	0	8	0	3

Qrs. in. long	In.	Pa.	S.	T.	Qrs. in. long	In.	Pa.	S.	T.	Qrs. in. long	In.	Pa.	S.	T.	Qrs. in. long	In.	Pa.	S.	T.
¼	0	2	0	0	¼	0	2	0	9	¼	0	2	1	6	¼	0	2	2	3
½	0	4	0	0	½	0	4	1	6	½	0	4	3	0	½	0	4	4	6
¾	0	6	0	0	¾	0	6	2	3	¾	0	6	4	6	¾	0	6	6	9

[TABLE III.

Ft. long	9 in. broad.			Ft. long	9¼ in. broad.			Ft. long	9½ in. broad.			Ft. long	9¾ in. broad.		
	Ft.	In.	Pa.		Ft.	In.	Pa.		Ft.	In.	Pa.		Ft.	In.	Pa.
1	0	9	0	1	0	9	3	1	0	9	6	1	0	9	9
2	1	6	0	2	1	6	6	2	1	7	0	2	1	7	6
3	2	3	0	3	2	3	9	3	2	4	6	3	2	5	3
4	3	0	0	4	3	1	0	4	3	2	0	4	3	3	0
5	3	9	0	5	3	10	3	5	3	11	6	5	4	0	9
6	4	6	0	6	4	7	6	6	4	9	0	6	4	10	6
7	5	3	0	7	5	4	9	7	5	6	6	7	5	8	3
8	6	0	0	8	6	2	0	8	6	4	0	8	6	6	0
9	6	9	0	9	6	11	3	9	7	1	6	9	7	3	9
10	7	6	0	10	7	8	6	10	7	11	0	10	8	1	6
11	8	3	0	11	8	5	9	11	8	8	6	11	8	11	3
12	9	0	0	12	9	3	0	12	9	6	0	12	9	9	0
13	9	9	0	13	10	0	3	13	10	3	6	13	10	6	9
14	10	6	0	14	10	9	6	14	11	1	0	14	11	4	6
15	11	3	0	15	11	6	9	15	11	10	6	15	12	2	3
16	12	0	0	16	12	4	0	16	12	8	0	16	13	0	0
17	12	9	0	17	13	1	3	17	13	5	6	17	13	9	9
18	13	6	0	18	13	10	6	18	14	3	0	18	14	7	6
19	14	3	0	19	14	7	9	19	15	0	6	19	15	6	3
20	15	0	0	20	15	5	0	20	15	10	0	20	16	5	0
21	15	9	0	21	16	2	3	21	16	7	6	21	17	0	9
22	16	6	0	22	16	11	6	22	17	5	0	22	17	10	6
23	17	3	0	23	17	8	9	23	18	2	6	23	18	8	3
24	18	0	0	24	18	6	0	24	19	0	0	24	19	6	0

In. long	Ft.	In.	Pa.	S.	In. long	Ft.	In.	Pa.	S.	In. long	Ft.	In.	Pa.	S.	In. long	Ft.	In.	Pa.	S.
1	0	0	9	0	1	0	0	9	3	1	0	0	9	6	1	0	0	9	9
2	0	1	6	0	2	0	1	6	6	2	0	1	7	0	2	0	1	7	6
3	0	2	3	0	3	0	2	3	9	3	0	2	4	6	3	0	2	5	3
4	0	3	0	0	4	0	3	1	0	4	0	3	2	0	4	0	3	3	0
5	0	3	9	0	5	0	3	10	3	5	0	3	11	6	5	0	4	0	9
6	0	4	6	0	6	0	4	7	6	6	0	4	9	0	6	0	4	10	6
7	0	5	3	0	7	0	5	4	9	7	0	5	6	6	7	0	5	8	3
8	0	6	0	0	8	0	6	2	0	8	0	6	4	0	8	0	6	6	0
9	0	6	9	0	9	0	6	11	3	9	0	7	1	6	9	0	7	3	9
10	0	7	6	0	10	0	7	8	6	10	0	7	11	0	10	0	8	1	6
11	0	8	3	0	11	0	8	5	9	11	0	8	8	6	11	0	8	11	3

Qrs. in. long	In.	Pa.	S.	T.	Qrs. in. long	In.	Pa.	S.	T.	Qrs. in. long	In.	Pa.	S.	T.	Qrs. in. long	In.	Pa.	S.	T.
¼	0	2	3	0	¼	0	2	3	9	¼	0	2	4	6	¼	0	2	5	3
½	0	4	6	0	½	0	4	7	6	½	0	4	9	0	½	0	4	10	6
¾	0	6	9	0	¾	0	6	11	3	¾	0	7	1	6	¾	0	7	3	9

[Table III.] 295

Ft. long	10 in. broad.			Ft. long	10¼ in. broad.			Ft. long	10½ in. broad.			Ft. long	10¾ in. broad.		
	Ft.	In.	Pa.		Ft.	In.	Pa.		Ft.	In.	Pa.		Ft.	In.	Pa.
1	0	10	0	1	0	10	3	1	0	10	6	1	0	10	9
2	1	8	0	2	1	8	6	2	1	9	0	2	1	9	6
3	2	6	0	3	2	6	9	3	2	7	6	3	2	8	3
4	3	4	0	4	3	5	0	4	3	6	0	4	3	7	0
5	4	2	0	5	4	3	3	5	4	4	6	5	4	5	9
6	5	0	0	6	5	1	6	6	5	3	0	6	5	4	6
7	5	10	0	7	5	11	9	7	6	1	6	7	6	3	3
8	6	8	0	8	6	10	0	8	7	0	0	8	7	2	0
9	7	6	0	9	7	8	3	9	7	10	6	9	8	0	9
10	8	4	0	10	8	6	6	10	8	9	0	10	8	11	6
11	9	2	0	11	9	4	9	11	9	7	6	11	9	10	3
12	10	0	0	12	10	3	0	12	10	6	0	12	10	9	0
13	10	10	0	13	11	1	3	13	11	4	6	13	11	7	9
14	11	8	0	14	11	11	6	14	12	3	0	14	12	6	6
15	12	6	0	15	12	9	9	15	13	1	6	15	13	5	3
16	13	4	0	16	13	8	0	16	14	0	0	16	14	4	0
17	14	2	0	17	14	6	3	17	14	10	6	17	15	2	9
18	15	0	0	18	15	4	6	18	15	9	0	18	16	1	6
19	15	10	0	19	16	2	9	19	16	7	6	19	17	0	3
20	16	8	0	20	17	1	0	20	17	6	0	20	17	11	0
21	17	6	0	21	17	11	3	21	18	4	6	21	18	9	9
22	18	4	0	22	18	9	6	22	19	3	0	22	19	8	6
23	19	2	0	23	19	7	9	23	20	1	6	23	20	7	3
24	20	0	0	24	20	6	0	24	21	0	0	24	21	6	0

In. long	Ft.	In.	Pa.	S.	In. long	Ft.	In.	Pa.	S.	In. long	Ft.	In.	Pa.	S.	In. long	Ft.	In.	Pa.	S.
1	0	0	10	0	1	0	0	10	3	1	0	0	10	6	1	0	0	10	9
2	0	1	8	0	2	0	1	8	6	2	0	1	9	0	2	0	1	9	6
3	0	2	6	0	3	0	2	6	9	3	0	2	7	6	3	0	2	8	3
4	0	3	4	0	4	0	3	5	0	4	0	3	6	0	4	0	3	7	0
5	0	4	2	0	5	0	4	3	3	5	0	4	4	6	5	0	4	5	9
6	0	5	0	0	6	0	5	1	6	6	0	5	3	0	6	0	5	4	6
7	0	5	10	0	7	0	5	11	9	7	0	6	1	6	7	0	6	3	3
8	0	6	8	0	8	0	6	10	0	8	0	7	0	0	8	0	7	2	0
9	0	7	6	0	9	0	7	8	3	9	0	7	10	6	9	0	8	0	9
10	0	8	4	0	10	0	8	6	6	10	0	8	9	0	10	0	8	11	6
11	0	9	2	0	11	0	9	4	9	11	0	9	7	6	11	0	9	10	3

Qrs. in. long	In.	Pa.	S.	T.	Qrs. in. long	In.	Pa.	S.	T.	Qrs. in. long	In.	Pa.	S.	T.	Qrs. in. long	In.	Pa.	S.	T.
¼	0	2	6	0	¼	0	2	6	9	¼	0	2	7	6	¼	0	2	8	3
½	0	5	0	0	½	0	5	1	6	½	0	5	3	0	½	0	5	4	6
¾	0	7	6	0	¾	0	7	8	3	¾	0	7	10	6	¾	0	8	0	9

[TABLE III.

Ft. long	11 in. broad.			Ft. long	11¼ in. broad.			Ft. long	11½ in. broad.			Ft. long	11¾ in. broad.		
	Ft.	In.	Pa.		Ft.	In.	Pa.		Ft.	In.	Pa.		Ft.	In.	Pa.
1	0	11	0	1	0	11	3	1	0	11	6	1	0	11	9
2	1	10	0	2	1	10	6	2	1	11	0	2	1	11	6
3	2	9	0	3	2	9	9	3	2	10	6	3	2	11	3
4	3	8	0	4	3	9	0	4	3	10	0	4	3	11	0
5	4	7	0	5	4	8	3	5	4	9	6	5	4	10	9
6	5	6	0	6	5	7	6	6	5	9	0	6	5	10	6
7	6	5	0	7	6	6	9	7	6	8	6	7	6	10	3
8	7	4	0	8	7	6	0	8	7	8	0	8	7	10	0
9	8	3	0	9	8	5	3	9	8	7	6	9	8	9	9
10	9	2	0	10	9	4	6	10	9	7	0	10	9	9	6
11	10	1	0	11	10	3	9	11	10	6	6	11	10	8	3
12	11	0	0	12	11	3	0	12	11	6	0	12	11	9	0
13	11	11	0	13	12	2	3	13	12	5	6	13	12	8	9
14	12	10	0	14	13	1	6	14	13	5	0	14	13	8	6
15	13	9	0	15	14	0	9	15	14	4	6	15	14	8	3
16	14	8	0	16	15	0	0	16	15	4	0	16	15	8	0
17	15	7	0	17	15	11	3	17	16	3	6	17	16	7	9
18	16	6	0	18	16	10	6	18	17	3	0	18	17	7	6
19	17	5	0	19	17	9	9	19	18	2	6	19	18	7	3
20	18	4	0	20	18	9	0	20	19	2	0	20	19	7	0
21	19	3	0	21	19	8	3	21	20	1	6	21	20	6	9
22	20	2	0	22	20	7	6	22	21	1	0	22	21	6	6
23	21	1	0	23	21	6	9	23	22	0	6	23	22	6	3
24	22	0	0	24	22	6	0	24	23	0	0	24	23	6	0

In. long	Ft.	In.	Pa.	S.	In. long	Ft.	In.	Pa.	S.	In. long	Ft.	In.	Pa.	S.	In. long	Ft.	In.	Pa.	S.
1	0	0	11	0	1	0	0	11	3	1	0	0	11	6	1	0	0	11	9
2	0	1	10	0	2	0	1	10	6	2	0	1	11	0	2	0	1	11	6
3	0	2	9	0	3	0	2	9	9	3	0	2	10	6	3	0	2	11	3
4	0	3	8	0	4	0	3	9	0	4	0	3	10	0	4	0	3	11	0
5	0	4	7	0	5	0	4	8	3	5	0	4	9	6	5	0	4	10	9
6	0	5	6	0	6	0	5	7	6	6	0	5	9	0	6	0	5	10	6
7	0	6	5	0	7	0	6	6	9	7	0	6	8	6	7	0	6	10	3
8	0	7	4	0	8	0	7	6	0	8	0	7	8	0	8	0	7	10	0
9	0	8	3	0	9	0	8	5	3	9	0	8	7	6	9	0	8	9	9
10	0	9	2	0	10	0	9	4	6	10	0	9	7	0	10	0	9	9	6
11	0	10	1	0	11	0	10	3	9	11	0	10	6	6	11	0	10	9	3

Qrs. in. long	In.	Pa.	S.	T.	Qrs. in. long	In.	Pa.	S.	T.	Qrs. in. long	In.	Pa.	S.	T.	Qrs. in. long	In.	Pa.	S.	T.
¼	0	2	9	0	¼	0	2	9	9	¼	0	2	10	6	¼	0	2	11	3
½	0	5	6	0	½	0	5	7	6	½	0	5	9	0	½	0	5	10	6
¾	0	8	3	0	¾	0	8	5	3	¾	0	8	7	6	¾	0	8	9	9

[TABLE III.] 297

Ft. long	12 in. broad.			Ft. long	12¼ in. broad.			Ft. long	12½ in. broad.			Ft. long	12¾ in. broad.		
	Ft.	In.	Pa.		Ft.	In.	Pa.		Ft.	In.	Pa.		Ft.	In.	Pa.
1	1	0	0	1	1	0	3	1	1	0	6	1	1	0	9
2	2	0	0	2	2	0	6	2	2	1	0	2	2	1	6
3	3	0	0	3	3	0	9	3	3	1	6	3	3	2	3
4	4	0	0	4	4	1	0	4	4	2	0	4	4	3	0
5	5	0	0	5	5	1	3	5	5	2	6	5	5	3	9
6	6	0	0	6	6	1	6	6	6	3	0	6	6	4	6
7	7	0	0	7	7	1	9	7	7	3	6	7	7	5	3
8	8	0	0	8	8	2	0	8	8	4	0	8	8	6	0
9	9	0	0	9	9	2	3	9	9	4	6	9	9	6	9
10	10	0	0	10	10	2	6	10	10	5	0	10	10	7	6
11	11	0	0	11	11	2	9	11	11	5	6	11	11	8	3
12	12	0	0	12	12	3	0	12	12	6	0	12	12	9	0
13	13	0	0	13	13	3	3	13	13	6	6	13	13	9	9
14	14	0	0	14	14	3	6	14	14	7	0	14	14	10	6
15	15	0	0	15	15	3	9	15	15	7	6	15	15	11	3
16	16	0	0	16	16	4	0	16	16	8	0	16	17	0	0
17	17	0	0	17	17	4	3	17	17	8	6	17	18	0	9
18	18	0	0	18	18	4	6	18	18	9	0	18	19	1	6
19	19	0	0	19	19	4	9	19	19	9	6	19	20	2	3
20	20	0	0	20	20	5	0	20	20	10	0	20	21	3	0
21	21	0	0	21	21	5	3	21	21	10	6	21	22	3	9
22	22	0	0	22	22	5	6	22	22	11	0	22	23	4	6
23	23	0	0	23	23	5	9	23	23	11	6	23	24	5	3
24	24	0	0	24	24	6	0	24	25	0	0	24	25	6	0

In. long	Ft.	In.	Pa.	S.	In. long	Ft.	In.	Pa.	S.	In. long	Ft.	In.	Pa.	S.	In. long	Ft.	In.	Pa.	S.
1	0	1	0	0	1	0	1	0	3	1	0	1	0	6	1	0	1	0	9
2	0	2	0	0	2	0	2	0	6	2	0	2	1	0	2	0	2	1	6
3	0	3	0	0	3	0	3	0	9	3	0	3	1	6	3	0	3	2	3
4	0	4	0	0	4	0	4	1	0	4	0	4	2	0	4	0	4	3	0
5	0	5	0	0	5	0	5	1	3	5	0	5	2	6	5	0	5	3	9
6	0	6	0	0	6	0	6	1	6	6	0	6	3	0	6	0	6	4	6
7	0	7	0	0	7	0	7	1	9	7	0	7	3	6	7	0	7	5	3
8	0	8	0	0	8	0	8	2	0	8	0	8	4	0	8	0	8	6	0
9	0	9	0	0	9	0	9	2	3	9	0	9	4	6	9	0	9	6	9
10	0	10	0	0	10	0	10	2	6	10	0	10	5	0	10	0	10	7	6
11	0	11	0	0	11	0	11	2	9	11	0	11	5	6	11	0	11	8	3

Qrs. in. long	In.	Pa.	S.	T.	Qrs. in. long	In.	Pa.	S.	T.	Qrs. in. long	In.	Pa.	S.	T.	Qrs. in. long	In.	Pa.	S.	T.
¼	0	3	0	0	¼	0	3	0	9	¼	0	3	1	6	¼	0	3	2	3
½	0	6	0	0	½	0	6	1	6	½	0	6	3	0	½	0	6	4	6
¾	0	9	0	0	¾	0	9	2	3	¾	0	9	4	6	¾	0	9	6	0

[TABLE III.

Ft. long	13 in. broad.			Ft. long	13¼ in. broad.			Ft. long	13½ in. broad.			Ft. long	13¾ in. broad.		
	Ft.	In.	Pa.		Ft.	In.	Pa.		Ft.	In.	Pa.		Ft.	In.	Pa.
1	1	1	0	1	1	1	3	1	1	1	6	1	1	1	9
2	2	2	0	2	2	2	6	2	2	3	0	2	2	3	6
3	3	3	0	3	3	3	9	3	3	4	6	3	3	5	3
4	4	4	0	4	4	5	0	4	4	6	0	4	4	7	0
5	5	5	0	5	5	6	3	5	5	7	6	5	5	8	9
6	6	6	0	6	6	7	6	6	6	9	0	6	6	10	6
7	7	7	0	7	7	8	9	7	7	10	6	7	8	0	3
8	8	8	0	8	8	10	0	8	9	0	0	8	9	2	0
9	9	9	0	9	9	11	3	9	10	1	6	9	10	3	9
10	10	10	0	10	11	0	6	10	11	3	0	10	11	5	6
11	11	11	0	11	12	1	9	11	12	4	6	11	12	7	3
12	13	0	0	12	13	3	0	12	13	6	0	12	13	9	0
13	14	1	0	13	14	4	3	13	14	7	6	13	14	10	9
14	15	2	0	14	15	5	6	14	15	9	0	14	16	0	6
15	16	3	0	15	16	6	9	15	16	10	6	15	17	2	3
16	17	4	0	16	17	8	0	16	18	0	0	16	18	4	0
17	18	5	0	17	18	9	3	17	19	1	6	17	19	5	9
18	19	6	0	18	19	10	6	18	20	3	0	18	20	7	6
19	20	7	0	19	20	11	9	19	21	4	6	19	21	9	3
20	21	8	0	20	22	1	0	20	22	6	0	20	22	11	0
21	22	9	0	21	23	2	3	21	23	7	6	21	24	0	9
22	23	10	0	22	24	3	6	22	24	9	0	22	25	2	6
23	24	11	0	23	25	4	9	23	25	10	6	23	26	4	3
24	26	0	0	24	26	6	0	24	27	0	0	24	27	6	0

In. long	Ft.	In.	Pa.	S.	In. long	Ft.	In.	Pa.	S.	In. long	Ft.	In.	Pa.	S.	In. long	Ft.	In.	Pa.	S.
1	0	1	1	0	1	0	1	1	3	1	0	1	1	6	1	0	1	1	9
2	0	2	2	0	2	0	2	2	6	2	0	2	3	0	2	0	2	3	6
3	0	3	3	0	3	0	3	3	9	3	0	3	4	6	3	0	3	5	3
4	0	4	4	0	4	0	4	5	0	4	0	4	6	0	4	0	4	7	0
5	0	5	5	0	5	0	5	6	3	5	0	5	7	6	5	0	5	8	9
6	0	6	6	0	6	0	6	7	6	6	0	6	9	0	6	0	6	10	6
7	0	7	7	0	7	0	7	8	9	7	0	7	10	6	7	0	8	0	3
8	0	8	8	0	8	0	8	10	0	8	0	9	0	0	8	0	9	2	0
9	0	9	9	0	9	0	9	11	3	9	0	10	1	6	9	0	10	3	9
10	0	10	10	0	10	0	11	0	6	10	0	11	3	0	10	0	11	5	6
11	0	11	11	0	11	0	12	1	9	11	1	0	4	6	11	1	0	7	3

Qrs. in. long	In.	Pa.	S.	T.	Qrs. in. long	In.	Pa.	S.	T.	Qrs. in. long	In.	Pa.	S.	T.	Qrs. in. long	In.	Pa.	S.	T.
¼	0	3	3	0	¼	0	3	3	9	¼	0	3	4	6	¼	0	3	5	3
½	0	6	6	0	½	0	6	7	6	½	0	6	9	0	½	0	6	10	6
¾	0	9	9	0	¾	0	9	11	3	¾	0	10	1	6	¾	0	10	3	9

[TABLE III.]

Ft. long	14 in. broad.			Ft. long	14¼ in. broad.			Ft. long	14½ in. broad.			Ft. long	14¾ in. broad.		
	Ft.	In.	Pa.		Ft.	In.	Pa.		Ft.	In.	Pa.		Ft.	In.	Pa.
1	1	2	0	1	1	2	3	1	1	2	6	1	1	2	9
2	2	4	0	2	2	4	6	2	2	5	0	2	2	5	6
3	3	6	0	3	3	6	9	3	3	7	6	3	3	8	3
4	4	8	0	4	4	9	0	4	4	10	0	4	4	11	0
5	5	10	0	5	5	11	3	5	6	0	6	5	6	1	9
6	7	0	0	6	7	1	6	6	7	3	0	6	7	4	6
7	8	2	0	7	8	3	9	7	8	5	6	7	8	7	3
8	9	4	0	8	9	6	0	8	9	8	0	8	9	10	0
9	10	6	0	9	10	8	3	9	10	10	6	9	11	0	9
10	11	8	0	10	11	10	6	10	12	1	0	10	12	3	6
11	12	10	0	11	13	0	9	11	13	3	6	11	13	6	3
12	14	0	0	12	14	3	0	12	14	6	0	12	14	9	0
13	15	2	0	13	15	5	3	13	15	8	6	13	15	11	9
14	16	4	0	14	16	7	6	14	16	11	0	14	17	2	6
15	17	6	0	15	17	9	9	15	18	1	6	15	18	5	3
16	18	8	0	16	19	0	0	16	19	4	0	16	19	8	0
17	19	10	0	17	20	2	3	17	20	6	6	17	20	10	9
18	21	0	0	18	21	4	6	18	21	9	0	18	22	1	6
19	22	2	0	19	22	6	9	19	22	11	6	19	23	4	3
20	23	4	0	20	23	9	0	20	24	2	0	20	24	7	0
21	24	6	0	21	24	11	3	21	25	4	6	21	25	9	9
22	25	8	0	22	26	1	6	22	26	7	0	22	27	0	6
23	26	10	0	23	27	3	9	23	27	9	6	23	28	3	3
24	28	0	0	24	28	6	0	24	29	0	0	24	29	6	0

In. long	Ft.	In.	Pa.	S.	In. long	Ft.	In.	Pa.	S.	In. long	Ft.	In.	Pa.	S.	In. long	Ft.	In.	Pa.	S.
1	0	1	2	0	1	0	1	2	3	1	0	1	2	6	1	0	1	2	9
2	0	2	4	0	2	0	2	4	6	2	0	2	5	0	2	0	2	5	6
3	0	3	6	0	3	0	3	6	9	3	0	3	7	6	3	0	3	8	3
4	0	4	8	0	4	0	4	9	0	4	0	4	10	0	4	0	4	11	0
5	0	5	10	0	5	0	5	11	3	5	0	6	0	6	5	0	6	1	9
6	0	7	0	0	6	0	7	1	6	6	0	7	3	0	6	0	7	4	6
7	0	8	2	0	7	0	8	3	9	7	0	8	5	6	7	0	8	7	3
8	0	9	4	0	8	0	9	6	0	8	0	9	8	0	8	0	9	10	0
9	0	10	6	0	9	0	10	8	3	9	0	10	10	6	9	0	11	0	9
10	0	11	8	0	10	0	11	10	6	10	1	0	1	0	10	1	0	3	6
11	1	0	10	0	11	1	1	0	9	11	1	1	3	6	11	1	1	6	3

Qrs. in. long	In.	Pa.	S.	T.	Qrs. in. long	In.	Pa.	S.	T.	Qrs. in. long	In.	Pa.	S.	T.	Qrs. in. long	In.	Pa.	S.	T.
¼	0	3	6	0	¼	0	3	6	9	¼	0	3	7	6	¼	0	3	8	3
½	0	7	0	0	½	0	7	1	6	½	0	7	3	0	½	0	7	4	6
¾	0	10	6	0	¾	0	10	8	3	¾	0	10	10	6	¾	0	11	0	9

300 [TABLE III.

Ft. long	15 in. broad.			Ft. long	15¼ in. broad.			Ft. long	15½ in. broad.			Ft. long	15¾ in. broad.		
	Ft.	In.	Pa.		Ft.	In.	Pa.		Ft.	In.	Pa.		Ft.	In.	Pa.
1	1	3	0	1	1	3	3	1	1	3	6	1	1	3	9
2	2	6	0	2	2	6	6	2	2	7	0	2	2	7	6
3	3	9	0	3	3	9	9	3	3	10	6	3	3	11	3
4	5	0	0	4	5	1	0	4	5	2	0	4	5	3	0
5	6	3	0	5	6	4	3	5	6	5	6	5	6	6	9
6	7	6	0	6	7	7	6	6	7	9	0	6	7	10	6
7	8	9	0	7	8	10	9	7	9	0	6	7	9	2	3
8	10	0	0	8	10	2	0	8	10	4	0	8	10	6	0
9	11	3	0	9	11	5	3	9	11	7	6	9	11	9	9
10	12	6	0	10	12	8	6	10	12	11	0	10	13	1	6
11	13	9	0	11	13	11	9	11	14	2	6	11	14	5	3
12	15	0	0	12	15	3	0	12	15	6	0	12	15	9	0
13	16	3	0	13	16	6	3	13	16	9	6	13	17	0	9
14	17	6	0	14	17	9	6	14	18	1	0	14	18	4	6
15	18	9	0	15	19	0	9	15	19	4	6	15	19	8	3
16	20	0	0	16	20	4	0	16	20	8	0	16	21	0	0
17	21	3	0	17	21	7	3	17	21	11	6	17	22	3	9
18	22	6	0	18	22	10	6	18	23	3	0	18	23	7	6
19	23	9	0	19	24	1	9	19	24	6	6	19	24	11	3
20	25	0	0	20	25	5	0	20	25	10	0	20	26	3	0
21	26	3	0	21	26	8	3	21	27	1	6	21	27	6	9
22	27	6	0	22	27	11	6	22	28	5	0	22	28	10	6
23	28	9	0	23	29	2	9	23	29	8	6	23	30	2	3
24	30	0	0	24	30	6	0	24	31	0	0	24	31	6	0

In. long	Ft.	In.	Pa.	S.	In. long	Ft.	In.	Pa.	S.	In. long	Ft.	In.	Pa.	S.	In. long	Ft.	In.	Pa.	S.
1	0	1	3	0	1	0	1	3	3	1	0	1	3	6	1	0	1	3	9
2	0	2	6	0	2	0	2	6	6	2	0	2	7	0	2	0	2	7	6
3	0	3	9	0	3	0	3	9	9	3	0	3	10	6	3	0	3	11	3
4	0	5	0	0	4	0	5	1	0	4	0	5	2	0	4	0	5	3	0
5	0	6	3	0	5	0	6	4	3	5	0	6	5	6	5	0	6	6	9
6	0	7	6	0	6	0	7	7	6	6	0	7	9	0	6	0	7	10	6
7	0	8	9	0	7	0	8	10	9	7	0	9	0	6	7	0	9	2	3
8	0	10	0	0	8	0	10	2	0	8	0	10	4	0	8	0	10	6	0
9	0	11	3	0	9	0	11	5	3	9	0	11	7	6	9	0	11	9	9
10	1	0	6	0	10	1	0	8	6	10	1	0	11	0	10	1	1	1	6
11	1	1	9	0	11	1	1	11	9	11	1	2	2	6	11	1	2	5	3

Qrs. in. long	In.	Pa.	S.	T.	Qrs. in. long	In.	Pa.	S.	T.	Qrs. in. long	In.	Pa.	S.	T.	Qrs. in. long	In.	Pa.	S.	T.
¼	0	3	9	0	¼	0	3	9	9	¼	0	3	10	6	¼	0	3	11	3
½	0	7	6	0	½	0	7	7	6	½	0	7	9	0	½	0	7	10	6
¾	0	11	3	0	¾	0	11	5	3	¾	0	11	7	6	¾	0	11	9	9

[TABLE III.] 301

Ft. long	16 in. broad.			Ft. long	16¼ in. broad.			Ft. long	16½ in. broad.			Ft. long	16¾ in. broad.		
	Ft.	In.	Pa.		Ft.	In.	Pa.		Ft.	In.	Pa.		Ft.	In.	Pa.
1	1	4	0	1	1	4	3	1	1	4	6	1	1	4	9
2	2	8	0	2	2	8	6	2	2	9	0	2	2	9	6
3	4	0	0	3	4	0	9	3	4	1	6	3	4	2	3
4	5	4	0	4	5	5	0	4	5	3	0	4	5	7	0
5	6	8	0	5	6	9	3	5	6	10	6	5	6	11	9
6	8	0	0	6	8	1	6	6	8	3	0	6	8	4	6
7	9	4	0	7	9	5	9	7	9	7	6	7	9	9	3
8	10	8	0	8	10	10	0	8	11	0	0	8	11	2	0
9	12	0	0	9	12	2	3	9	12	4	6	9	12	6	9
10	13	4	0	10	13	6	6	10	13	9	0	10	13	11	6
11	14	8	0	11	14	10	9	11	15	1	6	11	15	4	3
12	16	0	0	12	16	3	0	12	16	6	0	12	16	9	0
13	17	4	0	13	17	7	3	13	17	10	6	13	18	1	9
14	18	8	0	14	18	11	6	14	19	3	0	14	19	6	6
15	20	0	0	15	20	3	9	15	20	7	6	15	20	11	3
16	21	4	0	16	21	8	0	16	22	0	0	16	22	4	0
17	22	8	0	17	23	0	3	17	23	4	6	17	23	8	9
18	24	0	0	18	24	4	6	18	24	9	0	18	25	1	6
19	25	4	0	19	25	8	9	19	26	1	6	19	26	6	3
20	26	8	0	20	27	1	0	20	27	6	0	20	27	11	0
21	28	0	0	21	28	5	3	21	28	10	6	21	29	3	9
22	29	4	0	22	29	9	6	22	30	3	0	22	30	8	6
23	30	8	0	23	31	1	9	23	31	7	6	23	32	1	3
24	32	0	0	24	32	6	0	24	33	0	0	24	33	6	0

In. long	Ft.	In.	Pa.	S.	In. long	Ft.	In.	Pa.	S.	In. long	Ft.	In.	Pa.	S.	In. long	Ft.	In.	Pa.	S.
1	0	1	4	0	1	0	1	4	3	1	0	1	4	6	1	0	1	4	9
2	0	2	8	0	2	0	2	8	6	2	0	2	9	0	2	0	2	9	6
3	0	4	0	0	3	0	4	0	9	3	0	4	1	6	3	0	4	2	3
4	0	5	4	0	4	0	5	5	0	4	0	5	6	0	4	0	5	7	0
5	0	6	8	0	5	0	6	9	3	5	0	6	10	6	5	0	6	11	9
6	0	8	0	0	6	0	8	1	6	6	0	8	3	0	6	0	8	4	6
7	0	9	4	0	7	0	9	5	9	7	0	9	7	6	7	0	9	9	3
8	0	10	8	0	8	0	10	10	0	8	0	11	0	0	8	0	11	2	0
9	1	0	0	0	9	1	0	2	3	9	1	0	4	6	9	1	0	6	9
10	1	1	4	0	10	1	1	6	6	10	1	1	9	0	10	1	1	11	6
11	1	2	8	0	11	1	2	10	9	11	1	3	1	6	11	1	3	4	3

Qrs. in. long	In.	Pa.	S.	T.	Qrs. in. long	In.	Pa.	S.	T.	Qrs. in. long	In.	Pa.	S.	T.	Qrs. in. long	In.	Pa.	S.	T.
¼	0	4	0	0	¼	0	4	0	9	¼	0	4	1	6	¼	0	4	2	3
½	0	8	0	0	½	0	8	1	6	½	0	8	3	0	½	0	8	4	6
¾	1	0	0	0	¾	1	0	2	3	¾	1	0	4	6	¾	1	0	6	9

[TABLE III.

Ft. long	17 in. broad.			Ft. long	17¼ in. broad.			Ft. long	17½ in. broad.			Ft. long	17¾ in. broad.		
	Ft.	In.	Pa.		Ft.	In.	Pa.		Ft.	In.	Pa.		Ft.	In.	Pa.
1	1	5	0	1	1	5	3	1	1	5	6	1	1	5	9
2	2	10	0	2	2	10	6	2	2	11	0	2	2	11	6
3	4	3	0	3	4	3	9	3	4	4	6	3	4	5	3
4	5	8	0	4	5	9	0	4	5	10	0	4	5	11	0
5	7	1	0	5	7	2	3	5	7	3	6	5	7	4	9
6	8	6	0	6	8	7	6	6	8	9	0	6	8	10	6
7	9	11	0	7	10	0	9	7	10	2	6	7	10	4	3
8	11	4	0	8	11	6	0	8	11	8	0	8	11	10	0
9	12	9	0	9	12	11	3	9	13	1	6	9	13	3	9
10	14	2	0	10	14	4	6	10	14	7	0	10	14	9	6
11	15	7	0	11	15	9	9	11	16	0	6	11	16	3	3
12	17	0	0	12	17	3	0	12	17	6	0	12	17	9	0
13	18	5	0	13	18	8	3	13	18	11	6	13	19	2	9
14	19	10	0	14	20	1	6	14	20	5	0	14	20	8	6
15	21	3	0	15	21	6	9	15	21	10	6	15	22	2	3
16	22	8	0	16	23	0	0	16	23	4	0	16	23	8	0
17	24	1	0	17	24	5	3	17	24	9	6	17	25	1	9
18	25	6	0	18	25	10	6	18	26	3	0	18	26	7	6
19	26	11	0	19	27	3	9	19	27	8	6	19	28	1	3
20	28	4	0	20	28	9	0	20	29	2	0	20	29	7	0
21	29	9	0	21	30	2	3	21	30	7	6	21	31	0	9
22	31	2	0	22	31	7	6	22	32	1	0	22	32	6	6
23	32	7	0	23	33	0	9	23	33	6	6	23	34	0	3
24	34	0	0	24	34	6	0	24	35	0	0	24	35	6	0

In. long	Ft.	In.	Pa.	S.	In. long	Ft.	In.	Pa.	S.	In. long	Ft.	In.	Pa.	S.	In. long	Ft.	In.	Pa.	S.
1	0	1	5	0	1	0	1	5	3	1	0	1	5	6	1	0	1	5	9
2	0	2	10	0	2	0	2	10	6	2	0	2	11	0	2	0	2	11	6
3	0	4	3	0	3	0	4	3	9	3	0	4	4	6	3	0	4	5	3
4	0	5	8	0	4	0	5	9	0	4	0	5	10	0	4	0	5	11	0
5	0	7	1	0	5	0	7	2	3	5	0	7	3	6	5	0	7	4	9
6	0	8	6	0	6	0	8	7	6	6	0	8	9	0	6	0	8	10	6
7	0	9	11	0	7	0	10	0	9	7	0	10	2	6	7	0	10	4	3
8	0	11	4	0	8	0	11	6	0	8	0	11	8	0	8	0	11	10	0
9	1	0	9	0	9	1	0	11	3	9	1	1	1	6	9	1	1	3	9
10	1	2	2	0	10	1	2	4	6	10	1	2	7	0	10	1	2	9	6
11	1	3	7	0	11	1	3	9	9	11	1	4	0	6	11	1	4	3	3

Qrs. in. long	In.	Pa.	S.	T.	Qrs. in. long	In.	Pa.	S.	T.	Qrs. in. long	In.	Pa.	S.	T.	Qrs. in. long	In.	Pa.	S.	T.
¼	0	4	3	0	¼	0	4	3	9	¼	0	4	4	6	¼	0	4	5	3
½	0	8	6	0	½	0	8	7	6	½	0	8	9	0	½	0	8	10	6
¾	1	0	9	0	¾	1	0	11	3	¾	1	1	1	6	¾	1	1	3	9

TABLE III.]

Ft. long	18 in. broad.			Ft. long	18¼ in. broad.			Ft. long	18½ in. broad.			Ft. long	18¾ in. broad.		
	Ft.	In.	Pa.		Ft.	In.	Pa.		Ft.	In.	Pa.		Ft.	In.	Pa.
1	1	6	0	1	1	6	3	1	1	6	6	1	1	6	9
2	3	0	0	2	3	0	6	2	3	1	0	2	3	1	6
3	4	6	0	3	4	6	9	3	4	7	6	3	4	8	3
4	6	0	0	4	6	1	0	4	6	2	0	4	6	3	0
5	7	6	0	5	7	7	3	5	7	8	6	5	7	9	9
6	9	0	0	6	9	1	6	6	9	3	0	6	9	4	6
7	10	6	0	7	11	7	9	7	10	9	6	7	10	11	3
8	12	0	0	8	12	2	0	8	12	4	0	8	12	6	0
9	13	6	0	9	13	8	3	9	13	10	6	9	14	0	9
10	15	0	0	10	15	2	6	10	15	5	0	10	15	7	6
11	16	6	0	11	16	8	9	11	16	11	6	11	17	2	3
12	18	0	0	12	18	3	0	12	18	6	0	12	18	9	0
13	19	6	0	13	19	9	3	13	20	0	6	13	20	3	9
14	21	0	0	14	21	3	6	14	21	7	0	14	21	10	6
15	22	6	0	15	22	9	9	15	23	1	6	15	23	5	3
16	24	0	0	16	24	4	0	16	24	8	0	16	25	0	0
17	25	6	0	17	25	10	3	17	26	2	6	17	26	6	9
18	27	0	0	18	27	4	6	18	27	9	0	18	28	1	6
19	28	6	0	19	28	10	9	19	29	3	6	19	29	8	3
20	30	0	0	20	30	5	0	20	30	10	0	20	31	3	0
21	31	6	0	21	31	11	3	21	32	4	6	21	32	9	9
22	33	0	0	22	33	5	6	22	33	11	0	22	34	4	6
23	34	6	0	23	34	11	9	23	35	5	6	23	35	11	3
24	36	0	0	24	36	6	0	24	37	0	0	24	37	6	0

In. long	Ft.	In.	Pa.	S.	In. long	Ft.	In.	Pa.	S.	In. long	Ft.	In.	Pa.	S.	In. long	Ft.	In.	Pa.	S.
1	0	1	6	0	1	0	1	6	3	1	0	1	6	6	1	0	1	6	9
2	0	3	0	0	2	0	3	0	6	2	0	3	1	0	2	0	3	1	6
3	0	4	6	0	3	0	4	6	9	3	0	4	7	6	3	0	4	8	3
4	0	6	0	0	4	0	6	1	0	4	0	6	2	0	4	0	6	3	0
5	0	7	6	0	5	0	7	7	3	5	0	7	8	6	5	0	7	9	9
6	0	9	0	0	6	0	9	1	6	6	0	9	3	0	6	0	9	4	6
7	0	10	6	0	7	0	10	7	9	7	0	10	9	6	7	0	10	11	3
8	1	0	0	0	8	1	0	2	0	8	1	0	4	0	8	1	0	6	0
9	1	1	6	0	9	1	1	8	3	9	1	1	10	6	9	1	2	0	9
10	1	3	0	0	10	1	3	2	6	10	1	3	5	0	10	1	3	7	6
11	1	4	6	0	11	1	4	8	9	11	1	4	11	6	11	1	5	2	3

Qrs. in. long	In.	Pa.	S.	T.	Qrs. in. long	In.	Pa.	S.	T.	Qrs. in. long	In.	Pa.	S.	T.	Qrs. in. long	In.	Pa.	S.	T.
¼	0	4	6	0	¼	0	4	6	9	¼	0	4	7	6	¼	0	4	8	3
½	0	9	0	0	½	0	9	1	6	½	0	9	3	0	½	0	9	4	6
¾	1	1	6	0	¾	1	1	8	3	¾	1	1	10	6	¾	1	2	0	9

Ft. long	19 in. broad.			Ft. long	19¼ in. broad.			Ft. long	19½ in. broad.			Ft. long	19¾ in. broad.		
	Ft.	In.	Pa.		Ft.	In.	Pa.		Ft.	In.	Pa.		Ft.	In.	Pa.
1	1	7	0	1	1	7	3	1	1	7	6	1	1	7	9
2	3	2	0	2	3	2	6	2	3	3	0	2	3	3	6
3	4	9	0	3	4	9	9	3	4	10	6	3	4	11	3
4	6	4	0	4	6	5	0	4	6	6	0	4	6	7	0
5	7	11	0	5	8	0	3	5	8	1	6	5	8	2	9
6	9	6	0	6	9	7	6	6	9	9	0	6	9	10	6
7	11	1	0	7	11	2	9	7	11	4	6	7	11	6	3
8	12	8	0	8	12	10	0	8	13	0	0	8	13	2	0
9	14	3	0	9	14	5	3	9	14	7	6	9	14	9	9
10	15	10	0	10	16	0	6	10	16	3	0	10	16	5	6
11	17	5	0	11	17	7	9	11	17	10	6	11	18	1	3
12	19	0	0	12	19	3	0	12	19	6	0	12	19	9	0
13	20	7	0	13	20	10	3	13	21	1	6	13	21	4	9
14	22	2	0	14	22	5	6	14	22	9	0	14	23	0	6
15	23	9	0	15	24	0	9	15	24	4	6	15	24	8	3
16	25	4	0	16	25	8	0	16	26	0	0	16	26	4	0
17	26	11	0	17	27	3	3	17	27	7	6	17	27	11	9
18	28	6	0	18	28	10	6	18	29	3	0	18	29	7	6
19	30	1	0	19	30	5	9	19	30	10	6	19	31	3	3
20	31	8	0	20	32	1	0	20	32	6	0	20	32	11	0
21	33	3	0	21	33	8	3	21	34	1	6	21	34	6	9
22	34	10	0	22	35	3	6	22	35	9	0	22	36	2	6
23	36	5	0	23	36	10	9	23	37	4	6	23	37	10	3
24	38	0	0	24	38	6	0	24	39	0	0	24	39	6	0

In. long	Ft.	In.	Pa.	S.	In. long	Ft.	In.	Pa.	S.	In. long	Ft.	In.	Pa.	S.	In. long	Ft.	In.	Pa.	S.
1	0	1	7	0	1	0	1	7	3	1	0	1	7	6	1	0	1	7	9
2	0	3	2	0	2	0	3	2	6	2	0	3	3	0	2	0	3	3	6
3	0	4	9	0	3	0	4	9	9	3	0	4	10	6	3	0	4	11	3
4	0	6	4	0	4	0	6	5	0	4	0	6	6	0	4	0	6	7	0
5	0	7	11	0	5	0	8	0	3	5	0	8	1	6	5	0	8	2	9
6	0	9	6	0	6	0	9	7	6	6	0	9	9	0	6	0	9	10	6
7	0	11	1	0	7	0	11	2	9	7	0	11	4	6	7	0	11	6	3
8	1	0	8	0	8	1	0	10	0	8	1	1	0	0	8	1	1	2	0
9	1	2	3	0	9	1	2	5	3	9	1	2	7	6	9	1	2	9	9
10	1	3	10	0	10	1	4	0	6	10	1	4	3	0	10	1	4	5	6
11	1	5	5	0	11	1	5	7	9	11	1	5	10	6	11	1	6	1	3

Qrs. in. long	In.	Pa.	S.	T.	Qrs. in. long	In.	Pa.	S.	T.	Qrs. in. long	In.	Pa.	S.	T.	Qrs. in. long	In.	Pa.	S.	T.
¼	0	4	9	0	¼	0	4	9	9	¼	0	4	10	6	¼	0	4	11	3
½	0	9	6	0	½	0	9	7	6	½	0	9	9	0	½	0	9	10	6
¾	1	2	3	0	¾	1	2	5	3	¾	1	2	7	6	¾	1	2	9	9

Ft. long	20 in. broad.			Ft. long	20¼ in. broad.			Ft. long	20½ in. broad.			Ft. long	20¾ in. broad.		
	Ft.	In.	Pa.		Ft.	In.	Pa.		Ft.	In.	Pa.		Ft.	In.	Pa.
1	1	8	0	1	1	8	3	1	1	8	6	1	1	8	9
2	3	4	0	2	3	4	6	2	3	5	0	2	3	5	6
3	5	0	0	3	5	0	9	3	5	1	6	3	5	2	3
4	6	8	0	4	6	9	0	4	6	10	0	4	6	11	0
5	8	4	0	5	8	5	3	5	8	6	6	5	8	7	9
6	10	0	0	6	10	1	6	6	10	3	0	6	10	4	6
7	11	8	0	7	11	9	9	7	11	11	6	7	12	1	3
8	13	4	0	8	13	6	0	8	13	8	0	8	13	10	0
9	15	0	0	9	15	2	3	9	15	4	6	9	15	6	9
10	16	8	0	10	16	10	6	10	17	1	0	10	17	3	6
11	18	4	0	11	18	6	9	11	18	9	6	11	19	0	3
12	20	0	0	12	20	3	0	12	20	6	0	12	20	9	0
13	21	8	0	13	21	11	3	13	22	2	6	13	22	5	9
14	23	4	0	14	23	6	6	14	23	11	0	14	24	2	6
15	25	0	0	15	25	3	9	15	25	7	6	15	25	11	3
16	26	8	0	16	27	0	0	16	27	4	0	16	27	8	0
17	28	4	0	17	28	8	3	17	29	0	6	17	29	4	9
18	30	0	0	18	30	4	6	18	30	9	0	18	31	1	6
19	31	8	0	19	32	0	9	19	32	5	6	19	32	10	3
20	33	4	0	20	33	9	0	20	34	2	0	20	34	7	0
21	35	0	0	21	35	5	3	21	35	10	6	21	36	3	9
22	36	8	0	22	37	1	6	22	37	7	0	22	38	0	6
23	38	4	0	23	38	9	9	23	39	3	6	23	39	9	3
24	40	0	0	24	40	6	0	24	41	0	0	24	41	6	0

In. long	Ft.	In.	Pa.	S.	In. long	Ft.	In.	Pa.	S.	In. long	Ft.	In.	Pa.	S.	In. long	Ft.	In.	Pa.	S.
1	0	1	8	0	1	0	1	8	3	1	0	1	8	6	1	0	1	8	9
2	0	3	4	0	2	0	3	4	6	2	0	3	5	0	2	0	3	5	6
3	0	5	0	0	3	0	5	0	9	3	0	5	1	6	3	0	5	2	3
4	0	6	8	0	4	0	6	9	0	4	0	6	10	0	4	0	6	11	0
5	0	8	4	0	5	0	8	5	3	5	0	8	6	6	5	0	8	7	9
6	0	10	0	0	6	0	10	1	6	6	0	10	3	0	6	0	10	4	6
7	0	11	8	0	7	0	11	9	9	7	0	11	11	6	7	1	0	1	3
8	1	1	4	0	8	1	1	6	0	8	1	1	8	0	8	1	1	10	0
9	1	3	0	0	9	1	3	2	3	9	1	3	4	6	9	1	3	6	9
10	1	4	8	0	10	1	4	10	6	10	1	5	1	0	10	1	5	3	6
11	1	6	4	0	11	1	6	6	9	11	1	6	9	6	11	1	7	0	3

Qrs. in. long	In.	Pa.	S.	T.	Qrs. in. long	In.	Pa.	S.	T.	Qrs. in. long	In.	Pa.	S.	T.	Qrs. in. long	In.	Pa.	S.	T.
¼	0	5	0	0	¼	0	5	0	9	¼	0	5	1	6	¼	0	5	2	3
½	0	10	0	0	½	0	10	1	6	½	0	10	3	0	½	0	10	4	6
¾	1	3	0	0	¾	1	3	2	3	¾	1	3	4	6	¾	1	3	6	9

Ft. long	21 in. broad.			Ft. long	21¼ in. broad.			Ft. long	21½ in. broad.			Ft. long	21¾ in. broad.		
	Ft.	In.	Pa.		Ft.	In.	Pa.		Ft.	In.	Pa.		Ft.	In.	Pa.
1	1	9	0	1	1	9	3	1	1	9	6	1	1	9	9
2	3	6	0	2	3	6	6	2	3	7	0	2	3	7	6
3	5	3	0	3	5	3	9	3	5	4	6	3	5	5	3
4	7	0	0	4	7	1	0	4	7	2	0	4	7	3	0
5	8	9	0	5	8	10	3	5	8	11	6	5	9	0	9
6	10	6	0	6	10	7	6	6	10	9	0	6	10	10	6
7	12	3	0	7	12	4	9	7	12	6	6	7	12	8	3
8	14	0	0	8	14	2	0	8	14	4	0	8	14	6	0
9	15	9	0	9	15	11	3	9	16	1	6	9	16	3	9
10	17	6	0	10	17	8	6	10	17	11	0	10	18	1	6
11	19	3	0	11	19	5	9	11	19	8	6	11	19	11	3
12	21	0	0	12	21	3	0	12	21	6	0	12	21	9	0
13	22	9	0	13	23	0	3	13	23	3	6	13	23	6	9
14	24	6	0	14	24	9	6	14	25	1	0	14	25	4	6
15	26	3	0	15	26	6	9	15	26	10	6	15	27	2	3
16	28	0	0	16	28	4	0	16	28	8	0	16	29	0	0
17	29	9	0	17	30	1	3	17	30	5	6	17	30	9	9
18	31	6	0	18	31	10	6	18	32	3	0	18	32	7	6
19	33	3	0	19	33	7	9	19	34	0	6	19	34	5	3
20	35	0	0	20	35	5	0	20	35	10	0	20	36	3	0
21	36	9	0	21	37	2	3	21	37	7	6	21	38	0	9
22	38	6	0	22	38	11	6	22	39	5	0	22	39	10	6
23	40	3	0	23	40	8	9	23	41	2	6	23	41	8	3
24	42	0	0	24	42	6	0	24	43	0	0	24	43	6	0

In. long	Ft.	In.	Pa.	S.	In. long	Ft.	In.	Pa.	S.	In. long	Ft.	In.	Pa.	S.	In. long	Ft.	In.	Pa.	S.
1	0	1	9	0	1	0	1	9	3	1	0	1	9	6	1	0	1	9	9
2	0	3	6	0	2	0	3	6	6	2	0	3	7	0	2	0	3	7	6
3	0	5	3	0	3	0	5	3	9	3	0	5	4	6	3	0	5	5	3
4	0	7	0	0	4	0	7	1	0	4	0	7	2	0	4	0	7	3	0
5	0	8	9	0	5	0	8	10	3	5	0	8	11	6	5	0	9	0	9
6	0	10	6	0	6	0	10	7	6	6	0	10	9	0	6	0	10	10	6
7	1	0	3	0	7	1	0	4	9	7	1	0	6	6	7	1	0	8	3
8	1	2	0	0	8	1	2	2	0	8	1	2	4	0	8	1	2	6	0
9	1	3	9	0	9	1	3	11	3	9	1	4	1	6	9	1	4	3	9
10	1	5	6	0	10	1	5	8	6	10	1	5	11	0	10	1	6	1	6
11	1	7	3	0	11	1	7	5	9	11	1	7	8	6	11	1	7	11	3

Qrs. in. long	In.	Pa.	S.	T.	Qrs. in. long	In.	Pa.	S.	T.	Qrs. in. long	In.	Pa.	S.	T.	Qrs. in. long	In.	Pa.	S.	T.
¼	0	5	3	0	¼	0	5	3	9	¼	0	5	4	6	¼	0	5	5	3
½	0	10	6	0	½	0	10	7	6	½	0	10	9	0	½	0	10	10	6
¾	1	3	9	0	¾	1	3	11	3	¾	1	4	1	6	¾	1	4	3	9

[TABLE III.] 307

Ft. long	22 in. broad.			Ft. long	22¼ in. broad.			Ft. long	22½ in. broad.			Ft. long	22¾ in. broad.		
	Ft.	In.	Pa.		Ft.	In.	Pa.		Ft.	In.	Pa.		Ft.	In.	Pa.
1	1	10	0	1	1	10	3	1	1	10	6	1	1	10	9
2	3	8	0	2	3	8	6	2	3	9	0	2	3	9	6
3	5	6	0	3	5	6	9	3	5	7	6	3	5	8	3
4	7	4	0	4	7	5	0	4	7	6	0	4	7	7	0
5	9	2	0	5	9	3	3	5	9	4	6	5	9	5	9
6	11	0	0	6	11	1	6	6	11	3	0	6	11	4	6
7	12	10	0	7	12	11	9	7	13	1	6	7	13	3	3
8	14	8	0	8	14	10	0	8	15	0	0	8	15	2	0
9	16	6	0	9	16	8	3	9	16	10	6	9	17	0	9
10	18	4	0	10	18	6	6	10	18	9	0	10	18	11	6
11	20	2	0	11	20	4	9	11	20	7	6	11	20	10	3
12	22	0	0	12	22	3	0	12	22	6	0	12	22	9	0
13	23	10	0	13	24	1	3	13	24	4	6	13	24	7	9
14	25	8	0	14	25	11	6	14	26	3	0	14	26	6	6
15	27	6	0	15	27	9	9	15	28	1	6	15	28	5	3
16	29	4	0	16	29	8	0	16	30	0	0	16	30	4	0
17	31	2	0	17	31	6	3	17	31	10	6	17	32	2	9
18	33	0	0	18	33	4	6	18	33	9	0	18	34	1	6
19	34	10	0	19	35	2	9	19	35	7	6	19	36	0	3
20	36	8	0	20	37	1	0	20	37	6	0	20	37	11	0
21	38	6	0	21	38	11	3	21	39	4	6	21	39	9	9
22	40	4	0	22	40	9	6	22	41	3	0	22	41	8	6
23	42	2	0	23	42	7	9	23	43	1	6	23	43	7	3
24	44	0	0	24	44	6	0	24	45	0	0	24	45	6	0

In. long	Ft.	In.	Pa.	S.	In. long	Ft.	In.	Pa.	S.	In. long	Ft.	In.	Pa.	S.	In. long	Ft.	In.	Pa.	S.
1	0	1	10	0	1	0	1	10	3	1	0	1	10	6	1	0	1	10	9
2	0	3	8	0	2	0	3	8	6	2	0	3	9	0	2	0	3	9	6
3	0	5	6	0	3	0	5	6	9	3	0	5	7	6	3	0	5	8	3
4	0	7	4	0	4	0	7	5	0	4	0	7	6	0	4	0	7	7	0
5	0	9	2	0	5	0	9	3	3	5	0	9	4	6	5	0	9	5	9
6	0	11	0	0	6	0	11	1	6	6	0	11	3	0	6	0	11	4	6
7	1	0	10	0	7	1	0	11	9	7	1	1	1	6	7	1	1	3	3
8	1	2	8	0	8	1	2	10	0	8	1	3	0	0	8	1	3	2	0
9	1	4	6	0	9	1	4	8	3	9	1	4	10	6	9	1	5	0	9
10	1	6	4	0	10	1	6	6	6	10	1	6	9	0	10	1	6	11	6
11	1	8	2	0	11	1	8	4	9	11	1	8	7	6	11	1	8	10	3

Qrs. in. long	In.	Pa.	S.	T.	Qrs. in. long	In.	Pa.	S.	T.	Qrs. in. long	In.	Pa.	S.	T.	Qrs. in. long	In.	Pa.	S.	T.
¼	0	5	6	0	¼	0	5	6	9	¼	0	5	7	6	¼	0	5	8	3
½	0	11	0	0	½	0	11	1	6	½	0	11	3	0	½	0	11	4	6
¾	1	4	6	0	¾	1	4	8	3	¾	1	4	10	6	¾	1	5	0	9

x 2

Ft. long	23 in. broad.			Ft. long	23¼ in. broad.			Ft. long	23½ in. broad.			Ft. long	23¾ in. broad.		
	Ft.	In.	Pa.		Ft.	In.	Pa.		Ft.	In.	Pa.		Ft.	In.	Pa.
1	1	11	0	1	1	11	3	1	1	11	6	1	1	11	9
2	3	10	0	2	3	10	6	2	3	11	0	2	3	11	6
3	5	9	0	3	5	9	9	3	5	10	6	3	5	11	3
4	7	8	0	4	7	9	0	4	7	10	0	4	7	11	0
5	9	7	0	5	9	8	3	5	9	9	6	5	9	10	9
6	11	6	0	6	11	7	6	6	11	9	0	6	11	10	6
7	13	5	0	7	13	6	9	7	13	8	6	7	13	10	3
8	15	4	0	8	15	6	0	8	15	8	0	8	15	10	0
9	17	3	0	9	17	5	3	9	17	7	6	9	17	9	9
10	19	2	0	10	19	4	6	10	19	7	0	10	19	9	6
11	21	1	0	11	21	3	9	11	21	6	6	11	21	9	3
12	23	0	0	12	23	3	0	12	23	6	0	12	23	9	0
13	24	11	0	13	25	2	3	13	25	5	6	13	25	8	9
14	26	10	0	14	27	1	6	14	27	5	0	14	27	8	6
15	28	9	0	15	29	0	9	15	29	4	6	15	29	8	3
16	30	8	0	16	31	0	0	16	31	4	0	16	31	8	0
17	32	7	0	17	32	11	3	17	33	3	6	17	33	7	9
18	34	6	0	18	34	10	6	18	35	3	0	18	35	7	6
19	36	5	0	19	36	9	9	19	37	2	6	19	37	7	3
20	38	4	0	20	38	9	0	20	39	2	0	20	39	7	0
21	40	3	0	21	40	8	3	21	41	1	6	21	41	6	9
22	42	2	0	22	42	7	6	22	43	1	0	22	43	6	6
23	44	1	0	23	44	6	9	23	45	0	6	23	45	6	3
24	46	0	0	24	46	6	0	24	47	0	0	24	47	6	0

In. long					In. long					In. long					In. long				
	Ft.	In.	Pa.	S.		Ft.	In.	Pa.	S.		Ft.	In.	Pa.	S.		Ft.	In.	Pa.	S.
1	0	1	11	0	1	0	1	11	3	1	0	1	11	6	1	0	1	11	9
2	0	3	10	0	2	0	3	10	6	2	0	3	11	0	2	0	3	11	6
3	0	5	9	0	3	0	5	9	9	3	0	5	10	6	3	0	5	11	3
4	0	7	8	0	4	0	7	9	0	4	0	7	10	0	4	0	7	11	0
5	0	9	7	0	5	0	9	8	3	5	0	9	9	6	5	0	9	10	9
6	0	11	6	0	6	0	11	7	6	6	0	11	9	0	6	0	11	10	6
7	1	1	5	0	7	1	1	6	9	7	1	1	8	6	7	1	1	10	3
8	1	3	4	0	8	1	3	6	0	8	1	3	8	0	8	1	3	10	0
9	1	5	3	0	9	1	5	5	3	9	1	5	7	6	9	1	5	9	9
10	1	7	2	0	10	1	7	4	6	10	1	7	7	0	10	1	7	9	6
11	1	9	1	0	11	1	9	3	9	11	1	9	6	6	11	1	9	9	3

Qrs. in. long					Qrs. in. long					Qrs. in. long					Qrs. in. long				
	In.	Pa.	S.	T.		In.	Pa.	S.	T.		In.	Pa.	S.	T.		In.	Pa.	S.	T.
¼	0	5	9	0	¼	0	5	9	9	¼	0	5	10	6	¼	0	5	11	3
½	0	11	6	0	½	0	11	7	6	½	0	11	9	0	½	0	11	10	6
¾	1	5	3	0	¾	1	5	5	3	¾	1	5	7	6	¾	1	5	9	9

Feet long.	24 in. broad.			Feet long.	24 in. broad.			Inch long.	24 in. broad.				Qrs. in. long.	24 in. broad.			
	Ft.	In.	Pa.		Ft.	In.	Pa.		Ft.	In.	Pa.	S.		In.	Pa.	S.	T.
1	2	0	0	13	26	0	0	1	0	2	0	0	¼	0	6	C	0
2	4	0	0	14	28	0	0	2	0	4	0	0	½	1	0	0	0
3	6	0	0	15	30	0	0	3	0	6	0	0	¾	1	6	C	0
4	8	0	0	16	32	0	0	4	0	8	0	0	...		...		
5	10	0	0	17	34	0	0	5	0	10	0	0	...		...		
6	12	0	0	18	36	0	0	6	1	0	0	0	...		...		
7	14	0	0	19	38	0	0	7	1	2	0	0	...		...		
8	16	0	0	20	40	0	0	8	1	4	0	0	...		...		
9	18	0	0	21	42	0	0	9	1	6	0	0	...		...		
10	20	0	0	22	44	0	0	10	1	8	0	0	...		...		
11	22	0	0	23	46	0	0	11	1	10	0	0	...		...		
12	24	0	0	24	48	0	0	...		...			...		...		

TABLE IV.

WEIGHT OF VARIOUS BUILDING MATERIALS.

Kind of Material.	Per Cubic Foot in lbs.
Bell-metal	547·5
Brass	506·3
Brick, Red	135·5
Do. Stock	115·0
Brickwork	112·0
Concrete	140·0
Copper, Sheet	549·0
Earth, Compact	126·0
Do. Dry	120·0
Do. Shingle	112·0
Do. Sand (dry)	100·0
Do. Dry Clay	120·0
Do. Wet Clay	130·0
Do. Gravel	110·0
Glass, Crown	157·5
Do. Plate	172·5
Granite, Aberdeen	166·5
Do. Cornish	164·0
Do. Dublin	169·6
Do. Guernsey	187·5
Iron, Cast	444·0
Do. Wrought	480·0

Kind of Material.	Per Cubic Foot in lbs.
Lead, Milled	712·9
Limestone, Chalk	144·7
Do. Chilmark	157·4
Do. Kentish Rag	166·6
Do. Lias	154·2
Do. Purbeck	151·0
Magnesian Limestone, Anston	144·0
Do. Bolsover	151·7
Do. Huddlestone	137·8
Marble, Carrara	169·0
Mortar	109·0
Oolites, Ancaster	139·3
Do. Bath (Box)	123·0
Do. do. (Coombe Down)	116·0
Do. Doulting	134·3
Do. Ketton	127·8
Do. Portland	135·6
Sandstone, Calverley	118·1
Do. Cragleith	141·6
Do. Darley-dale	148·3
Do. Dundee	160·1
Do. Heddon	130·7
Do. Huddersfield	153·5
Do. Leeds	142·2
Do. Mansfield, red	148·7
Do. Reigate	103·1
Do. Tisbury	111·1
Slate, Cornish	157·0
Do. Welsh	180·5
Do. Westmoreland	173·0
Steel	490·0
Water	62·5
Wood, Beech, dry	43·1
Do. Elm, seasoned	34·6
Do. Fir, American Spruce	29·1
Do. Fir, Norway Spruce	32·0
Do. Fir, Memel, dry	34·0
Do. Fir, Riga, dry	29·1
Do. Mahogany, Spanish	53·3
Do. Mahogany, Honduras	35·0
Do. Oak, American, red	47·0
Do. Oak, English, seasoned	48·6
Do. Oak, Riga	43·0
Do. Pine, American, dry	23·0
Do. Pine, American Pitch, seasoned	46·3
Do. Teak	46·6
Zinc	439·3

APPENDIX II.

TAKING OFF QUANTITIES FROM PLANS.

It will be seen from the "Preliminary Observations," page 3, that this Treatise was originally intended to teach the art of measuring up Artificers' Works that had actually been executed; it having been a common practice in former days for the Builder to give a "Schedule of prices" from which the work was to be valued when finished, instead of contracting to erect the building for a lump sum. Of late years, however, this system has been almost entirely superseded by that of contract based upon a "Bill of Quantities" taken off the plans and specification either by the Architect himself or by a Surveyor appointed by him. These "Quantities" are sometimes made to form a part of the contract, so that whatever work is done on the building that is not mentioned in the Quantities is charged as an "extra," and whatever is included in the "Quantities" but is not carried out in the building becomes a "deduction." In either case the difference is valued by the Architect and added to or deducted from the amount to be received by the Builder, as the case may be.

The *principles* of measurement are the same whether taken from the plans or from the work actually executed, but there is a wide difference in their practical application to the two cases. When measuring the work done in a building the Surveyor has it all (or nearly all) before his eyes, and has only to measure the work as he finds it executed. But in taking off "Quantities" from the plans and specification, he requires to have a thorough understanding of the various modes of construction, and to be able to correctly interpret the meaning of the drawings and specification; and as it is impossible to show everything on the plans, he must be capable of filling up the deficiencies and providing for all contingencies.

In taking off the measurements of a building of moderate dimensions which is in a single block, the whole of the work of each trade should be completely taken off before proceeding with the next trade. But in large buildings consisting of more than one block it is advisable to take the whole of the several trades in each block together, so that in case of any dispute arising hereafter the Surveyor may know to which part of the building any item may apply. In bringing into Bill, however, the quantities for the different blocks may be brought together, unless it is desired to have a separate estimate for each block.

With a view to assist the student in acquiring a sufficient knowledge of construction to enable him to take off "Quantities" from plans, a large amount of practical information on the technicalities and operations of the several trades was added to the fourth and fifth editions by the Editor. In the present edition a few corrections have been made in the text, and it has been considered advisable to give a complete "Bill of Quantities" to further assist the student in measuring and "bringing into bill" from plans and specification. Portions of such "Bill" have already been given under the various trades, and the following "Bill" is only an amplification of those that appear in the text. The *quantities* themselves are of course only imaginary, and are only inserted to show when the measurement is to be cubic, square, or lineal, or in rods, yards, squares, feet, or inches.

It is frequently the case that the Surveyor charges for taking off the Quantities in the Summary, and he is then to be paid for them by the Contractor out of the first instalment at the rate of $1\frac{1}{2}$ per cent. for large and 2 per cent. for small jobs on the amount of the contract. It is far better and more straightforward for the client to pay the Surveyor himself and not through the Builder, if it can be so arranged; as in any case the cost of taking out the quantities must come out of the client's pocket, in whatever way they are paid for.

BILL OF QUANTITIES of Works to be done in the erection and completion of sundry Buildings at A——, in the County of B——, for X—— Y——, Esq., from Plans and Specification prepared by C—— D——, Esq., Architect.

SUNDRIES.

	£	s.	d.
Hoarding for enclosing the works during their progress			
Tarpaulins for covering up the works to protect them from weather			
Water supply during progress of works . .			
Office for Clerk of Works			
Fees to District Surveyor, &c.			
Fire Insurance until roof is covered in .			
Clearing away rubbish, &c., &c. . . .			
To summary £			

EXCAVATING AND WELL-SINKING.

[The digging for different kinds of soil is to be kept separate. Strutting and planking to trenches may be taken either by foot run, or by yard super., measuring each side, and the ramming to bottom by the yard super., although they are frequently included in the cube digging. Trenches for small drains may be taken by foot run, and the depth stated.]

Yds.	ft.	in.		£	s.	d.
45	1	4	Cube digging and throwing out loose stuff to basement story and cellars .			
105	3	1	Ditto, ditto, in stiff clay (or other kind of earth), not exceeding 5 ft. in depth			
105	3	1	Ditto, ditto, ditto, from 5 to 10 ft. deep			
45	11	4	Ditto, ditto, ditto, to trenches for foundations and drains			
8	10	3	Ditto, filling in only and ramming earth on top of foundations and drains . .			
41	21	11	Ditto, wheeling only, 20 yds. distance			
232	24	6	Ditto, carting only, 1 mile distance .			
21	23	0	Ditto, digging and throwing out to form well, 30 ft. deep and carting 1 mile			
10	25	0	Ditto, ditto, ditto, above 30 ft. deep, 1 mile			
7	4	6	Super. clay tempered and laid over vaults, 6 in. thick and puddled . .			
			Carry forward £			

Yds.	ft.	in.		£	s.	d.
			Brought forward .			
14	4	0	Super. removing top soil 12 in. deep and levelling ground . .			
58	4	3	Ditto, ramming only to bottom of trenches			
25	7	6	Ditto, planking and strutting to trenches			
	95	0	Run planking and strutting to trenches 18 in. wide and 4 ft. 9 in. deep . .			
	30	0	Ditto, well-sinking, including steining in ½ brick, 5 ft. diameter, under 30 ft. deep			
	15	0	Ditto, ditto, steining in 1 brick, 5 ft. diameter, 30 ft. to 45 ft. deep . .			
	10	0	Ditto, boring in clay (or other soil), including tools and tackle, with 3½ in. auger, down to 10 ft. deep .			
	10	0	Ditto, ditto, ditto, 10 to 20 ft. deep . .			
			To summary £			

FOUNDATIONS AND BRICKWORK.

The *Concrete* used in the foundations to be composed of the best freshly-ground lime (or selenitic lime or Portland cement) mixed with broken stones or bricks that will pass through a 2 in. mesh, and clean sharp sand, in the proportion of 1 part lime to 6 of ballast and sand, to be well rammed and levelled after being filled into the trenches.

The *mortar* to be composed of the best stone lime freshly burnt and free from core, mixed with twice its bulk of clean sharp sand. Mortar for pargetting to be mixed with fresh cow dung. *Cement* to be best quality of Portland. *Stock bricks* of best quality to be used in the foundations and walls, except *facings*, which are to be of best *red* or *yellow* malms. No 4 courses to rise more than 12 in. and well flushed up with mortar.

Yds.	ft.	in.		£	s.	d.
16	19	3	Cube lime concrete, as above, filled into trenches, rammed and levelled .			
50	1	9	Super. 6 in. of cement concrete laid over surface of ground and levelled			
Rods	ft.	in.				
20	260	0	Super. reduced stock brickwork as above, in mortar			
5	20	0	Ditto, ditto, ditto, in garden walls, joints struck on both sides . .			
			Carry forward £			

BILL OF QUANTITIES.

				£	s.	d.
Rods	ft.	in.	Brought forward .			
1	15	0	Super. reduced stock brickwork in Portland cement with 2 parts sand .			
	50	6	Ditto, sleeper walls in mortar, 9 in. thick			
Yds.	ft.	in.				
18	5	0	Ditto, bricknogging in mortar, laid flat			
6	3	0	Ditto, paving with malm paviors on edge in mortar			
14	7	0	Ditto, tile paving in cement, ½ in. tiles 6 in. by 6 in.			
115	5	0	Ditto, tuck pointing			
	28	6	Ditto, rough arches in stocks . . .			
	64	9	Ditto, gauged ditto in best malms, set in putty			
	1040	8	Ditto, best red brick facings, extra only on the stock brickwork . .			
	15	6	Ditto, extra only to arches in cement			
	42	0	Ditto, half-brick trimmers . . .			
	80	0	Ditto, cutting to splays			
	214	0	Ditto, damp proof course . . .			
	156	0	Ditto, extra to battering face of wall .			
	90	6	Run cutting to narrow splays . . .			
	170	0	Ditto, ditto, birdsmouth . . .			
	48	9	Ditto, ditto, skewback 4 in. wide . .			
	15	3	Ditto, ditto, and pinning into wall in cement			
	45	0	Ditto, cement filleting			
	40	0	Ditto, raking out and pointing to flashings, ditto			
	665	0	Ditto, stout iron hooping 1¼ in. wide, tarred and sanded and laid in walls as bond			
	8	0	Ditto, 4 in. pipe drain jointed in cement			
	25	6	Ditto, 6 in. ditto, ditto, ditto			
	12	0	Ditto, 9 in. ditto, ditto, ditto			
Nos.						
8			Terra-cotta chimney moulds, 3 ft. high and setting in cement . .			
18			Door and window frames bedded and pointed			
8			Flues cored			
1			9 in. syphon trap to drain . .			
1			6 in. to 9 in. junction ditto . . .			
1			4 in. to 6 in. ditto ditto . .			
2			4 in. bends ditto . . .			
1			6 in. ditto ditto . . .			

Carry forward £

Nos.		£	s.	d.
	Brought forward .			
	Labour and materials to setting . .			
1	4 ft. 6 range			
1	2 ft. copper			
4	Register stoves			
2	Elliptic ditto			
	Cutting away and making good for fixing smith's, gas-fitter's, and bell-hanger's work			
	To summary £			

TILING AND SLATING.

Sqrs.	ft.	in.		£	s.	d.
10	40	0	Super. pantiling bedded in hair and ash mortar			
3	12	0	Ditto plain tiling on double fir lath and wrought nails			
	47	6	Run ridge and hip tiles in mortar .			
9	2	0	Super. Bangor Countess slating to $2\frac{1}{2}$ in. lap, on stout fir laths, two $1\frac{1}{2}$ in. copper nails to each slate . .			
8	58	0	Ditto asphalte felting laid under slates with 1 in. lap, and metal nails . .			
	21	0	Ditto inch sawn slate cistern . . .			
	8	9	Ditto $1\frac{1}{4}$ in. ditto, ditto . . .			
	26	0	Run sawn slate ridge in cement . .			
	18	0	Ditto, ditto, $1\frac{1}{2}$ in. steps, 11 in. wide, with rounded nosing and in. risers 6 in. wide			
Nos.						
4			$\frac{1}{2}$ in. galvanized iron bolts, 2 ft. 6 in. long with nuts and screws . .			
			To summary £			

CARPENTRY.

All fir timber to be best Memel, Dantzic, or Riga, free from sap, shakes, large or loose knots, and well seasoned.

All oak to be of English growth, well seasoned.

[In taking off Quantities, the timber is taken as cube fir (or oak), including labour and nails.]

Ft.	in.		£	s.	d.
50	0	Cube fir (or oak) no labour . . .			
120	0	Ditto, ditto, framed rough in shoring, use and waste			
		Carry forward £			

BILL OF QUANTITIES.

Sqrs.	Ft.	in.		£	s.	d.
			Brought forward .			
	30	0	Cube fir in bond, wall, pole and curb plates			
	65	6	Ditto, ditto, framed			
	45	8	Ditto, ditto, in trusses			
	25	3	Ditto, ditto, ditto, and wrought one (or more) face			
	16	9	Ditto, ditto, ditto, and rebated . .			
	12	7	Ditto, ditto, ditto, and beaded . . .			
	9	9	Ditto, ditto, proper door and window cases			
Sqrs.	ft.	in.				
5	25	6	Super. inch battening plugged to walls, 2¼ in. wide, 9 in. centre to centre .			
9	12	8	Ditto, battening for slates (Countess) .			
3	55	9	Ditto, ¾ in. sound boarding and fillets .			
9	17	5	Ditto, ditto, rough boarding for slates, edges shot			
7	15	6	Ditto, inch ditto, ditto, ploughed and tongued			
5	42	7	Ditto, ditto, wrought batten weather boarding			
1	55	0	Ditto, centering to vaults . . .			
	12	6	Ditto, ditto, trimmer arch . . .			
	36	6	Ditto, inch deal gutter and bearers . .			
	75	0	Ditto, bracketing to cornices . .			
	8	9	Ditto, ditto, ditto, circular on plan . .			
	16	8	Run circular head to door and window head			
	52	6	Ditto, tilting fillet			
	165	9	Ditto, herring-bone strutting . .			
	56	8	Ditto, hip and ridge roll . . .			
	18	0	Ditto, centres to 4 in. gauged flat arches			
	6	6	Ditto, ditto, ditto, circular ditto . .			
	52	3	Ditto, inch feather edge eaves board .			
			To summary £			

JOINERY AND IRONMONGERY.

All Deals to be of best Christiania growth, well seasoned, free from sap, large or loose knots, and other imperfections.

Sqrs.	ft.	in.		£	s.	d.
4	28	9	Super. ¾ in. yellow batten matched and beaded linings, plugged to walls .			
5	36	8	Ditto, inch white deal wrought floor laid folding			
			Carry forward £			

Sqrs.	ft.	in.		£	s.	d.
			Brought forward			
11	22	6	Super. 1¼ in. yellow batten straight joint floor, edge nailed, splayed headings			
6	15	4	Ditto, ditto, ditto, ploughed and tongued with hoop-iron tongues, tongued headings			
7	48	6	Ditto inch straight joint wainscot floor, with tongued headings, edge nailed			
	55	8	Ditto inch solid parquet floor, oak and walnut			
	62	9	Ditto, ditto, square skirting, plugged			
	38	6	Ditto, ditto, torus ditto, ditto			
	17	8	Ditto, ditto, raking, scribed to steps			
	5	6	Ditto, ditto, ramped			
	3	9	Ditto, ditto, writhed			
	28	7	Ditto, moulded skirting, rebated for double plinth			
	75	8	Ditto, ditto, deal keyed and tongued dado			
	26	9	Ditto, ditto, framed and beaded closet front, ploughed for plastering			
	11	9	Ditto, ditto, ditto, ditto, circular on plan			
	29	6	Ditto, 1¼ inch moulded and square spandril framing			
	34	6	Ditto, ditto, framed and beaded grounds ploughed for plastering			
	56	8	Ditto, ditto, rebated and beaded door linings			
	25	7	Ditto, 1¼ in. framed and moulded door linings, 2 panel jambs, and 1 panel soffit			
	17	6	Ditto, 1¼ in. deal tongued, beaded, and ledged door			
	36	9	Ditto, 1½ in. moulded and square door			
	38	4	Ditto, 2 in. ditto, ditto, both sides, ditto			
	21	6	Ditto, 2¼ in. ditto, external door, bolection moulded and bead flush			
	55	9	Ditto, 1¼ in. architrave moulding			
	63	0	Ditto, 2½ in. deal coach-house doors, framed, ledged, braced and filled in with inch deal, ploughed, tongued and beaded			
	39	6	Ditto, 2 in. oak 6 panel bead flush and moulded door			

Carry forward £

BILL OF QUANTITIES. 319

ft.	in.		£	s.	d.
		Brought forward .			
36	0	Super. 2 in. mahogany astragal folding casement			
75	8	Ditto, deal cased frames, oak sunk sills, 2 in. ovolo sash, double hung with brass-cased pulleys, patent lines and iron weights			
82	6	Ditto, ditto, ditto, 2 in. lamb's tongue mahogany sash, double-hung, brass axle-pulleys, brass chains, and lead weights			
9	8	Ditto, inch deal window back in one panel moulded, plain keyed . .			
11	7	Ditto, ditto, square framed backs, elbows and soffits, splayed . .			
13	9	Ditto, 1¼ in. framed, rebated, and beaded boxings, on splay . .			
35	0	Ditto, ditto, 2 panel bead butt and square folding shutters . .			
48	6	Ditto, ditto, treads and inch risers, rounded nosings, glued, blocked and bracketed on 2 fir carriages, mitred to cut string with returned nosings, risers tongued to steps . .			
15	0	Ditto, ditto, ditto, ditto, winders circular at one end . . .			
9	6	Ditto, 1¼ in. rebated and beaded outer framed string, cut and mitred to riser			
3	6	Ditto, ditto, ditto, writhed, glued upright, ditto, ditto . .			
10	8	Ditto, ditto, plain wall string and plugging			
4	6	Ditto, ditto, ditto, ramped . .			
85	4	Run ¾ in. wrought 1 side grounds 2¼ in. wide			
125	9	Ditto, inch deal moulding 3½ in. girt .			
68	6	Ditto, ditto, wrought and framed fillet 2 in. wide			
47	10	Ditto, ditto, square skirting 4½ in. wide			
62	0	Ditto, ditto, ditto, bar balusters .			
5	6	Ditto, 2¼ in. framed newel . .			
12	6	Ditto, 2 in. deal moulded handrail 2¼ in. wide			
2	6	Ditto, ditto, ditto, ramped . .			
10	9	Ditto, 2¼ in. mahogany moulded ditto grooved and beaded . . .			
5	3	Ditto, ditto, ditto, writhe . .			

Carry forward £

Ft.	in.		£	s.	d.
		Brought forward			
11	9	Run external mitre to dado			
15	6	Ditto, beaded and tongued capping to window back			
Nos.					
1		Mahogany seat flap frame skirting and riser to W. C.			
1		Deal, ditto, ditto, ditto			
22		1½ in. deal turned balusters			
1		Curtail end to step			
46		Housings of balusters to steps			
1		Curtail end to mahogany handrail			
1		Deal newel cap, turned			
12		Mitres to 9 in. skirting			
4		Ditto, to 11 in. ditto			
		Cutting away and making good for other trades			

IRONMONGERY AND FIXING.

Nos.				
2	¾ in. brass flush bolts 4 in. long			
2	6 in. iron barrel ditto			
2	⅝ in. brass espagnolette bolts, 5 ft. 6 in. long			
6	Pair 2 in. back-flap hinges and screws			
6	Ditto, 2½ in. wrought iron butts and screws			
8	Ditto, 3½ in. ditto, ditto			
4	Ditto, 4 in. ditto, ditto			
1½	Ditto, 4 in. brass ditto, ditto			
1	Ditto, 10 in. cross garnet hinges			
1	Ditto, 3 ft. cup and ball, coach-house hinges			
1	Wrought-iron swing bar fastening to coach-house doors, with staples, &c.			
2	3 in. cupboard lock			
4	6 in. rimmed lock and brass furniture			
4	6 in. mortice lock			
1	Norfolk thumb-latch			
4	Sets of china and gold furniture			
2	Wrought iron shutter bar and spring catch			
2	China knob and rose			
12	Brass sash fastening			

To summary £

MASONRY.

All the stone to be of the best quality of its kind, perfectly sound in every respect, free from cracks, vents, sandholes, or other defects, and set on its natural or quarry bed in fine mortar.

Yds.	ft.	in.		£	s.	d
36	5	0	Super. rough stone walling 2 ft. 6 in. thick in random (or regular) courses in mortar			
125	7	6	Ditto, ditto, ditto, 2 ft. 3 in. thick, hammer dressed			
	66	8	Ditto, extra labour to external quoins .			
			Ditto, ditto, ditto, internal ditto .			
	75	8	Cube Portland (or Bath) stone and hoisting			
	31	9	Ditto, ditto, ditto, above 30 ft. high .			
	112	6	Super. sawing to Portland (or Bath) .			
	94	8	Ditto, plain work to beds and joints .			
	47	3	Ditto, tooled ditto			
	53	2	Ditto, rubbed ditto			
	43	6	Ditto, sunk ditto			
	29	8	Ditto, moulded ditto			
	23	9	Ditto, circular ditto . . .			
	15	6	Ditto, circular moulded work . .			
	27	3	Ditto, plain circular to columns . .			
	8	6	Ditto, spherical work			
	9	4	4 in. rubbed landing			
	4	6	Run moulded edge to landing . .			
	65	4	Ditto, cutting to groove . . .			
Nos.						
161			Extra only to through or bond stones in rough walling			
24			Holes drilled for iron bars . . .			

YORKSHIRE STONE.

Yds.	ft.	in.				
47	4	6	Super. 2½ in. tooled paving bedded in mortar			
	18	6	Ditto, 4 in. rubbed landing . . .			

Carry forward £

Y

APPENDIX II.

Ft.	in.	Brought forward	£	s.	d.
5	3	Run of moulded edge to 4 in. landing .			
5	3	Ditto, joggled joint to ditto . .			
26	0	Ditto, 2 in. core 18 in. wide . . .			
24	8	Ditto, 3 in. tooled, weathered, and throated sill			
58	6	Ditto, 2½ in. ditto coping 14 in. wide .			
36	9	Ditto, ditto, ditto treads			
40	3	Ditto, 2 in. ditto risers			
Nos.					
18		Holes drilled in York for iron bars .			
2		Bath stone boxed chimney-pieces . .			
4		Veined marble ditto, value 5 gs. ea. p. c.			
2		Statuary ditto, ditto, 20 gs. each ditto			
		To summary £			

PLASTERING.

External Work.

Yds.	ft.	in.		£	s.	d.
108	7	6	Super rendering in Portland cement gauged with 3 parts sand, to outside front of building, and jointing same as masonry			
	55	6	Ditto, moulded cornice . . .			
	23	8	Ditto, ditto, ditto, ditto, circular . .			
	94	4	Ditto, cement skirting			
	80	9	Run 4½in. cement reveals . . .			
	95	8	Ditto, arris in cement			
	75	0	Ditto, moulded string course 6 in. girt do.			
Nos.						
4			External angles to cement skirting .			
10			Cement moulded balusters 18 in. high			

Internal Work.

Yds.	ft.	in.				
45	8	6	Super. rough render on brick . .			
96	3	2	Ditto, render and set ditto . . .			
75	4	7	Ditto, ditto, float and set . . .			
53	6	2	Ditto, lath plaster and set . . .			
78	8	3	Ditto, ditto, ditto, float and set . .			
127	8	6	Ditto, whiting or distemper to ceiling.			
62	5	8	Ditto, lime-whiting to walls . .			
	65	9	Ditto, moulded cornice			
			Carry forward £			

BILL OF QUANTITIES.

Sqrs.	ft.	in.			£	s.	d.
			Brought forward				
4	56	0	Super. pugging under floor . .				
	54	6	Run quirk to angle bead . . .				
	Nos.						
	6		Mitres to cornice 12 in. girt . .				
	6		Ditto, ditto, 18 in. girt . . .				
	2		Centre flowers 3 ft. diameter . .				
			Cutting away and making good for other trades				
			To summary £				

SMITH'S WORK AND BELL-HANGING.

Cwts.	qr.	lb.		£	s.	d.
15	2	8	Rolled iron joists and fixing . .			
7	3	15	Cast iron columns, ditto . . .			
1	2	9	Ditto railings, ditto			
		64	Wrought iron flat bar			
		15	Ditto, ditto, chimney bar, turned up and down			
		18	Lead and running same into stone			
		56	Bolts and straps in timber framing .			
2	1	13	Hoop iron in bond, tarred and sanded			

Yds.	ft.	in.				
18	2	6	Run 6 in. cast-iron half-round eaves gutter, and fixing			
36	1	9	Ditto, 4 in. ditto rain-water pipe, ditto.			
	Nos.					
	4		Stopt ends to eaves-gutter . . .			
	2		Nozzles to ditto			
	2		Rain-water pipe heads, 4 in. diameter.			
	6		Wrought-iron straps to eaves-gutter, bolted to feet of rafters . .			
	1		4 ft. 6 in. kitchener . . .			
	1		2 ft. copper, bearing bars, &c. . .			
	4		Register stoves value 3 gs. ea. p. c. .			
	2		Elliptic, ditto			
	8		Bells hung with cranks, wire, springs, carriages, tubing, &c. . . .			
	2		Lever pulls with china knobs . .			
			To summary £			

APPENDIX II.

PLUMBING AND GAS-FITTING.

Cwts.	qr.	lb.		£	s.	d.
2	3	14	Milled lead cut to dimensions			
14	1	18	Ditto, ditto, in gutters and flats			
3	2	7	Ditto, ditto, hips and ridges			
2	3	9	Ditto, ditto, flashings			
	ft.	in.				
	29	6	Run $\frac{1}{2}$ in. stout lead pipe and fixing			
	22	9	Ditto, $\frac{3}{4}$ in. ditto, ditto			
	15	8	Ditto, 4 in. soil pipe			
	15	9	Ditto $\frac{1}{2}$ in. composition pipe for gas			
	12	6	Ditto $\frac{3}{4}$ in. wrought iron welded gas tubing			
	45	4	Ditto $\frac{2}{5}$ in. block-tin pipe			
Nos.						
3			Solder joints to $\frac{1}{2}$ in. pipe			
2			Ditto, $\frac{3}{4}$ in. ditto			
2			Lead traps 4 in. diameter			
2			W. C. apparatus			
1			$\frac{3}{4}$ in. stop-cock			
2			$\frac{1}{2}$ in. bib-cock			
1			Brass bell trap			
1			Iron bend to $\frac{3}{4}$ in. gas pipe			
3			Ditto, tees, $\frac{1}{2}$ in. to $\frac{3}{8}$ in.			
4			Ditto, $\frac{3}{8}$ in. elbows			
2			Doz. wall hooks			
			To summary £			

PAINTING.

Yds.	ft.	in.		£	s.	d
76	5	6	Super. knotting, stopping, priming, and painting woodwork 3 times in oil and lead colour			
49	3	8	Ditto, ditto, ditto, 4 times and flat on plaster			
97	6	3	Ditto, ditto, 5 times on cement work			
24	7	9	Ditto, ditto, 4 times balusters of stairs			
55	8	2	Ditto, ditto, 5 times iron railing			
33	4	9	Ditto, graining wainscot and twice varnishing			
22	5	7	Run of $4\frac{1}{2}$ in. reveals 5 times in oil			
17	8	3	Ditto, cutting-in party colours to architraves			
11	3	6	Ditto, painting 5 oils to rain-water pipe			
16	5	5	Ditto, ditto, eaves-gutter			
			Carry forward £			

BILL OF QUANTITIES.

Yds.	ft.	in.		£	s.	d.
			Brought forward			
21	2	8	Run painting narrow skirting 3 oils			
	16	0	Ditto, French-polishing handrail . .			
Nos.						
15			Sashes and frames, 4 oils . . .			
9			Dozen squares, ditto			
2			Stone chimney-pieces, 4 oils, grained marble and varnished . . .			
			To summary £			

GLAZING.

Ft.	in.		£	s.	d.
85	6	Super. glazing with best 21 oz. sheet glass			
36	9	Ditto, ditto, best British plate in ordinary squares			
48	0	Ditto, ditto, ditto, in squares, 12 sq. ft.			
13	3	Ditto, ditto, with 26 oz. sheet glass, bent to sweep			
7	6	Run cutting to circular edge of plate glass			
		To summary £			

PAPER-HANGING AND DECORATING.

Yds.	ft.	in.		£	s.	d.
62	7	6	Super. distempering in 2 tints . .			
22	5	8	Ditto, ditto, cornices in 3 tints . .			
	112	6	Ditto, marble mosaic paving, value 7s. per ft.			
	68	9	Run gold moulding and fixing with needle points			
	25	4	Ditto, gilding to ornaments of cornice			
Nos.						
88			Pieces lining paper and hanging walls, previously rubbed down and sized			
32			Ditto, paper, value 1s. per piece, p. c., and hanging			
26			Ditto, ditto, 2s. ditto			
28			Ditto, ditto, 5s. ditto			
16			Ditto, ditto, 6s. ditto			
18			Ditto, marbled paper, value 4s. per piece, hung in blocks, sized and varnished .			
3			Dozen yards borders, value 1d. per yd.			
			To summary £			

SUMMARY.

	£	s.	d.
Sundries			
Excavating and Well-sinking			
Foundations and Brickwork			
Tiling and Slating			
Carpentry			
Joinery and Ironmongery			
Masonry			
Plastering			
Smith's work and Bell-hanging			
Plumbing and Gas-fitting			
Painting			
Glazing			
Paper-hanging and Decorating			
Estimate £			

Surveyor's charge for Quantities (unless paid for by client) 1½ to 2 per cent. on the amount of estimate . . . .

Total £

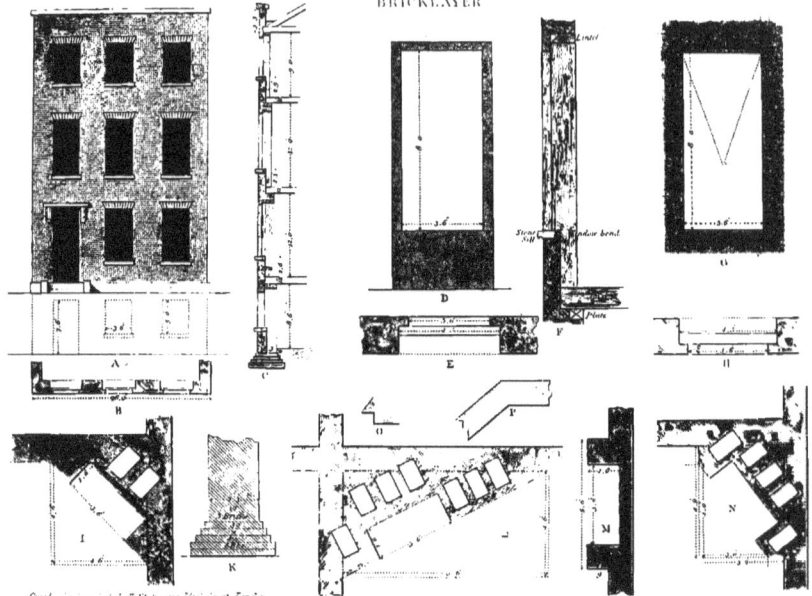

h.va

oa

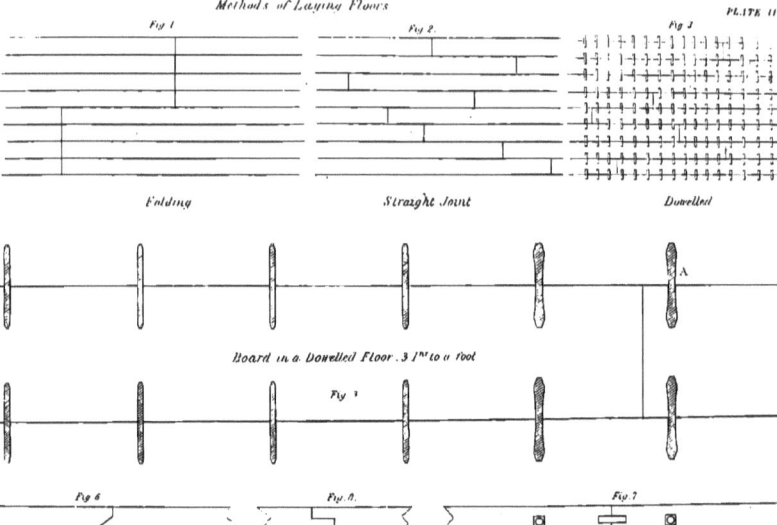

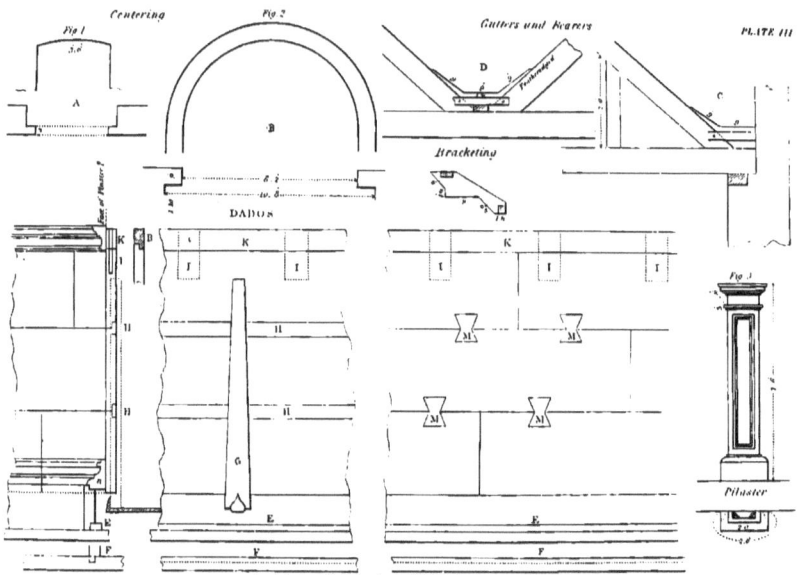

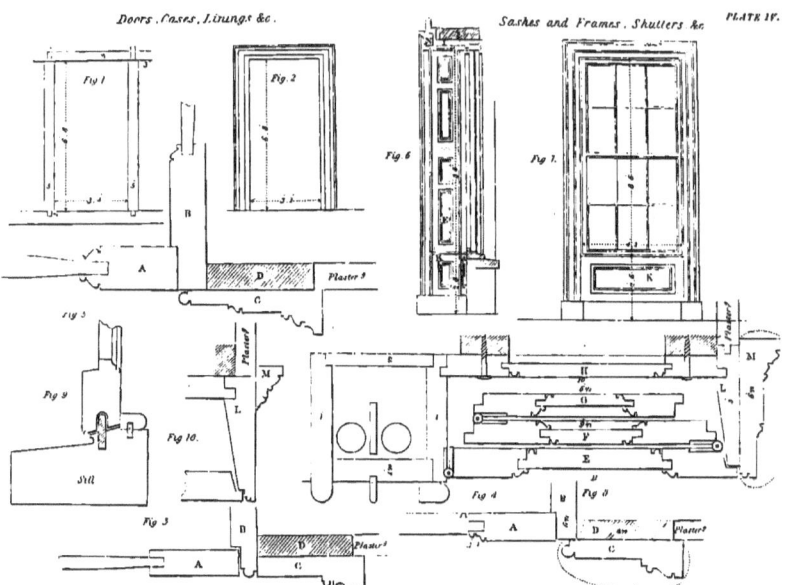

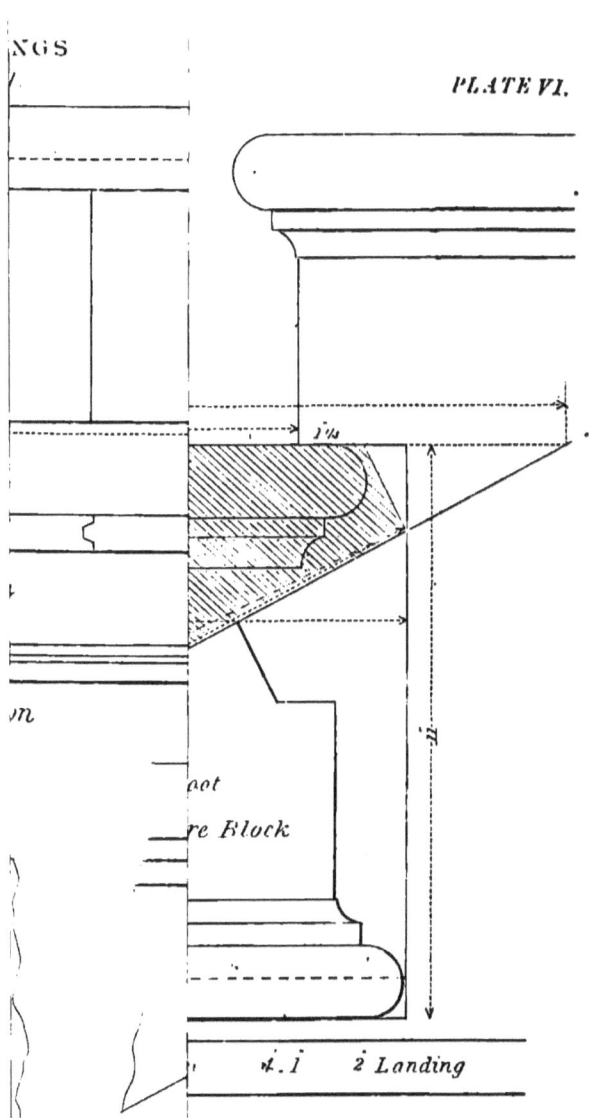

PLATE VI.

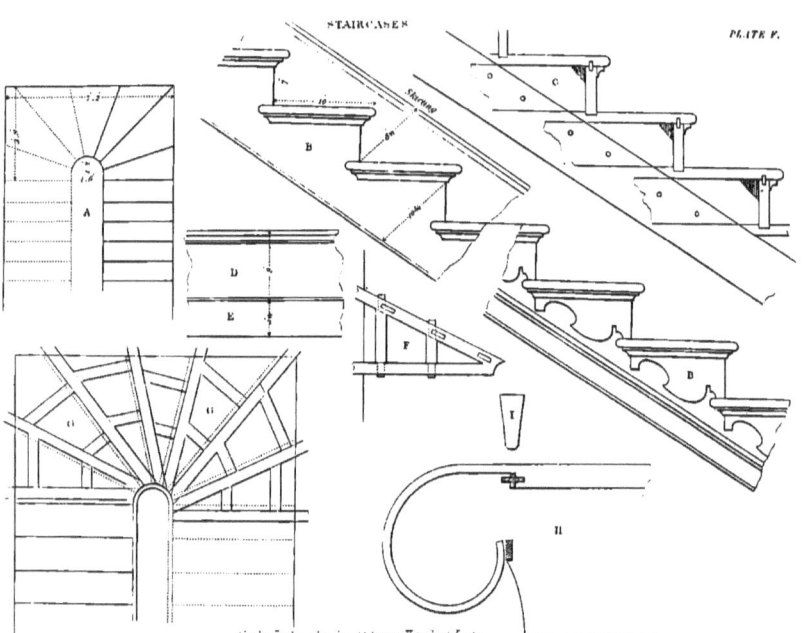

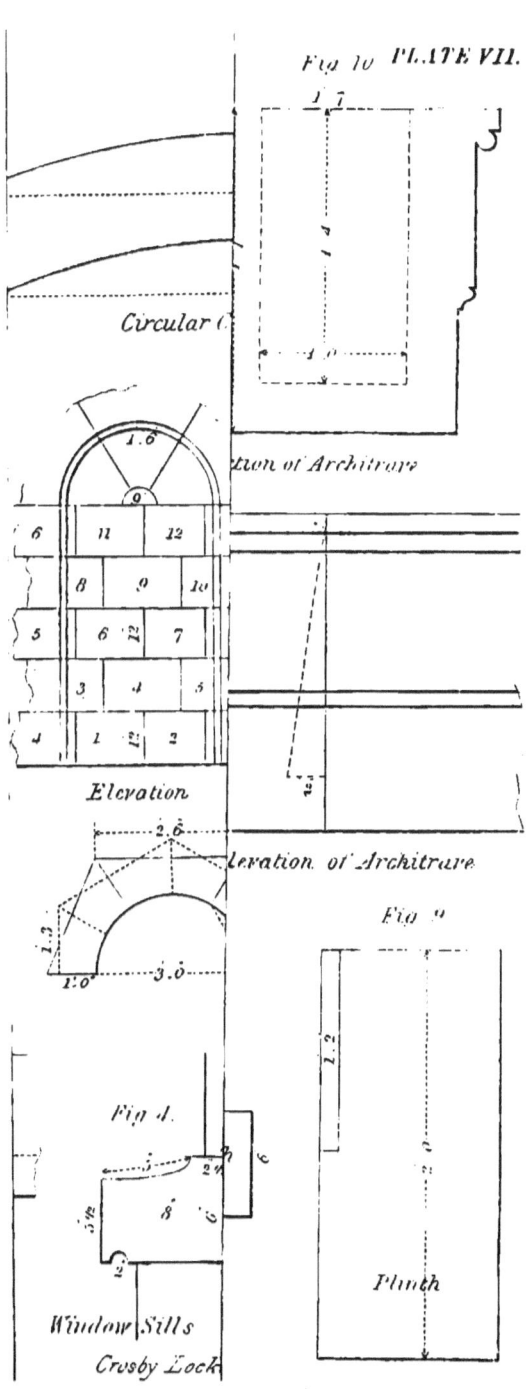

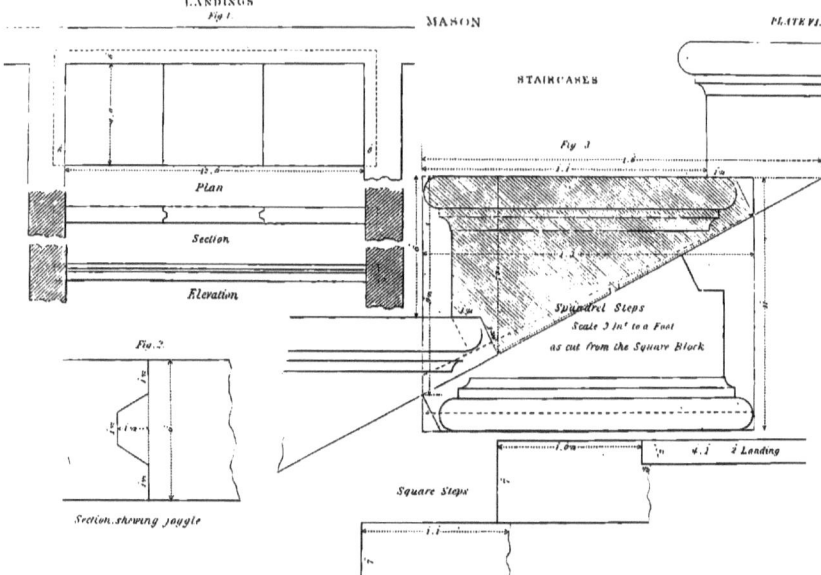

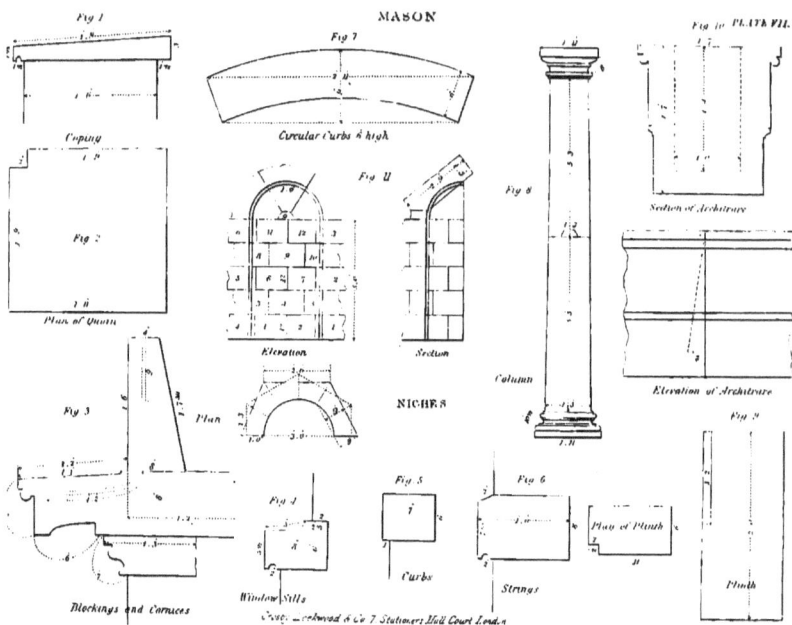

INDEX.

	PAGE
ABBREVIATION	6
——— in Carpenters' and Joiners' work	118
——————— Masons' work	157
————————— Plasterers' work	175
————————— Painters' work	227
Abstracting	7
Abstract of Bricklayers' work	66
——————— Carpenters' and Joiners' work	128
——————— Plasterers' work	177
——————— Painters' work	230
Abutment of arches	150
Acre, a measure	24
Act of Parliament	43
Adulteration of white lead	226
Air	197
Air-bricks	186
Alhambra	171
Ancaster stone	145
Ancient lights, obstruction of	255
Angle chimney	56
Angle-iron, weight of	192
Angle staff	108
Anston stone	146
Apron	107
Apron, lead	214
Arbitration	254
Arc of a circle	18
Arch, stone	149
Architraves	108
——— over columns	163
Area of circles, to find	17
——— table of	26
——— geometrical figures	13

	PAGE
Arris	165
Arris gutters	104
——— rails	92
Artificial stone	51
Ashlar-facing	148
Ashlering to attics	87
————————— measured	103
Asphalte	53
Aubigny stone	146
BACK of a window	108
Back-flaps	108
Balconies	196
Balusters	107
Ball-cocks	217
Banker, masons'	152
Barge-board	90
Bar iron, flat, weight of	193
——— round, ditto	193
Basement story, digging to	27
Bastard stucco	174
Bath, hot water to	220
Bath-stone	145
Batten floors	112
Battens for slating	74
——— widths of	104
Battening walls for plaster	170
Batter to stone walls	148
Battering to brick walls	47
Baulk of timber	93
Bead, in brickwork	58
Bead-butt, bead-flush	114
Bead-and-flat	114
Beam, cast-iron	180
——— wrought-iron	188
Beam-filling	152

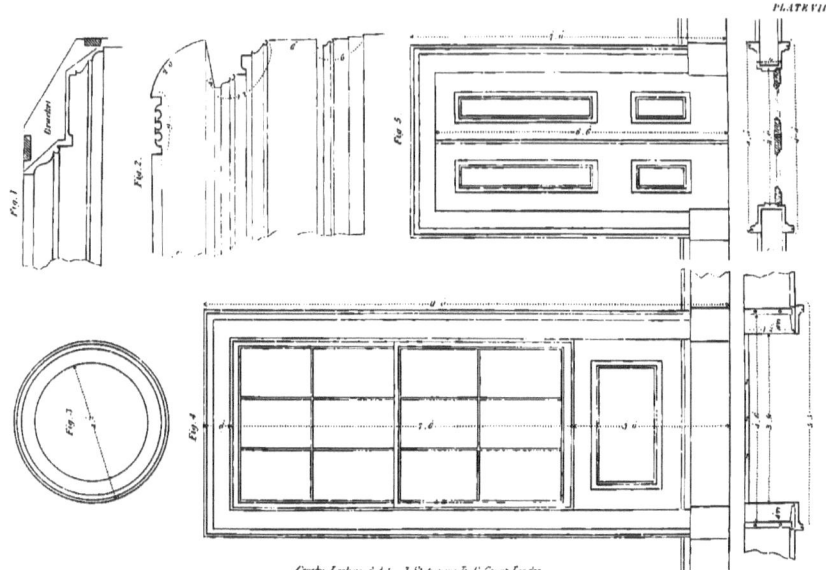

BEA

	PAGE
Bearers to gutters	86
Bearing of joists	78
Beds of brickwork	40
Bell-hanging	209
Bells	209
—— electric	210
—— pneumatic	211
Bending glass	233
Bevil	105
Bib-cocks	217
Binders	80
Bitumen paving	54
Bird's-mouth	49
Blinds	116
Blockings	106
Blue-lias lime	173
Bolsover stone	146
Bolts	206
Bond, in brickwork	41
Bond-timber	50
Borders to paper-hanging	236
Boring for water	32
Box-girders	188
Box stone	145
Boxings to shutters	108
—————— measured	125
Braces to partitions	82
Bracketing to cornices	103
Brass sash-bars for shops	111
Breaking-joint	41
Breaking-weight of timber	94
——————— cast-iron	181
——————— wrought-iron	188
Breast of a chimney	51
Bressummer	81
Brick drains	57
Bricklayers'-work	39
——————— measured	55
Brick-nogging	51
——————— measured	58
Brick-on-edge coping	58
Brick-paving	51
——————— labour on	70
——————— measured	58
Bricks, load of	24
—— sizes of	68
Bridging-joists	80
Bringing-forward painters' work	226
British-plate glass	233
Building Act	28
Building-stones	142
Bulk of earth	29

	PAGE
Bundle of laths	25
Bushel, measure	24

CLI

	PAGE
CABIN hooks	209
Caen stone	146
Canting a corner	111
Canvas	236
Carpenters'-work	78
Carriages to stairs	107
Carting-earth	29
Casements	109
——————— measured	123
Cast-iron	180
——— beams and pillars	181
Cast-lead traps	217
Cathedral-glass	233
Ceiling-joists	80
Ceilings, plastered	170
Cellars, digging to	27
Cement-concrete	37
Cement, Portland	172
Centering for arches	91
Centre of a circle	20
Cesspools	31
Chain, a measure	24
Chamfer, in wood	105
——— in stone	149
Chase	49
Chilmark stone	145
Chimney-bars	191
——— breast	51
——— flues	46
——— shafts, measured	57
Chord of an arc	18
Church bells	209
Circle	18
——— table of area and circumference of	26
Circular iron stairs	196
Cisterns, iron	191
——— lead	214
——— slate	75
Claircolle	173
Clamp of bricks	40
Clamping	106
Claying to vaults	29
Cleaning paint	227
Clearcole	226
Clinch	213
Clinker-bricks	40

CLO	PAGE
Close-boarding	86
Close-string	106
Closers	41
Closet, earth	219
Coach-house doors	113
Cocks	217
Coiling-shutters	202
Collar-beam	84
Collar-roof, Gothic	89
Columns, cast-iron	181
——— stone	161
Concrete	36
Conductor, lightning	203
Cone, solidity of	21
——— surface of	23
Constants of labour	8
Contract	249
Coping, stone	151
——— measured	159
Copper to roofs	220
Coppers	202
Core for handrails	191
——— cornices	172
Coring chimney-flues	49
Cornice, brick	58
——— plaster	170
Corrugated-iron	190
——— zinc	221
Corsehill stone	145
Countess slates	74
Course of bricks	40
Coursed-rubble-work	147
Coves	176
Cowhouses	196
Cradling	103
Cramps	152
Creasing, tile	59
Cross-wall	46
Crown-glass	232
Cubical-content	11
Cubical foot	23
Cubical-measure, table of	265
Cubit, a measure	23
Curb-roof	81
Curbs, stone	161
Curtail-step	107
Cutting bricks	57
Cutting and pinning	151
Cylinder, solidity of	21
——— cast-iron, weight of	184
Cylindrical roof	88

ELL	PAGE
DADO	122
Damp-proof course	47
Dead-knots	93
Deals, hundred of	24
——— width of	104
Decagon, area of	17
Decimals	12
Decorating	235
Delorme's roofing	88
Depeter, depretor	175
Diagonal	14
Digging	27
Dilapidations	241
Dishing	111
Distemper	227
Dodecagon, area of	17
Dog-legged stairs	107
Domical roofing	88
Doorcases	120
Doors	113
——— measured	120
Dormer windows	111
Double-floor	80
Doulting stone	146
Dovetail	106
Dowels for masonry	152
——— floors	112
Dowelled floors	112
Drains	57
Drawn lead pipe	216
Dripping-eaves	58
Drips	213
Dry-steening	31
Dry walling	147
Dubbing-out	172
Duchess slates	75
Dumfries stone	145
Duodecimals	10
Dwelling-houses, walls of	43
EARTH closets	219
Eaves	58
Eaves-board	86
Eaves-gutters	185
Ecclesiastical dilapidations	242
Elbows to windows	108
——— rain-water pipes and gutters	185
Electric-bells	210
——— lights	223
Ell, a measure	24

ELL	PAGE
Ellipse, area of	20
Elliptic-stove	201
Enamelled-slate	238
English-bond	41
Escutcheon	207
Espagnolette bolts	206
Excavating	27
Extrados	150
FACIA, in joinery	111
———— brickwork, measured	58
Facing-bricks	40
Facings of stone	165
———— brickwork, measured	58
Factories, walls of	45
Fall-pipes	184
Fanlight	110
Fathom, a measure	24
Feather-edge fencing	92
Feather-tongue	105
Felt under slating	74
Fen, a measure	24
Fences	92
Fender-walls	48
Filleting	60
Filling-in earth	28
Fir-beams, strength of	94
Fir-framed	99
Fir-pillars, strength of	94
Fire-bricks	40
Fireproof-floors	187
Fished-joint	91
Flagstones	151
Flank-wall	47
Flashing	214
Flat-joint pointing	50
Flatting	226
Flemish bond	42
Flint-walling	148
Flitch-girders	81
Flooring, naked	78
Floors, boarded	111
———— measured	120
Floreutine blinds	117
Flues, size of	46
Flush-bolts	206
Flush-panels	113
Flush-rings	208
Flyers	107
Fodder of lead, a measure	25
Folding floors	112

GRO	PAGE
Folding shutters	115
Foot, lineal, square, cubic	23
Footings of walls	47
Force-pump	218
Forest-pole, a measure	24
Foundations	36
———— of stone walls	147
Framed doors	113
———— floors	80
———— trusses	96
French-polishing	227
———— sashes	109
Fresco-painting	237
Frieze	176
Frustum of a cone	21
Furlong, a measure	24
Furniture	207
Furring-up	79
Furrings	102
GABLE	59
Galvanising	191
Gantry	99
Gas-fitting	222
———— pipes	223
———— stoves	223
Gate hinges	207
Gauge, of slating	74
Gauged-arches, measured	58
———— work	49
General conditions for building contracts	250
Geometrical-pace, a measure	24
Gilding	239
Girders, wood	80
———— wrought-iron	188
Glass, British-plate	233
———— cathedral	233
———— crown	232
———— patent-plate	233
———— rough-plate	233
———— sheet	232
Glaziers'-work	232
Gothic arches and vaulting	150
Graining	227
Granite	142
———— polished	239
Grits, sandstones	144
Groins, measured	58
Grooving	104
Grounds	108

INDEX.

GRO	PAGE
Grouting	51
Gutter-boards	86
Gutters, iron	185
—— lead	214
Gypsum concrete	51
HACKING earth	28
Hacking-out glass	234
Half-space of stairs	107
Halving collar to rafters	84
—— wall plates	91
Hammer-beam roof	89
Hammer-dressed stone	147
Hand, a measure	23
Handrail	107
Haunches of arch	150
Head of fall-pipe	185
Header in brickwork	41
Heading-joints	112
Helioscene	117
Herring-bone strutting	79
——————walling	47
Hexagon	16
Hinges	206
Hip-hooks	60
—— roof	84
—— tiles	59
Hips	58
Hod of bricks	68
Hoisting stone	151
Hollow cylinder, solidity of	22
Hollow iron cylinders	39
Hollow-walls	47
Hoop-iron bond	50
Hoppus' tables	261
Horizontal-thrust of arches	150
Hot-water apparatus	198
———— to baths	220
House-bells	210
Housing	106
Hundred, a measure	24
Hung-sashes	109
Hydraulic lifts	204
I-JOISTS	192
Imitations of columns and pilasters	140
Inch, lineal, square, cubic	23
Intrados	150
Inverts	48

LIF	PAGE
Iron	180
Ironmongery	205
Irregular polygon, area of	15
JACK-rafters	84
Jalousies	117
Joggled-joints	149
Joiners'-work	104
Joists, wood	78
———— rolled-iron	186
———————— weight of	192
KEY, master	208
Key, for paint	225
———— plaster	170
Key-stone of arch	150
Kiln, brick	40
King-post roof	84
Kitchener	201
Knotting	225
LABOUR, constants of	8
———— calculation of, to bricklayers' work	70
———————— carpenters' work	133
———————— joiners' work	135
———————— masons' work	166
———————— painters' work	232
———————— plasterers' work	179
Lamb's-tongue moulding	110
Landings, stone	151
———— measured	158
Lap of slates	74
Latches	208
Lath-and-plaster	174
Lathing	170
Laths, bundle of	25
Lead gutters	214
—— hundred of	25
—— lights	234
—— milled	212
—— pipes	216
Lean-to roof	82
Ledged-doors	113
———— valued	137
Letters, valued	228
Lewis	152
Lifting shutters	115
Lifts	204

INDEX.

LIG

	PAGE
Lightning-conductor	203
Lime, load of	24
Lime-concrete	37
Limestone	145
Lime-whiting	173
Lineal inch, foot, yard	23
Linings of doors	115
———————— measured	121
Linseed oil	226
Lintels	81
Load of materials	24
Locks	207
Looking-glasses	239
Loose box	195

	PAGE
MAGNESIAN limestone	146
Malm-bricks	39
Mansarde-roof	87
Mansfield-stone	145
Marble in decoration	238
——— labour to	166
Marble-cement	171
Masonry	142
Mastic	226
Matched-boarding	109
Measurement of artificers' work	5
——— ——— brickwork	55
——— ——— carpentry	99
——— ——— digging	33
——— ——— glazing	234
——— ——— joinery	119
——— ——— masonry	153
——— ——— obstruction of ancient lights	255
——— ——— painting	228
——— ——— plastering	175
——— ——— slating	76
——— ——— tiling	60
Mensuration	10
Meters, gas	223
Mile, a measure	24
Milled-lead	212
——— weight of	215
Mitre	105
Mitred border	105
——— measured	120
Models for iron castings	180
Mortice and tenon in carpentry	79
——— joinery	104
Mosaic decoration	237
Moulded bricks	40

PAV

	PAGE
Moulded work to stone	149
Mouldings in joinery, measured	120

	PAGE
NAILS, hundred of	25
——— varieties of	205
Naked-flooring	78
Natural-bed of sandstones	144
Needling	98
Newel-post	107
Niches, measuring	164
Nogging to partitions	51
——————— measured	69
Nosing	106
Nozzles to iron eaves-gutters	185

	PAGE
OAK-beams, strength of	94
— pillars, ditto	94
— trusses	102
Oblong, area of	13
Obstruction of light	255
Octagon	16
Offsets to footings	47
Ogee moulding	114
Oil, linseed	225
Oolites	145
Ornamental tiling	60
Ornaments, plaster	175
Outer-string of stairs	106
Ovens	57
Ovolo-moulding	110

	PAGE
PACE, a measure	24
Painswick stone	146
Painters'-work	225
Paling for fences	92
Palm, a measure	23
Panel-doors and framing	113
Pantiling	59
Paper-hanging	235
Parallelogram, area of	14
Parallelopiped	21
Pargeting	48
Parian-cement	171
Parquetry floors	113
Parting-slip	109
Partitions, quartered	82
Party-walls	46
Patent plate-glass	233
Paving bricks	40

INDEX.

PAV

	PAGE
Paving bricks, sizes of	68
Pedestal stoves	198
Perch, a measure	24
Permanent-load on wood beams	94
——————————— pillars	94
——————————— cast-iron beams	181
——————————— pillars	182
——————————— wrought-iron beams	188
Piggeries	196
Pilasters, wood, measured	122
Piling	38
Pillars, iron, strength of	181
——— timber, strength of	94
Pinning into walls	151
Pipe-drains	57
Pipes, lead	216
——— gas	223
Pitch of a roof	87
Pitching-piece to stairs	107
Place-bricks	40
Plain-tiles, load of	24
Plain-tiling	59
Planking	29
Planks, load of	24
——— width of	104
Plaster-of-paris	171
Plastering	170
Plates, wall	79
——— pole	86
Plate-glass, British	233
——————— patent	233
Plinth, wood, measured	122
——— stone, ditto	160
Ploughing	104
Plumbers'-work	212
Pneumatic bells	211
Pointing	50
Pole, a measure	24
Pole-plate	86
Polished granite	239
——— marble	238
Polishing, French	227
Polygons	15
Portland-cement	172
——— stone	145
Preservation of stone from decay	146
Pricking-up	173
Priming	225
Principal	85
Prism	21

ROD

	PAGE
Prison-cells, warming	199
Puddling with clay	29
Pugging	175
Pulley-pieces	109
Pulling down old walls	58
Pumicing	226
Pumping water from wells	32
Pumps	218
Purbeck stone	145
Purlins	86
Putty	233
QUANTITIES, taking off from Plans	311
Quarter-space of stairs	107
Quartered partitions	82
Quarters	82
Queen-post roof	85
Quirked mouldings in joinery	114
Quirks in brickwork	58
——— mouldings	114
Quoin-stones	143
RAFTERS	83
Rag-stone	145
Railings, painting	228
Rails of framing	113
Rain-water-pipes	184
Raking out joints	58
Ramming earth	33
Ramp	107
Ranges, cooking	201
Rebating	104
Red bricks	40
——— sandstones	145
Reduced-brickwork	55
Register-stoves	201
Regular polygon	15
Rendering	173
Repairs	243
Retaining walls	52
Reveals of windows	49
Revolving-shutters	202
Ribs of vaulting	150
Ridge-piece	83
——— roll	86
——— tiles	59
Right-solid, solidity of	21
Ring, area of	18
Riveted beams	188
Risers	106
Rod of brickwork	55

ROD		SQU	
	PAGE		PAGE
Rod of fencing	92	Semi-circular arch	150
Rolled glass	233	Setting to plaster	174
Roller-blinds	116	—— stoves	58
Rood, a measure	24	Sham-fronts	201
Roofs, timber	82	Shed-roof	82
—— iron	188	Sheet-glass	232
Rotation in measuring	6	—— iron, weight of	193
———————— brickwork	67	Shoes to rain-water-pipes	185
———————— carpentry	130	Shooting edges	105
———————— digging	35	Shop fronts	111
———————— masonry	168	—— shutters	115
———————— painting	228	Shoring	97
———————— plastering	178	Shutter-bars and lifts	208
Rough-boarding, valuation of	136	Shutters	115
Rough-cast	174	Sideling-earth, measured	30
Rough plate-glass	233	Silicate of lime	147
Rough timber, load of	24	Sills of window-frames	109
Rubbing-down painters' work	226	—— stone	152
Rubble-walling	147	Silvering glass	239
Run of wheeling earth	28	Single-joisted floor	78
		Sinks	152
		Skew-backs	49
SADDLE-backed coping	151	—— to stone arch	150
—— bars	192	Sketches	117
Sand, load of	24	Skirtings	108
Sandstones	144	———— measured	122
Sap-wood	93	———— valued	137
Sash-bars	110	Skylights	111
Sash fastenings	209	Slate, enamelled	238
Sash-lines	209	Slates, thousand of	25
Sashes and frames	109	Slating	73
———————— measured	123	Sleepers	79
———————— valued	139	Sleeper-walls	47
Sawing stone	149	Smiths'-work	180
Sawn slate	75	Soakers	214
Sawyer	141	Solder	214
Scaffolding	99	Solid measure, table of	265
Scagliola	237	Solidity	21
Scantlings of binders	80	Sound-boarding	99
———————— girders	81	Span, a measure	23
———————— joists	79	Spandril-steps	153
———————— timbers of roofs	97	Spanish blinds	117
Scarfing timber	91	Sphere, surface and solidity of	23
Scribing	122	Splays in brickwork	50
Screw piling	39	—— of stone	149
Scroll to curtail step	107	Spring hinges	207
Seamless soil pipes	216	Square, a measure	24
Sector of circle	18	Square, area of	14
Segment, ditto	19	Square inch, foot, yard	23
Segmental arch	150	Square of unequal-sided timber,	
Selenitic cement	172	table of	261
Self-coiling shutters	202	Square-framing	114

INDEX. 335

SQU	PAGE
Squaring dimensions	10
Stable-fittings	193
Staining woodwork	227
Staircase, wood	106
———————— measured	125
——————— iron, ditto	196
——————— stone, ditto	157
Stall-board to shops	111
Steining to wells	31
Stepped-flashing	214
Steps, square stone, measuring	158
———— spandril	154
Stiles of framing	113
Stock-bricks	39
Stop-cocks	217
Stones, building	142
Stopped-chamfer	114
Story-posts, iron	183
Stoves	201
———— open	201
———— close	198
Straight-joint floor	112
Straining-beam	85
Strength of timber	93
Stretcher-bricks	40
Stretching-force in timber	96
———————————— wrought-iron	186
Striking joints of brickwork	50
String-courses, measuring	160
String of stairs	106
Struts	85
Strutting to digging	29
Stucco, external	172
———— trowelled, internal	174
———— bastard, internal	174
Suction-pump	218
Summary	240
Sunburner, ventilator	200
Sunk-work in stone, valuation of	166
Superficial measure, table of	234
Syphon-ventilator	200
———— trap	217

	PAGE
TABLE of area and circumference of circles	25
——————— cast-iron cylinders	184
——————— flat measure	284
——————— glass	25
——————— measures	23
——————— slaters' work	75
——————— solid measure	265

VAL	PAGE
Table of square of unequal-sided timber	261
——————— value of brickwork	73
——————— weight of various building materials	309
Tanks, iron	197
Tar paper	236
Tar paving	54
Tee-iron	192
Tenon	104
Terra-cotta	54
Thickness of brick walls	43
——————— floor-boards	111
Thousand of slates, &c.	25
Throating	149
Through-stones	148
Throwing-out earth	27
Thumb-latches	208
Tie-beam	84
Tiling	58
Tile-creasing	59
———— paving	60
Tiles, per square	70
Tilting-fillet	86
Timber	92
Tinfoil	236
Tin-lined pipes	216
Ton of iron	25
Tongued-floors	112
Tongueing	105
Transverse strength of stone	168
Trap-doors	215
Trapezium, area of	14
Traps	217
Treads	106
Trenches for foundations	27
Triangle	13
Trimmer-arch	48
Trimming-joist	79
Trowelled-stucco	174
Truss	85
Tuck-pointing	50
Turning-pieces	91

	PAGE
UNCOURSED walling	147
Under-pinning	51
Unequal-sided timber, table of	261

	PAGE
VALLEY of a roof	59
———— boards	84

VAL

	PAGE
Valley-tiles	60
Valuation	7
———— of brickwork	68
———— ———— carpenters' work	132
———— ———— masonry	166
———— ———— painting	231
———— ———— plastering	178
———— ———— repairs	248
———— ———— slating	75
Varnishing	227
Vaults, measured	56
Venetian-blinds	116
———— windows	110
Ventilation through ceilings	171
———— of rooms	199
Versed-sine of an arc	19
Voussoirs of an arch	149
V-roof	83

	PAGE
WAINSCOT floors	112
Walling, stone	147
Wall-plates	79
Wall-strings	106
Walls, brick	39
———— thickness of	43
Warehouse-walls, thickness of	45
Warming apparatus	198
———— rooms	197
Waste preventers	219
Water-closet apparatus	218
———— fittings	117
———— trunks	104
Weather-boarding	98
———— valued	135
Weight of cast-iron	183
———— timber	132
———— wrought-iron	191
———— various building materials	309

ZIN

	PAGE
Well-hole of stairs	107
Well-sinking	31
Welsh slates	73
Westmoreland slates	74
Wheeling-earth	28
White-lead	226
Whitening	173
Winders of stairs	107
Window-back	108
Withe of a flue	46
Wood bricks	99
Woodland-pole, a measure	24
Wreathe of a handrail	107
Wrought-iron	186
———— joists	187
———— ———— weight of	192
———— girders	188
———— roofs	188
———— tanks	191
———— angle-iron, weight of	192
———— flat-bar iron, ditto	193
———— round ditto, ditto	193
———— sheet ditto, ditto	193
———— tee-iron, ditto	192

	PAGE
YARD, a measure	23
Yorkshire stone	151

	PAGE
ZINC	221
———— corrugated	221
———— gutters	222
———— perforated	222
———— rain-water pipes	222
———— roofs	221
———— white	226
———— working	221

THE END.

7, Stationers' Hall Court, London, E.C.
February, 1893.

A CATALOGUE OF BOOKS
INCLUDING NEW AND STANDARD WORKS IN
ENGINEERING: CIVIL, MECHANICAL, AND MARINE;
ELECTRICITY AND ELECTRICAL ENGINEERING;
MINING, METALLURGY; ARCHITECTURE,
BUILDING, INDUSTRIAL AND DECORATIVE ARTS;
SCIENCE, TRADE AND MANUFACTURES;
AGRICULTURE, FARMING, GARDENING;
AUCTIONEERING, VALUING AND ESTATE AGENCY;
LAW AND MISCELLANEOUS.

PUBLISHED BY

CROSBY LOCKWOOD & SON.

MECHANICAL ENGINEERING, etc.

D. K. Clark's Pocket-Book for Mechanical Engineers.
THE MECHANICAL ENGINEER'S POCKET-BOOK OF TABLES, FORMULÆ, RULES AND DATA. A Handy Book of Reference for Daily Use in Engineering Practice. By D. KINNEAR CLARK, M.Inst.C.E., Author of "Railway Machinery," "Tramways," &c. Second Edition, Revised and Enlarged. Small 8vo, 700 pages, 9s. bound in flexible leather covers, with rounded corners and gilt edges. [*Just published.*

SUMMARY OF CONTENTS.

MATHEMATICAL TABLES.—MEASUREMENT OF SURFACES AND SOLIDS.—ENGLISH WEIGHTS AND MEASURES.—FRENCH METRIC WEIGHTS AND MEASURES.—FOREIGN WEIGHTS AND MEASURES.—MONEYS.—SPECIFIC GRAVITY. WEIGHT AND VOLUME—MANUFACTURED METALS.—STEEL PIPES.—BOLTS AND NUTS.—SUNDRY ARTICLES IN WROUGHT AND CAST IRON, COPPER, BRASS, LEAD, TIN, ZINC.—STRENGTH OF MATERIALS.—STRENGTH OF TIMBER.—STRENGTH OF CAST IRON.—STRENGTH OF WROUGHT IRON.—STRENGTH OF STEEL.—TENSILE STRENGTH OF COPPER, LEAD, ETC.—RESISTANCE OF STONES AND OTHER BUILDING MATERIALS.—RIVETED JOINTS IN BOILER PLATES.—BOILER SHELLS—WIRE ROPES AND HEMP ROPES.—CHAINS AND CHAIN CABLES.—FRAMING.—HARDNESS OF METALS, ALLOYS AND STONES.—LABOUR OF ANIMALS.—MECHANICAL PRINCIPLES.—GRAVITY AND FALL OF BODIES.—ACCELERATING AND RETARDING FORCES.—MILL GEARING, SHAFTING, ETC.—TRANSMISSION OF MOTIVE POWER.—HEAT.—COMBUSTION: FUELS.—WARMING, VENTILATION, COOKING STOVES.—STEAM.—STEAM ENGINES AND BOILERS.—RAILWAYS—TRAMWAYS.—STEAM SHIPS.—PUMPING STEAM ENGINES AND PUMPS.—COAL GAS, GAS ENGINES, ETC.—AIR IN MOTION.—COMPRESSED AIR.—HOT AIR ENGINES.—WATER POWER.—SPEED OF CUTTING TOOLS.—COLOURS.—ELECTRICAL ENGINEERING.

* OPINIONS OF THE PRESS.

"Mr. Clark manifests what is an innate perception of what is likely to be useful in a pocket-book, and he is really unrivalled in the art of condensation. . . . It is very difficult to hit upon any mechanical engineering subject concerning which this work supplies no information, and the excellent index at the end adds to its utility. In one word, it is an exceedingly handy and efficient tool, possessed of which the engineer will be saved many a wearisome calculation, or yet more wearisome hunt through various text-books and treatises, and, as such, we can heartily recommend it to our readers, who must not run away with the idea that Mr. Clark's Pocket-book is only Molesworth in another form. On the contrary, each contains what is not to be found in the other; and Mr. Clark takes more room and deals at more length with many subjects than Molesworth possibly could."—*The Engineer*, Sept 16th, 1892.

"Just the kind of work that practical men require to have near to them."—*English Mechanic.*

MR. HUTTON'S PRACTICAL HANDBOOKS.

Handbook for Works' Managers.

THE WORKS' MANAGER'S HANDBOOK OF MODERN RULES, TABLES, AND DATA. For Engineers, Millwrights, and Boiler Makers; Tool Makers, Machinists, and Metal Workers; Iron and Brass Founders, &c. By W. S. HUTTON, Civil and Mechanical Engineer, Author of "The Practical Engineer's Handbook." Fourth Edition, carefully Revised and partly Re-written. In One handsome Volume, medium 8vo, price 15s. strongly bound.

☞ *The Author having compiled Rules and Data for his own use in a great variety of modern engineering work, and having found his notes extremely useful, decided to publish them—revised to date—believing that a practical work, suited to the* DAILY REQUIREMENTS OF MODERN ENGINEERS, *would be favourably received.*
In the Fourth Edition the First Section has been re-written and improved by the addition of numerous Illustrations and new matter relating to STEAM ENGINES *and* GAS ENGINES. *The Second Section has been enlarged and Illustrated, and throughout the book a great number of emendations and alterations have been made, with the object of rendering the book more generally useful.*

*** OPINIONS OF THE PRESS.

"The author treats every subject from the point of view of one who has collected workshop notes for application in workshop practice, rather than from the theoretical or literary aspect. The volume contains a great deal of that kind of information which is gained only by practical experience, and is seldom written in books."—*Engineer.*

"The volume is an exceedingly useful one, brimful with engineers' notes, memoranda, and rules, and well worthy of being on every mechanical engineer's bookshelf."—*Mechanical World.*

"A formidable mass of facts and figures, readily accessible through an elaborate index ... Such a volume will be found absolutely necessary as a book of reference in all sorts of 'works' connected with the metal trades."—*Ryland's Iron Trades Circular.*

"Brimful of useful information, stated in a concise form, Mr. Hutton's books have met a pressing want among engineers. The book must prove extremely useful to every practical man possessing a copy."—*Practical Engineer.*

New Manual for Practical Engineers.

THE PRACTICAL ENGINEER'S HAND-BOOK. Comprising a Treatise on Modern Engines and Boilers: Marine, Locomotive and Stationary. And containing a large collection of Rules and Practical Data relating to recent Practice in Designing and Constructing all kinds of Engines, Boilers, and other Engineering work. The whole constituting a comprehensive Key to the Board of Trade and other Examinations for Certificates of Competency in Modern Mechanical Engineering. By WALTER S. HUTTON, Civil and Mechanical Engineer, Author of "The Works' Manager's Handbook for Engineers," &c. With upwards of 370 Illustrations. Fourth Edition, Revised, with Additions. Medium 8vo, nearly 500 pp., price 18s. Strongly bound.

☞ *This work is designed as a companion to the Author's "*WORKS' MANAGER'S HAND-BOOK.*" It possesses many new and original features, and contains, like its predecessor, a quantity of matter not originally intended for publication, but collected by the author for his own use in the construction of a great variety of* MODERN ENGINEERING WORK.
The information is given in a condensed and concise form, and is illustrated by upwards of 370 Woodcuts; and comprises a quantity of tabulated matter of great value to all engaged in designing, constructing, or estimating for ENGINES, BOILERS, *and* OTHER ENGINEERING WORK.

*** OPINIONS OF THE PRESS.

"We have kept it at hand for several weeks, referring to it as occasion arose, and we have not on a single occasion consulted its pages without finding the information of which we were in quest."—*Athenæum.*

"A thoroughly good practical handbook, which no engineer can go through without learning something that will be of service to him."—*Marine Engineer.*

"The author has collected together a surprising quantity of rules and practical data, and has shown much judgment in the selections he has made. . . . There is no doubt that this book is one of the most useful of its kind published, and will be a very popular compendium."—*Engineer.*

"A mass of information, set down in simple language, and in such a form that it can be easily referred to at any time. The matter is uniformly good and well chosen and is greatly elucidated by the illustrations. The book will find its way on to most engineers' shelves, where it will rank as one of the most useful books of reference."—*Practical Engineer.*

"Full of useful information and should be found on the office shelf of all practical engineers."—*English Mechanic.*

MR. HUTTON'S PRACTICAL HANDBOOKS—continued.

Practical Treatise on Modern Steam-Boilers.

STEAM-BOILER CONSTRUCTION. A Practical Handbook for Engineers, Boiler-Makers, and Steam Users. Containing a large Collection of Rules and Data relating to Recent Practice in the Design, Construction, and Working of all Kinds of Stationary, Locomotive, and Marine Steam-Boilers. By WALTER S. HUTTON, Civil and Mechanical Engineer, Author of "The Works' Manager's Handbook," "The Practical Engineer's Handbook," &c. With upwards of 300 Illustrations. Second Edition. Medium 8vo, 19s. cloth. [*Just published.*

☞ THIS WORK is issued in continuation of the Series of Handbooks written by the Author, viz :—"THE WORKS' MANAGERS' HANDBOOK" and "THE PRACTICAL ENGINEER'S HANDBOOK," which are so highly appreciated by Engineers for the practical nature of their information; and is consequently written in the same style as those works.

The Author believes that the concentration, in a convenient form for easy reference, of such a large amount of thoroughly practical information on Steam-Boilers, will be of considerable service to those for whom it is intended, and he trusts the book may be deemed worthy of as favourable a reception as has been accorded to its predecessors.

*** OPINIONS OF THE PRESS.

"Every detail, both in boiler design and management, is clearly laid before the reader. The volume shows that boiler construction has been reduced to the condition of one of the most exact sciences; and such a book is of the utmost value to the *fin de siècle* Engineer and Works' Manager."—*Marine Engineer.*

"There has long been room for a modern handbook on steam boilers; there is not that room now, because Mr. Hutton has filled it. It is a thoroughly practical book for those who are occupied in the construction, design, selection, or use of boilers."—*Engineer.*

"The book is of so important and comprehensive a character that it must find its way into the libraries of everyone interested in boiler using or boiler manufacture if they wish to be thoroughly informed. We strongly recommend the book for the intrinsic value of its contents."—*Machinery Market.*

"The value of this book can hardly be over-estimated. The author's rules, formulæ, &c., are all very fresh, and it is impossible to turn to the work and not find what you want. No practical engineer should be without it."—*Colliery Guardian.*

Hutton's "Modernised Templeton."

THE PRACTICAL MECHANICS' WORKSHOP COMPANION. Comprising a great variety of the most useful Rules and Formulæ in Mechanical Science, with numerous Tables of Practical Data and Calculated Results for Facilitating Mechanical Operations. By WILLIAM TEMPLETON, Author of "The Engineer's Practical Assistant," &c. &c. Sixteenth Edition, Revised, Modernised, and considerably Enlarged by WALTER S. HUTTON, C.E., Author of "The Works' Manager's Handbook," &c. Fcap. 8vo, nearly 500 pp., with 8 Plates and upwards of 250 Illustrative Diagrams, 6s., strongly bound for workshop or pocket wear and tear.

*** OPINIONS OF THE PRESS.

"In its modernised form Hutton's 'Templeton' should have a wide sale, for it contains much valuable information which the mechanic will often find of use, and not a few tables and notes which he might look for in vain in other works. This modernised edition will be appreciated by all who have learned to value the original editions of 'Templeton.'"—*English Mechanic.*

"It has met with great success in the engineering workshop, as we can testify; and there are a great many men who, in a great measure, owe their rise in life to this little book."—*Building News.*

"This familiar text-book—well known to all mechanics and engineers—is of essential service to the every-day requirements of engineers, millwrights, and the various trades connected with engineering and building. The new modernised edition is worth its weight in gold."—*Building News.* (Second Notice.)

"This well-known and largely used book contains information, brought up to date, of the sort so useful to the foreman and draughtsman. So much fresh information has been introduced as to constitute it practically a new book. It will be largely used in the office and workshop."—*Mechanical World.*

Templeton's Engineer's and Machinist's Assistant.

THE ENGINEER'S, MILLWRIGHT'S, and MACHINIST'S PRACTICAL ASSISTANT. A collection of Useful Tables, Rules and Data. By WILLIAM TEMPLETON. 7th Edition, with Additions. 18mo, 2s. 6d. cloth.

"Occupies a foremost place among books of this kind. A more suitable present to an apprentice to any of the mechanical trades could not possibly be made."—*Building News.*

"A deservedly popular work. It should be in the 'drawer' of every mechanic."—*English Mechanic.*

Foley's Reference Book for Mechanical Engineers.

THE MECHANICAL ENGINEER'S REFERENCE BOOK, for Machine and Boiler Construction. In Two Parts. Part I. GENERAL ENGINEERING DATA. Part II. BOILER CONSTRUCTION. With 51 Plates and numerous Illustrations. By NELSON FOLEY, M.I.N.A. Folio, £5 5s. half-bound. [*Just published.*

SUMMARY OF CONTENTS.

PART I.

MEASURES.—CIRCUMFERENCES AND AREAS, &c., SQUARES, CUBES, FOURTH POWERS.—SQUARE AND CUBE ROOTS.—SURFACE OF TUBES—RECIPROCALS.—LOGARITHMS.—MENSURATION.—SPECIFIC GRAVITIES AND WEIGHTS.—WORK AND POWER.—HEAT.—COMBUSTION.—EXPANSION AND CONTRACTION.—EXPANSION OF GASES.—STEAM.—STATIC FORCES.—GRAVITATION AND ATTRACTION.—MOTION AND COMPUTATION OF RESULTING FORCES.—ACCUMULATED WORK.—CENTRE AND RADIUS OF GYRATION.—MOMENT OF INERTIA.—CENTRE OF OSCILLATION.—ELECTRICITY.—STRENGTH OF MATERIALS.—ELASTICITY. — TEST SHEETS OF METALS. — FRICTION. — TRANSMISSION OF POWER.—FLOW OF LIQUIDS.—FLOW OF GASES.—AIR PUMPS, SURFACE CONDENSERS, &c.—SPEED OF STEAMSHIPS.—PROPELLERS. — CUTTING TOOLS.—FLANGES. — COPPER SHEETS AND TUBES.—SCREWS, NUTS, BOLT HEADS, &c.—VARIOUS RECIPES AND MISCELLANEOUS MATTER.

WITH DIAGRAMS FOR VALVE-GEAR, BELTING AND ROPES, DISCHARGE AND SUCTION PIPES, SCREW PROPELLERS, AND COPPER PIPES.

PART II.

TREATING OF, POWER OF BOILERS.—USEFUL RATIOS.—NOTES ON CONSTRUCTION.—CYLINDRICAL BOILER SHELLS. — CIRCULAR FURNACES. — FLAT PLATES.—STAYS.—GIRDERS.—SCREWS. — HYDRAULIC TESTS. — RIVETING.—BOILER SETTING, CHIMNEYS, AND MOUNTINGS.—FUELS, &c.—EXAMPLES OF BOILERS AND SPEEDS OF STEAMSHIPS.—NOMINAL AND NORMAL HORSE POWER.

WITH DIAGRAMS FOR ALL BOILER CALCULATIONS AND DRAWINGS OF MANY VARIETIES OF BOILERS.

⁎ OPINIONS OF THE PRESS.

"This appears to be a work for which there should be a large demand on the part of mechanical engineers. It is no easy matter to compile a book of this class, and the labour involved is enormous, particularly when—as the author informs us—the majority of the tables and diagrams have been specially prepared for the work. The diagrams are exceptionally well executed, and generally constructed on the method adopted in a previous work by the same author. . . . The tables are very numerous, and deal with a greater variety of subjects than will generally be found in a work of this kind; they have evidently been compiled with great care and are unusually complete. All the information given appears to be well up to date. . . . It would be quite impossible within the limits at our disposal to even enumerate all the subjects treated; it should, however, be mentioned that the author does not confine himself to a mere bald statement of formulæ and laws, but in very many instances shows succinctly how these are derived. . . . The latter part of the book is devoted to diagrams relating to Boiler Construction, and to nineteen beautifully-executed plates of working drawings of boilers and their details. As samples of how such drawings should be got out, they may be cordially recommended to the attention of all young, and even some elderly, engineers. . . . Altogether the book is one which every mechanical engineer may, with advantage to himself add to his library."—*Industries.*

"Mr. Foley is well fitted to compile such a work. . . . The diagrams are a great feature of the work. . . . Regarding the whole work, it may be very fairly stated that Mr. Foley has produced a volume which will undoubtedly fulfil the desire of the author and become indispensable to all mechanical engineers."—*Marine Engineer.*

"We have carefully examined this work, and pronounce it a most excellent reference book for the use of marine engineers."—*Journal of American Society of Naval Engineers.*

"A veritable monument of industry on the part of Mr. Foley, who has succeeded in producing what is simply invaluable to the engineering profession."—*Steamship.*

Coal and Speed Tables.

A POCKET BOOK OF COAL AND SPEED TABLES, for Engineers and Steam-users. By NELSON FOLEY, Author of "The Mechanical Engineer's Reference Book." Pocket-size, 3s. 6d. cloth.

"These tables are designed to meet the requirements of every-day use; they are of sufficient scope for most practical purposes, and may be commended to engineers and users of steam."—*Iron.*

"This pocket-book well merits the attention of the practical engineer. Mr. Foley has compiled a very useful set of tables, the information contained in which is frequently required by engineers, coal consumers and users of steam."—*Iron and Coal Trades Review.*

Steam Engine.

TEXT-BOOK ON THE STEAM ENGINE. With a Supplement on Gas Engines, and PART II. ON HEAT ENGINES. By T. M. GOODEVE, M.A., Barrister-at-Law, Professor of Mechanics at the Normal School of Science and the Royal School of Mines; Author of "The Principles of Mechanics," "The Elements of Mechanism," &c. Eleventh Edition, Enlarged. With numerous Illustrations. Crown 8vo, 6s. cloth.

"Professor Goodeve has given us a treatise on the steam engine which will bear comparison with anything written by Huxley or Maxwell, and we can award it no higher praise."—*Engineer*.
" Mr. Goodeve's text-book is a work of which every young engineer should possess himself."—*Mining Journal*.
"Essentially practical in its aim. The manner of exposition leaves nothing to be desired."—*Scotsman*.

Gas Engines.

ON GAS-ENGINES. Being a Reprint, with some Additions, of the Supplement to the *Text-book on the Steam Engine*, by T. M. GOODEVE, M.A. Crown 8vo, 2s. 6d. cloth.

"Like all Mr. Goodeve's writings, the present is no exception in point of general excellence. It is a valuable little volume."—*Mechanical World*.

Steam Engine Design.

THE STEAM ENGINE: A Practical Manual for Draughtsmen, Designers, and Constructors. Translated from the German of HERMANN HAEDER; Revised and Adapted to English Practice by H. H. P. POWLES, A.M.I.C.E., Translator of Kick's Treatise on "Flour Manufacture." Upwards of 1,000 Diagrams. Crown 8vo, cloth. [*In the press*.

Steam Boilers.

A TREATISE ON STEAM BOILERS: Their Strength, Construction, and Economical Working. By ROBERT WILSON, C.E. Fifth Edition. 12mo, 6s. cloth.

"The best treatise that has ever been published on steam boilers."—*Engineer*.
"The author shows himself perfect master of his subject, and we heartily recommend all employing steam power to possess themselves of the work."—*Ryland's Iron Trade Circular*.

Boiler Chimneys.

BOILER AND FACTORY CHIMNEYS; Their Draught-Power and Stability. With a Chapter on *Lightning Conductors*. By ROBERT WILSON, A.I.C.E., Author of "A Treatise on Steam Boilers," &c. Second Edition. Crown 8vo, 3s. 6d. cloth.

"Full of useful information, definite in statement, and thoroughly practical in treatment.'—*The Local Government Chronicle*.
"A valuable contribution to the literature of scientific building."—*The Builder*.

Boiler Making.

THE BOILER-MAKER'S READY RECKONER & ASSISTANT. With Examples of Practical Geometry and Templating, for the Use of Platers, Smiths and Riveters. By JOHN COURTNEY, Edited by D. K. CLARK, M.I.C.E. Third Edition, 480 pp., with 140 Illusts. Fcap. 8vo, 7s. half-bound.

"A most useful work. . . . No workman or apprentice should be without this book."—*Iron Trade Circular*.
"Boiler-makers will readily recognise the value of this volume. . . . The tables are clearly printed, and so arranged that they can be referred to with the greatest facility, so that it cannot be doubted that they will be generally appreciated and much used."—*Mining Journal*.

Locomotive Engine Development.

THE LOCOMOTIVE ENGINE AND ITS DEVELOPMENT. A Popular Treatise on the Gradual Improvements made in Railway Engines between the Years 1803 and 1892. By CLEMENT E. STRETTON, C.E., Author of "Safe Railway Working," &c. Second Edition, Revised and much Enlarged. With 94 Illustrations. Crown 8vo, 3s. 6d. cloth. [*Just published*.

"Students of railway history and all who are interested in the evolution of the modern locomotive will find much to attract and entertain in this volume."—*The Times*.
"The volume cannot fail to be popular, because it contains, in a condensed and readable form, a great deal of just the kind of information that multitudes of people want."—*Engineer*.
"The author of this work is well known to the railway world as one who has long taken a great interest in everything pertaining thereto. No one probably has a better knowledge of the history and development of the locomotive. It is with much pleasure we welcome the volume before us . . . . which, taken as a whole, is most interesting, and should be of value to all connected with the railway system of this country, as a book of reference."—*Nature*.

Fire Engineering.

FIRES, FIRE-ENGINES, AND FIRE-BRIGADES. With a History of Fire-Engines, their Construction, Use, and Management; Remarks on Fire-Proof Buildings, and the Preservation of Life from Fire; Statistics of the Fire Appliances in English Towns; Foreign Fire Systems; Hints on Fire-Brigades, &c. &c. By CHARLES F. T. YOUNG, C.E. With numerous Illustrations. 544 pp., demy 8vo, £1 4s. cloth.

"To such of our readers as are interested in the subject of fires and fire apparatus, we can most heartily commend this book. It is really the only English work we now have upon the subject"—*Engineering.*

Estimating for Engineering Work, &c.

ENGINEERING ESTIMATES, COSTS AND ACCOUNTS: A Guide to Commercial Engineering. With numerous Examples of Estimates and Costs of Millwright Work, Miscellaneous Productions, Steam Engines and Steam Boilers; and a Section on the Preparation of Costs Accounts. By A GENERAL MANAGER. Demy 8vo, 12s. cloth.

"This is an excellent and very useful book, covering subject-matter in constant requisition in every factory and workshop. . . . The book is invaluable, not only to the young engineer, but also to the estimate department of every works."—*Builder.*

"We accord the work unqualified praise. The information is given in a plain, straightforward manner, and bears throughout evidence of the intimate practical acquaintance of the author with every phase of commercial engineering"—*Mechanical World.*

Engineering Construction.

PATTERN-MAKING: A Practical Treatise, embracing the Main Types of Engineering Construction, and including Gearing, both Hand and Machine made, Engine Work, Sheaves and Pulleys, Pipes and Columns, Screws, Machine Parts, Pumps and Cocks, the Moulding of Patterns in Loam and Greensand, &c., together with the methods of Estimating the weight of Castings; to which is added an Appendix of Tables for Workshop Reference. By a FOREMAN PATTERN MAKER. With upwards of 370 Illustrations. Crown 8vo, 7s. 6d. cloth.

"A well-written technical guide, evidently written by a man who understands and has practised what he has written about. . . . We cordially recommend it to engineering students, young journeymen, and others desirous of being initiated into the mysteries of pattern-making."—*Builder.*

"More than 370 illustrations help to explain the text, which is, however, always clear and explicit, thus rendering the work an excellent *vade mecum* for the apprentice who desires to become master of his trade."—*English Mechanic.*

Dictionary of Mechanical Engineering Terms.

LOCKWOOD'S DICTIONARY OF TERMS USED IN THE PRACTICE OF MECHANICAL ENGINEERING, embracing those current in the Drawing Office, Pattern Shop, Foundry, Fitting, Turning, Smith's and Boiler Shops, &c. &c. Comprising upwards of 6,000 Definitions. Edited by A FOREMAN PATTERN-MAKER, Author of "Pattern Making." Second Edition, Revised, with Additions. Crown 8vo, 7s. 6d. cloth. [*Just published.*

"Just the sort of handy dictionary required by the various trades engaged in mechanical engineering. The practical engineering pupil will find the book of great value in his studies, and every foreman engineer and mechanic should have a copy."—*Building News.*

"One of the most useful books which can be presented to a mechanic or student."—*English Mechanic.*

"Not merely a dictionary, but, to a certain extent, also a most valuable guide. It strikes us as a happy idea to combine with a definition of the phrase useful information on the subject of which it treats."—*Machinery Market.*

Mill Gearing.

TOOTHED GEARING: A Practical Handbook for Offices and Workshops. By A FOREMAN PATTERN MAKER, Author of "Pattern Making," "Lockwood's Dictionary of Mechanical Engineering Terms," &c. With 184 Illustrations. Crown 8vo, 6s. cloth. [*Just published.*

SUMMARY OF CONTENTS.

CHAP. I. PRINCIPLES.—II. FORMATION OF TOOTH PROFILES.—III. PROPORTIONS OF TEETH.—IV. METHODS OF MAKING TOOTH FORMS.—V. INVOLUTE TEETH.—VI. SOME SPECIAL TOOTH FORMS.—VII. BEVEL WHEELS.—VIII. SCREW GEARS.—IX. WORM GEARS.—X. HELICAL WHEELS.—XI. SKEW BEVELS.—XII. VARIABLE AND OTHER GEARS.—XIII. DIAMETRICAL PITCH.—XIV. THE ODONTOGRAPH.—XV. PATTERN GEARS.—XVI. MACHINE MOULDING GEARS.—XVII. MACHINE CUT GEARS.—XVIII. PROPORTION OF WHEELS.

"We must give the book our unqualified praise for its thoroughness of treatment, and we can heartily recommend it to all interested as the most practical book on the subject yet written."—*Mechanical World.*

Stone-working Machinery.

STONE-WORKING MACHINERY, and the Rapid and Economical Conversion of Stone. With Hints on the Arrangement and Management of Stone Works. By M. POWIS BALE, M.I.M.E. With Illusts. Crown 8vo, 9s.

"The book should be in the hands of every mason or student of stone-work."—*Colliery Guardian.*

"A capital handbook for all who manipulate stone for building or ornamental purposes."—*Machinery Market.*

Pump Construction and Management.

PUMPS AND PUMPING: A Handbook for Pump Users. Being Notes on Selection, Construction and Management. By M. POWIS BALE, M.I.M.E., Author of "Woodworking Machinery," "Saw Mills," &c. Second Edition, Revised. Crown 8vo, 2s. 6d. cloth. [*Just published.*

"The matter is set forth as concisely as possible. In fact, condensation rather than diffuseness has been the author's aim throughout; yet he does not seem to have omitted anything likely to be of use."—*Journal of Gas Lighting.*

"Thoroughly practical and simply and clearly written."—*Glasgow Herald.*

Milling Machinery, etc.

MILLING MACHINES AND PROCESSES: A Practical Treatise on Shaping Metals by Rotary Cutters, including Information on Making and Grinding the Cutters. By PAUL N. HASLUCK, Author of "Lathe-work," "Handybooks for Handicrafts," &c. With upwards of 300 Engravings, including numerous Drawings by the Author. Large crown 8vo, 352 pages, 12s. 6d. cloth.

"A new departure in engineering literature. . . . We can recommend this work to all interested in milling machines; it is what it professes to be—a practical treatise."—*Engineer.*

"A capital and reliable book, which will no doubt be of considerable service, both to those who are already acquainted with the process as well as to those who contemplate its adoption."—*Industries.*

Turning.

LATHE-WORK: A Practical Treatise on the Tools, Appliances, and Processes employed in the Art of Turning. By PAUL N. HASLUCK. Fourth Edition, Revised and Enlarged. Cr. 8vo, 5s. cloth.

"Written by a man who knows, not only how work ought to be done, but who also knows how to do it, and how to convey his knowledge to others. To all turners this book would be valuable."—*Engineering.*

"We can safely recommend the work to young engineers. To the amateur it will simply be invaluable. To the student it will convey a great deal of useful information."—*Engineer.*

Screw-Cutting.

SCREW THREADS: And Methods of Producing Them. With Numerous Tables, and complete directions for using Screw-Cutting Lathes. By PAUL N. HASLUCK, Author of "Lathe-Work," &c. With Seventy-four Illustrations. Third Edition, Revised and Enlarged. Waistcoat-pocket size, 1s. 6d. cloth.

"Full of useful information, hints and practical criticism. Taps, dies and screwing-tools generally are illustrated and their action described."—*Mechanical World.*

"It is a complete compendium of all the details of the screw cutting lathe; in fact a *multum in parvo* on all the subjects it treats upon."—*Carpenter and Builder.*

Smith's Tables for Mechanics, etc.

TABLES, MEMORANDA, AND CALCULATED RESULTS, FOR MECHANICS, ENGINEERS, ARCHITECTS, BUILDERS, etc. Selected and Arranged by FRANCIS SMITH. Fifth Edition, thoroughly Revised and Enlarged, with a New Section of ELECTRICAL TABLES, FORMULÆ, and MEMORANDA. Waistcoat-pocket size, 1s. 6d. limp leather.

"It would, perhaps, be as difficult to make a small pocket-book selection of notes and formulæ to suit ALL engineers as it would be to make a universal medicine; but Mr. Smith's waistcoat-pocket collection may be looked upon as a successful attempt."—*Engineer.*

"The best example we have ever seen of 270 pages of useful matter packed into the dimensions of a card-case."—*Building News.* "A veritable pocket treasury of knowledge."—*Iron.*

French-English Glossary for Engineers, etc.

A POCKET GLOSSARY of TECHNICAL TERMS: ENGLISH-FRENCH, FRENCH-ENGLISH; with Tables suitable for the Architectural, Engineering, Manufacturing and Nautical Professions. By JOHN JAMES FLETCHER, Engineer and Surveyor. Second Edition, Revised and Enlarged, 200 pp. Waistcoat-pocket size, 1s. 6d. limp leather.

"It is a very great advantage for readers and correspondents in France and England to have so large a number of the words relating to engineering and manufacturers collected in a lilliputian volume. The little book will be useful both to students and travellers."—*Architect.*

"The glossary of terms is very complete, and many of the tables are new and well arranged. We cordially commend the book."—*Mechanical World.*

Portable Engines.

THE PORTABLE ENGINE; ITS CONSTRUCTION AND MANAGEMENT. A Practical Manual for Owners and Users of Steam Engines generally. By WILLIAM DYSON WANSBROUGH. With 90 Illustrations. Crown 8vo, 3s. 6d. cloth.

"This is a work of value to those who use steam machinery. . . . Should be read by everyone who has a steam engine, on a farm or elsewhere."—*Mark Lane Express.*

"We cordially commend this work to buyers and owners of steam engines, and to those who have to do with their construction or use."—*Timber Trades Journal.*

"Such a general knowledge of the steam engine as Mr. Wansbrough furnishes to the reader should be acquired by all intelligent owners and others who use the steam engine."—*Building News.*

"An excellent text-book of this useful form of engine, which describes with all necessary minuteness the details of the various devices. . . 'The Hints to Purchasers' contain a good deal of commonsense and practical wisdom.'—*English Mechanic.*

Iron and Steel.

"IRON AND STEEL": A Work for the Forge, Foundry, Factory, and Office. Containing ready, useful, and trustworthy Information for Ironmasters and their Stock-takers; Managers of Bar, Rail, Plate, and Sheet Rolling Mills; Iron and Metal Founders; Iron Ship and Bridge Builders; Mechanical, Mining, and Consulting Engineers; Architects, Contractors, Builders, and Professional Draughtsmen. By CHARLES HOARE, Author of "The Slide Rule," &c. Eighth Edition, Revised throughout and considerably Enlarged. 32mo, 6s. leather.

"For comprehensiveness the book has not its equal."—*Iron.*
"One of the best of the pocket books."—*English Mechanic.*
"We cordially recommend this book to those engaged in considering the details of all kinds o iron and steel works."—*Naval Science.*

Elementary Mechanics.

CONDENSED MECHANICS. A Selection of Formulæ, Rules, Tables, and Data for the Use of Engineering Students, Science Classes, &c. In Accordance with the Requirements of the Science and Art Department. By W. G. CRAWFORD HUGHES, A.M.I.C.E. Crown 8vo, 2s. 6d. cloth.

"The book is well fitted for those who are either confronted with practical problems in their work, or are preparing for examination and wish to refresh their knowledge by going through their formulæ again."—*Marine Engineer.*

"It is well arranged, and well adapted to meet the wants of those for whom it is intended.'—*Railway News.*

Steam.

THE SAFE USE OF STEAM. Containing Rules for Unprofessional Steam-users. By an ENGINEER. Sixth Edition. Sewed, 6d.

"If steam-users would but learn this little book by heart, boiler explosions would become sensations by their rarity."—*English Mechanic.*

Warming.

HEATING BY HOT WATER; with Information and Suggestions on the best Methods of Heating Public, Private and Horticultural Buildings. By WALTER JONES. With upwards of 50 Illustrations. Crown 8vo, 2s. cloth.

"We confidently recommend all interested in heating by hot water to secure a copy of this valuable little treatise."—*The Plumber and Decorator.*

THE POPULAR WORKS OF MICHAEL REYNOLDS
("THE ENGINE DRIVER'S FRIEND").

Locomotive-Engine Driving.

LOCOMOTIVE-ENGINE DRIVING: *A Practical Manual for Engineers in charge of Locomotive Engines*. By MICHAEL REYNOLDS, Member of the Society of Engineers, formerly Locomotive Inspector L. B. and S. C. R. Eighth Edition. Including a KEY TO THE LOCOMOTIVE ENGINE. With Illustrations and Portrait of Author. Crown 8vo, 4s. 6d. cloth.

"Mr. Reynolds has supplied a want, and has supplied it well. We can confidently recommend the book, not only to the practical driver, but to everyone who takes an interest in the performance of locomotive engines."—*The Engineer.*

"Mr. Reynolds has opened a new chapter in the literature of the day. This admirable practical treatise, of the practical utility of which we have to speak in terms of warm commendation."—*Athenæum.*

"Evidently the work of one who knows his subject thoroughly."—*Railway Service Gazette.*

"Were the cautions and rules given in the book to become part of the every-day working of our engine-drivers, we might have fewer distressing accidents to deplore."—*Scotsman.*

Stationary Engine Driving.

STATIONARY ENGINE DRIVING: *A Practical Manual for Engineers in charge of Stationary Engines*. By MICHAEL REYNOLDS. Fourth Edition, Enlarged. With Plates and Woodcuts. Crown 8vo, 4s. 6d. cloth.

"The author is thoroughly acquainted with his subjects, and his advice on the various points treated is clear and practical. . . . He has produced a manual which is an exceedingly useful one for the class for whom it is specially intended."—*Engineering.*

"Our author leaves no stone unturned. He is determined that his readers shall not only know something about the stationary engine, but all about it."—*Engineer.*

"An engineman who has mastered the contents of Mr. Reynolds's book will require but little actual experience with boilers and engines before he can be trusted to look after them."—*English Mechanic.*

The Engineer, Fireman, and Engine-Boy.

THE MODEL LOCOMOTIVE ENGINEER, FIREMAN, and ENGINE-BOY. Comprising a Historical Notice of the Pioneer Locomotive Engines and their Inventors. By MICHAEL REYNOLDS. With numerous Illustrations and a fine Portrait of George Stephenson. Crown 8vo, 4s. 6d. cloth.

"From the technical knowledge of the author it will appeal to the railway man of to-day more forcibly than anything written by Dr. Smiles. . . . The volume contains information of a technical kind, and facts that every driver should be familiar with."—*English Mechanic.*

"We should be glad to see this book in the possession of everyone in the kingdom who has ever laid, or is to lay, hands on a locomotive engine."—*Iron.*

Continuous Railway Brakes.

CONTINUOUS RAILWAY BRAKES: *A Practical Treatise on the several Systems in Use in the United Kingdom; their Construction and Performance*. With copious Illustrations and numerous Tables. By MICHAEL REYNOLDS. Large crown 8vo, 9s. cloth.

"A popular explanation of the different brakes. It will be of great assistance in forming public opinion, and will be studied with benefit by those who take an interest in the brake."—*English Mechanic.*

"Written with sufficient technical detail to enable the principle and relative connection of the various parts of each particular brake to be readily grasped."—*Mechanical World.*

Engine-Driving Life.

ENGINE-DRIVING LIFE: *Stirring Adventures and Incidents in the Lives of Locomotive-Engine Drivers*. By MICHAEL REYNOLDS. Second Edition, with Additional Chapters. Crown 8vo, 2s. cloth.

"From first to last perfectly fascinating. Wilkie Collins's most thrilling conceptions are thrown into the shade by true incidents, endless in their variety, related in every page."—*North British Mail.*

"Anyone who wishes to get a real insight into railway life cannot do better than read 'Engine-Driving Life' for himself; and if he once take it up he will find that the author's enthusiasm and real love of the engine-driving profession will carry him on till he has read every page."—*Saturday Review.*

Pocket Companion for Enginemen.

THE ENGINEMAN'S POCKET COMPANION AND PRACTICAL EDUCATOR FOR ENGINEMEN, BOILER ATTENDANTS, AND MECHANICS. By MICHAEL REYNOLDS. With Forty-five Illustrations and numerous Diagrams. Second Edition, Revised. Royal 18mo, 3s. 6d., strongly bound for pocket wear.

"This admirable work is well suited to accomplish its object, being the honest workmanship of a competent engineer."—*Glasgow Herald.*

"A most meritorious work, giving in a succinct and practical form all the information an engine-minder desirous of mastering the scientific principles of his daily calling would require."—*The Miller.*

"A boon to those who are striving to become efficient mechanics."—*Daily Chronicle.*

CIVIL ENGINEERING, SURVEYING, etc.

MR. HUMBER'S VALUABLE ENGINEERING BOOKS.

The Water Supply of Cities and Towns.
A COMPREHENSIVE TREATISE on the WATER-SUPPLY OF CITIES AND TOWNS. By WILLIAM HUMBER, A-M.Inst.C.E., and M. Inst. M.E., Author of "Cast and Wrought Iron Bridge Construction," &c. &c. Illustrated with 50 Double Plates, 1 Single Plate, Coloured Frontispiece, and upwards of 250 Woodcuts, and containing 400 pages of Text. Imp. 4to, £6 6s. elegantly and substantially half-bound in morocco.

List of Contents.

I. Historical Sketch of some of the means that have been adopted for the Supply of Water to Cities and Towns.—II. Water and the Foreign Matter usually associated with it.—III. Rainfall and Evaporation.—IV. Springs and the water-bearing formations of various districts.—V. Measurement and Estimation of the flow of Water—VI. On the Selection of the Source of Supply.—VII. Wells.—VIII. Reservoirs.—IX. The Purification of Water.—X. Pumps. — XL Pumping Machinery. — XII. Conduits.—XIII. Distribution of Water.—XIV. Meters, Service Pipes, and House Fittings.— XV. The Law and Economy of Water Works. XVI. Constant and Intermittent Supply.— XVII. Description of Plates. — Appendices, giving Tables of Rates of Supply, Velocities, &c. &c., together with Specifications of several Works Illustrated, among which will be found: Aberdeen, Bideford, Canterbury, Dundee, Halifax, Lambeth, Rotherham, Dublin, and others.

"The most systematic and valuable work upon water supply hitherto produced in English, or in any other language. . . . Mr. Humber's work is characterised almost throughout by an exhaustiveness much more distinctive of French and German than of English technical treatises."—*Engineer.*

"We can congratulate Mr. Humber on having been able to give so large an amount of information on a subject so important as the water supply of cities and towns. The plates, fifty in number, are mostly drawings of executed works, and alone would have commanded the attention of every engineer whose practice may lie in this branch of the profession."—*Builder.*

Cast and Wrought Iron Bridge Construction.
A COMPLETE AND PRACTICAL TREATISE ON CAST AND WROUGHT IRON BRIDGE CONSTRUCTION, including Iron Foundations. In Three Parts—Theoretical, Practical, and Descriptive. By WILLIAM HUMBER, A.M.Inst.C.E., and M.Inst.M.E. Third Edition, Revised and much improved, with 115 Double Plates (20 of which now first appear in this edition), and numerous Additions to the Text. In Two Vols., imp. 4to, £6 16s. 6d. half-bound in morocco.

"A very valuable contribution to the standard literature of civil engineering. In addition to elevations, plans and sections, large scale details are given which very much enhance the instructive worth of those illustrations."—*Civil Engineer and Architect's Journal.*

"Mr. Humber's stately volumes, lately issued—in which the most important bridges erected during the last five years, under the direction of the late Mr. Brunel, Sir W. Cubitt, Mr. Hawkshaw, Mr. Page, Mr. Fowler, Mr. Hemans, and others among our most eminent engineers, are drawn and specified in great detail."—*Engineer.*

Strains, Calculation of.
A HANDY BOOK FOR THE CALCULATION OF STRAINS IN GIRDERS AND SIMILAR STRUCTURES, AND THEIR STRENGTH. Consisting of Formulæ and Corresponding Diagrams, with numerous details for Practical Application. &c. By WILLIAM HUMBER, A-M.Inst.C.E., &c. Fifth Edition. Crown 8vo, nearly 100 Woodcuts and 3 Plates, 7s. 6d. cloth.

"The formulæ are neatly expressed, and the diagrams good."—*Athenæum.*
"We heartily commend this really *handy* book to our engineer and architect readers."—*English Mechanic.*

Barlow's Strength of Materials, enlarged by Humber.
A TREATISE ON THE STRENGTH OF MATERIALS; with Rules for Application in Architecture, the Construction of Suspension Bridges, Railways, &c. By PETER BARLOW, F.R.S. A New Edition, revised by his Sons, P. W. BARLOW, F.R.S., and W. H. BARLOW, F.R.S.; to which are added, Experiments by HODGKINSON, FAIRBAIRN, and KIRKALDY; and Formulæ for Calculating Girders, &c. Arranged and Edited by WM. HUMBER, A-M.Inst.C.E. Demy 8vo, 400 pp., with 19 large Plates and numerous Woodcuts, 18s. cloth.

"Valuable alike to the student, tyro, and the experienced practitioner, it will always rank in future, as it has hitherto done, as the standard treatise on that particular subject."—*Engineer.*
"There is no greater authority than Barlow."—*Building News.*
"As a scientific work of the first class, it deserves a foremost place on the bookshelves of every civil engineer and practical mechanic."—*English Mechanic.*

CIVIL ENGINEERING, SURVEYING, etc.

MR. HUMBER'S GREAT WORK ON MODERN ENGINEERING.

Complete in Four Volumes, imperial 4to, price £12 12s., half-morocco. Each Volume sold separately as follows:—

A RECORD OF THE PROGRESS OF MODERN ENGINEERING. FIRST SERIES.
Comprising Civil, Mechanical, Marine, Hydraulic, Railway, Bridge, and other Engineering Works, &c. By WILLIAM HUMBER, A-M.Inst.C.E., &c. Imp. 4to, with 36 Double Plates, drawn to a large scale, Photographic Portrait of John Hawkshaw, C.E., F.R.S., &c., and copious descriptive Letterpress, Specifications, &c., £3 3s. half-morocco.

List of the Plates and Diagrams.

Victoria Station and Roof, L. B. & S. C. R. (8 plates); Southport Pier (2 plates); Victoria Station and Roof, L. C. & D. and G. W. R. (6 plates); Roof of Cremorne Music Hall; Bridge over G. N. Railway; Roof of Station, Dutch Rhenish Rail (2 plates); Bridge over the Thames, West London Extension Railway (5 plates); Armour Plates; Suspension Bridge, Thames (4 plates); The Allen Engine; Suspension Bridge, Avon (3 plates); Underground Railway (3 plates).

"Handsomely lithographed and printed. It will find favour with many who desire to preserve in a permanent form copies of the plans and specifications prepared for the guidance of the contractors for many important engineering works."—*Engineer*.

HUMBER'S PROGRESS OF MODERN ENGINEERING.
SECOND SERIES. Imp. 4to, with 36 Double Plates, Photographic Portrait of Robert Stephenson, C.E., M.P., F.R.S., &c., and copious descriptive Letterpress, Specifications, &c., £3 3s. half-morocco.

List of the Plates and Diagrams.

Birkenhead Docks, Low Water Basin (15 plates); Charing Cross Station Roof, C. C. Railway (3 plates); Digswell Viaduct, Great Northern Railway; Robbery Wood Viaduct, Great Northern Railway; Iron Permanent Way; Clydach Viaduct, Merthyr, Tredegar, and Abergavenny Railway; Ebbw Viaduct, Merthyr, Tredegar, and Abergavenny Railway; College Wood Viaduct, Cornwall Railway; Dublin Winter Palace Roof (3 plates); Bridge over the Thames, L. C. & D. Railway (6 plates); Albert Harbour, Greenock (4 plates).

"Mr. Humber has done the profession good and true service, by the fine collection of examples he has here brought before the profession and the public."—*Practical Mechanic's Journal*.

HUMBER'S PROGRESS OF MODERN ENGINEERING.
THIRD SERIES. Imp. 4to, with 40 Double Plates, Photographic Portrait of J. R. M'Clean, late Pres. Inst. C.E., and copious descriptive Letterpress, Specifications, &c., £3 3s. half-morocco.

List of the Plates and Diagrams.

MAIN DRAINAGE, METROPOLIS.—*North Side*.—Map showing Interception of Sewers; Middle Level Sewer (2 plates); Outfall Sewer, Bridge over River Lea (3 plates); Outfall Sewer, Bridge over Marsh Lane, North Woolwich Railway, and Bow and Barking Railway Junction; Outfall Sewer, Bridge over Bow and Barking Railway (3 plates); Outfall Sewer, Bridge over East London Waterworks' Feeder (2 plates); Outfall Sewer, Reservoir and Outlet; Outfall Sewer, Tumbling Bay and Outlet; Outfall Sewer, Penstocks. *South Side*.—Outfall Sewer, Bermondsey Branch (2 plates); Outfall Sewer, Reservoir and Outlet (4 plates); Outfall Sewer, Filth Hoist; Sections of Sewers (North and South Sides). THAMES EMBANKMENT.—Section of River Wall; Steamboat Pier, Westminster (2 plates); Landing Stairs between Charing Cross and Waterloo Bridges; York Gate (2 plates); Overflow and Outlet at Savoy Street Sewer (3 plates); Steamboat Pier, Waterloo Bridge (3 plates); Junction of Sewers, Plans and Sections; Gullies, Plans and Sections; Rolling Stock; Granite and Iron Forts.

"The drawings have a constantly increasing value, and whoever desires to possess clear representations of the two great works carried out by our Metropolitan Board will obtain Mr. Humber's volume."—*Engineer*.

HUMBER'S PROGRESS OF MODERN ENGINEERING.
FOURTH SERIES. Imp. 4to, with 36 Double Plates, Photographic Portrait of John Fowler, late Pres. Inst. C.E., and copious descriptive Letterpress, Specifications, &c., £3 3s. half-morocco.

List of the Plates and Diagrams.

Abbey Mills Pumping Station, Main Drainage, Metropolis (4 plates); Barrow Docks (5 plates); Manquis Viaduct, Santiago and Valparaiso Railway (2 plates); Adam's Locomotive, St. Helen's Canal Railway (2 plates); Cannon Street Station Roof, Charing Cross Railway (3 plates); Road Bridge over the River Moka (2 plates); Telegraphic Apparatus for Mesopotamia; Viaduct over the River Wye, Midland Railway (3 plates); St. Germans Viaduct, Cornwall Railway (2 plates); Wrought-Iron Cylinder for Diving Bell; Millwall Docks (6 plates); Milroy's Patent Excavator; Metropolitan District Railway (6 plates); Harbours, Ports, and Breakwaters (3 plates).

"We gladly welcome another year's issue of this valuable publication from the able pen of Mr. Humber. The accuracy and general excellence of this work are well known, while its usefulness in giving the measurements and details of some of the latest examples of engineering, as carried out by the most eminent men in the profession, cannot be too highly prized."—*Artisan*.

Statics, Graphic and Analytic.

GRAPHIC AND ANALYTIC STATICS, in their Practical Application to the Treatment of Stresses in Roofs, Solid Girders, Lattice, Bowstring and Suspension Bridges, Braced Iron Arches and Piers, and other Frameworks. By R. HUDSON GRAHAM, C.E. Containing Diagrams and Plates to Scale. With numerous Examples, many taken from existing Structures. Specially arranged for Class-work in Colleges and Universities. Second Edition, Revised and Enlarged. 8vo, 16s. cloth.

"Mr. Graham's book will find a place wherever graphic and analytic statics are used or studied."—*Engineer.*

"The work is excellent from a practical point of view, and has evidently been prepared with much care. The directions for working are ample, and are illustrated by an abundance of well-selected examples. It is an excellent text-book for the practical draughtsman."—*Athenæum.*

Practical Mathematics.

MATHEMATICS FOR PRACTICAL MEN: Being a Commonplace Book of Pure and Mixed Mathematics. Designed chiefly for the use of Civil Engineers, Architects and Surveyors. By OLINTHUS GREGORY, LL.D., F.R.A.S., Enlarged by HENRY LAW, C.E. 4th Edition, carefully Revised by J. R. YOUNG, formerly Professor of Mathematics, Belfast College. With 13 Plates. 8vo, £1 1s. cloth.

"The engineer or architect will here find ready to his hand rules for solving nearly every mathematical difficulty that may arise in his practice. The rules are in all cases explained by means of examples, in which every step of the process is clearly worked out."—*Builder.*

"One of the most serviceable books for practical mechanics. . . It is an instructive book for the student, and a text-book for him who, having once mastered the subjects it treats of, needs occasionally to refresh his memory upon them."—*Building News.*

Hydraulic Tables.

HYDRAULIC TABLES, CO-EFFICIENTS, and FORMULÆ for finding the Discharge of Water from Orifices, Notches, Weirs, Pipes, and Rivers. With New Formulæ, Tables, and General Information on Rainfall, Catchment-Basins, Drainage, Sewerage, Water Supply for Towns and Mill Power. By JOHN NEVILLE, Civil Engineer, M.R.I.A. Third Ed., carefully Revised, with considerable Additions. Numerous Illusts. Cr. 8vo, 14s. cloth.

"Alike valuable to students and engineers in practice; its study will prevent the annoyance of avoidable failures, and assist them to select the readiest means of successfully carrying out any given work connected with hydraulic engineering."—*Mining Journal.*

"It is, of all English books on the subject, the one nearest to completeness. . . . From the good arrangement of the matter, the clear explanations, and abundance of formulæ, the carefully calculated tables, and, above all, the thorough acquaintance with both theory and construction, which is displayed from first to last, the book will be found to be an acquisition."—*Architect.*

Hydraulics.

HYDRAULIC MANUAL. Consisting of Working Tables and Explanatory Text. Intended as a Guide in Hydraulic Calculations and Field Operations. By LOWIS D'A. JACKSON, Author of "Aid to Survey Practice," "Modern Metrology," &c. Fourth Edition, Enlarged. Large cr. 8vo, 16s. cl.

"The author has had a wide experience in hydraulic engineering and has been a careful observer of the facts which have come under his notice, and from the great mass of material at his command he has constructed a manual which may be accepted as a trustworthy guide to this branch of the engineer's profession. We can heartily recommend this volume to all who desire to be acquainted with the latest development of this important subject."—*Engineering.*

"The standard-work in this department of mechanics."—*Scotsman.*

"The most useful feature of this work is its freedom from what is superannuated, and its thorough adoption of recent experiments; the text is, in fact, in great part a short account of the great modern experiments."—*Nature.*

Drainage.

ON THE DRAINAGE OF LANDS, TOWNS, AND BUILDINGS. By G. D. DEMPSEY, C.E., Author of "The Practical Railway Engineer," &c. Revised, with large Additions on RECENT PRACTICE IN DRAINAGE ENGINEERING, by D. KINNEAR CLARK, M.Inst.C.E. Author of "Tramways: Their Construction and Working," "A Manual of Rules, Tables, and Data for Mechanical Engineers," &c. Second Edition, Corrected. Fcap. 8vo, 5s. cloth.

"The new matter added to Mr. Dempsey's excellent work is characterised by the comprehensive grasp and accuracy of detail for which the name of Mr. D. K. Clark is a sufficient voucher."—*Athenæum.*

"As a work on recent practice in drainage engineering, the book is to be commended to all who are making that branch of engineering science their special study."—*Iron.*

"A comprehensive manual on drainage engineering, and a useful introduction to the student."—*Building News.*

Water Storage, Conveyance, and Utilisation.

WATER ENGINEERING: A Practical Treatise on the Measurement, Storage, Conveyance, and Utilisation of Water for the Supply of Towns, for Mill Power, and for other Purposes. By CHARLES SLAGG, Water and Drainage Engineer, A.M.Inst.C.E., Author of "Sanitary Work in the Smaller Towns, and in Villages," &c. With numerous Illusts. Cr. 8vo, 7s. 6d. cloth.

"As a small practical treatise on the water supply of towns, and on some applications of water-power, the work is in many respects excellent."—*Engineering*.

"The author has collated the results deduced from the experiments of the most eminent authorities, and has presented them in a compact and practical form, accompanied by very clear and detailed explanations. . . . The application of water as a motive power is treated very carefully and exhaustively."—*Builder*.

"For anyone who desires to begin the study of hydraulics with a consideration of the practical applications of the science there is no better guide."—*Architect*.

River Engineering.

RIVER BARS: The Causes of their Formation, and their Treatment by "Induced Tidal Scour;" with a Description of the Successful Reduction by this Method of the Bar at Dublin. By I. J. MANN, Assist. Eng. to the Dublin Port and Docks Board. Royal 8vo, 7s. 6d. cloth.

"We recommend all interested in harbour works—and, indeed, those concerned in the improvements of rivers generally—to read Mr. Mann's interesting work on the treatment of river bars."—*Engineer*.

Trusses.

TRUSSES OF WOOD AND IRON. Practical Applications of Science in Determining the Stresses, Breaking Weights, Safe Loads, Scantlings, and Details of Construction, with Complete Working Drawings. By WILLIAM GRIFFITHS, Surveyor, Assistant Master, Tranmere School of Science and Art. Oblong 8vo, 4s. 6d. cloth.

"This handy little book enters so minutely into every detail connected with the construction of roof trusses, that no student need be ignorant of these matters."—*Practical Engineer*.

Railway Working.

SAFE RAILWAY WORKING. A Treatise on Railway Accidents: Their Cause and Prevention; with a Description of Modern Appliances and Systems. By CLEMENT E. STRETTON, C.E., Vice-President and Consulting Engineer, Amalgamated Society of Railway Servants. With Illustrations and Coloured Plates. Third Edition, Enlarged. Crown 8vo, 3s. 6d. cloth.

"A book for the engineer, the directors, the managers; and, in short, all who wish for information on railway matters will find a perfect encyclopædia in 'Safe Railway Working.'"—*Railway Review*.

"We commend the remarks on railway signalling to all railway managers, especially where a uniform code and practice is advocated."—*Herepath's Railway Journal*.

"The author may be congratulated on having collected, in a very convenient form, much valuable information on the principal questions affecting the safe working of railways."—*Railway Engineer*.

Oblique Bridges.

A PRACTICAL AND THEORETICAL ESSAY ON OBLIQUE BRIDGES. With 13 large Plates. By the late GEORGE WATSON BUCK, M.I.C.E. Third Edition, revised by his Son, J. H. WATSON BUCK, M.I.C.E.; and with the addition of Description to Diagrams for Facilitating the Construction of Oblique Bridges, by W. H. BARLOW, M.I.C.E. Royal 8vo, 12s. cloth.

"The standard text-book for all engineers regarding skew arches is Mr. Buck's treatise, and it would be impossible to consult a better."—*Engineer*.

"Mr. Buck's treatise is recognised as a standard text-book, and his treatment has divested the subject of many of the intricacies supposed to belong to it. As a guide to the engineer and architect, on a confessedly difficult subject, Mr. Buck's work is unsurpassed."—*Building News*.

Tunnel Shafts.

THE CONSTRUCTION OF LARGE TUNNEL SHAFTS: A Practical and Theoretical Essay. By J. H. WATSON BUCK, M.Inst.C.E., Resident Engineer, London and North-Western Railway. Illustrated with Folding Plates. Royal 8vo, 12s. cloth.

"Many of the methods given are of extreme practical value to the mason; and the observations on the form of arch, the rules for ordering the stone, and the construction of the templates will be found of considerable use. We commend the book to the engineering profession."—*Building News*.

"Will be regarded by civil engineers as of the utmost value, and calculated to save much time and obviate many mistakes."—*Colliery Guardian*.

Student's Text-Book on Surveying.

PRACTICAL SURVEYING: A Text-Book for Students preparing for Examination or for Survey-work in the Colonies. By GEORGE W. USILL, A.M.I.C.E., Author of "The Statistics of the Water Supply of Great Britain." With Four Lithographic Plates and upwards of 330 Illustrations. Second Edition, Revised. Crown 8vo, 7s. 6d. cloth.

"The best forms of instruments are described as to their construction, uses and modes of employment, and there are innumerable hints on work and equipment such as the author, in his experience as surveyor, draughtsman, and teacher, has found necessary, and which the student in his inexperience will find most serviceable."—*Engineer.*

"The latest treatise in the English language on surveying, and we have no hesitation in saying that the student will find it a better guide than any of its predecessors . . . . Deserves to be recognised as the first book which should be put in the hands of a pupil of Civil Engineering, and every gentleman of education who sets out for the Colonies would find it well to have a copy."—*Architect.*

Survey Practice.

AID TO SURVEY PRACTICE, for Reference in Surveying, Levelling, and Setting-out; and in Route Surveys of Travellers by Land and Sea. With Tables, Illustrations, and Records. By LOWIS D'A. JACKSON, A.M.I.C.E., Author of "Hydraulic Manual," "Modern Metrology," &c. Second Edition, Enlarged. Large crown 8vo, 12s. 6d. cloth.

"Mr. Jackson has produced a valuable *vade-mecum* for the surveyor. We can recommend this book as containing an admirable supplement to the teaching of the accomplished surveyor."—*Athenæum.*

"As a text-book we should advise all surveyors to place it in their libraries, and study well the matured instructions afforded in its pages."—*Colliery Guardian.*

"The author brings to his work a fortunate union of theory and practical experience which, aided by a clear and lucid style of writing, renders the book a very useful one."—*Builder.*

Surveying, Land and Marine.

LAND AND MARINE SURVEYING, in Reference to the Preparation of Plans for Roads and Railways; Canals, Rivers, Towns' Water Supplies; Docks and Harbours. With Description and Use of Surveying Instruments. By W. D. HASKOLL, C.E., Author of "Bridge and Viaduct Construction," &c. Second Edition, Revised, with Additions. Large cr. 8vo, 9s. cl.

"This book must prove of great value to the student. We have no hesitation in recommending it, feeling assured that it will more than repay a careful study."—*Mechanical World.*

"A most useful and well arranged book for the aid of a student. We can strongly recommend it as a carefully-written and valuable text-book. It enjoys a well-deserved repute among surveyors."—*Builder.*

"This volume cannot fail to prove of the utmost practical utility. It may be safely recommended to all students who aspire to become clean and expert surveyors."—*Mining Journal.*

Field-Book for Engineers.

THE ENGINEER'S, MINING SURVEYOR'S, AND CONTRACTOR'S FIELD-BOOK. Consisting of a Series of Tables, with Rules, Explanations of Systems, and use of Theodolite for Traverse Surveying and Plotting the Work with minute accuracy by means of Straight Edge and Set Square only; Levelling with the Theodolite, Casting-out and Reducing Levels to Datum, and Plotting Sections in the ordinary manner; setting-out Curves with the Theodolite by Tangential Angles and Multiples, with Right and Left-hand Readings of the Instrument: Setting-out Curves without Theodolite, on the System of Tangential Angles by sets of Tangents and Offsets; and Earthwork Tables to 80 feet deep, calculated for every 6 inches in depth. By W. DAVIS HASKOLL, C.E. With numerous Woodcuts. Fourth Edition, Enlarged. Crown 8vo, 12s. cloth.

"The book is very handy; the separate tables of sines and tangents to every minute will make it useful for many other purposes, the genuine traverse tables existing all the same."—*Athenæum.*

"Every person engaged in engineering field operations will estimate the importance of such a work and the amount of valuable time which will be saved by reference to a set of reliable tables prepared with the accuracy and fulness of those given in this volume."—*Railway News.*

Levelling.

A TREATISE ON THE PRINCIPLES AND PRACTICE OF LEVELLING. Showing its Application to purposes of Railway and Civil Engineering, in the Construction of Roads; with Mr. TELFORD'S Rules for the same. By FREDERICK W. SIMMS, F.G.S., M.Inst.C.E. Seventh Edition, with the addition of LAW's Practical Examples for Setting-out Railway Curves, and TRAUTWINE's Field Practice of Laying-out Circular Curves. With 7 Plates and numerous Woodcuts. 8vo, 8s. 6d. cloth. **** TRAUTWINE on Curves may be had separate, 5s.

"The text-book on levelling in most of our engineering schools and colleges."—*Engineer.*

"The publishers have rendered a substantial service to the profession, especially to the younger members, by bringing out the present edition of Mr. Simms's useful work."—*Engineering.*

Trigonometrical Surveying.

AN OUTLINE OF THE METHOD OF CONDUCTING A TRIGONOMETRICAL SURVEY, for the Formation of Geographical and Topographical Maps and Plans, Military Reconnaissance, Levelling, &c., with Useful Problems, Formulæ, and Tables. By Lieut.-General FROME, R.E. Fourth Edition, Revised and partly Re-written by Major General Sir CHARLES WARREN, G.C.M.G., R.E. With 19 Plates and 115 Woodcuts. Royal 8vo, 16s. cloth.

"The simple fact that a fourth edition has been called for is the best testimony to its merits. No words of praise from us can strengthen the position so well and so steadily maintained by this work. Sir Charles Warren has revised the entire work, and made such additions as were necessary to bring every portion of the contents up to the present date."—*Broad Arrow.*

Field Fortification.

A TREATISE ON FIELD FORTIFICATION, THE ATTACK OF FORTRESSES, MILITARY MINING, AND RECONNOITRING. By Colonel I. S. MACAULAY, late Professor of Fortification in the R.M.A., Woolwich. Sixth Edition. Crown 8vo, with separate Atlas of 12 Plates, 12s. cloth.

Tunnelling.

PRACTICAL TUNNELLING. Explaining in detail the Setting-out of the works, Shaft-sinking and Heading-driving, Ranging the Lines and Levelling underground, Sub-Excavating, Timbering, and the Construction of the Brickwork of Tunnels, with the amount of Labour required for, and the Cost of, the various portions of the work. By FREDERICK W. SIMMS, F.G.S., M.Inst.C.E. Third Edition, Revised and Extended by D. KINNEAR CLARK, M.Inst.C.E. Imperial 8vo, with 21 Folding Plates and numerous Wood Engravings, 30s. cloth.

"The estimation in which Mr. Simms's book on tunnelling has been held for over thirty years cannot be more truly expressed than in the words of the late Prof. Rankine:—'The best source of information on the subject of tunnels is Mr. F.W. Simms's work on Practical Tunnelling.'"—*Architect.*

"It has been regarded from the first as a text-book of the subject. . . . Mr. Clark has added immensely to the value of the book."—*Engineer.*

Tramways and their Working.

TRAMWAYS: THEIR CONSTRUCTION AND WORKING. Embracing a Comprehensive History of the System; with an exhaustive Analysis of the various Modes of Traction, including Horse-Power, Steam, Compressed Air, Electric Traction, &c.; a Description of the Varieties of Rolling Stock; and ample Details of Cost and Working Expenses. New Edition, Thoroughly Revised, and Including the Progress recently made in Tramway Construction, &c. &c. By D. KINNEAR CLARK, M.Inst.C.E. With numerous Illustrations. In One Volume, 8vo. [*In preparation.*

"All interested in tramways must refer to it, as all railway engineers have turned to the author's work 'Railway Machinery.'"—*Engineer.*

"An exhaustive and practical work on tramways, in which the history of this kind of locomotion, and a description and cost of the various modes of laying tramways, are to be found."—*Building News.*

"The best form of rails, the best mode of construction, and the best mechanical appliances are so fairly indicated in the work under review, that any engineer about to construct a tramway will be enabled at once to obtain the practical information which will be of most service to him."—*Athenæum.*

Curves, Tables for Setting-out.

TABLES OF TANGENTIAL ANGLES AND MULTIPLES for Setting-out Curves from 5 to 200 Radius. By ALEXANDER BEAZELEY, M.Inst.C.E. Fourth Edition. Printed on 48 Cards, and sold in a cloth box, waistcoat-pocket size, 3s. 6d.

"Each table is printed on a small card, which, being placed on the theodolite, leaves the hands free to manipulate the instrument—no small advantage as regards the rapidity of work."—*Engineer.*

"Very handy; a man may know that all his day's work must fall on two of these cards, which he puts into his own card-case, and leaves the rest behind."—*Athenæum.*

Earthwork.

EARTHWORK TABLES. Showing the Contents in Cubic Yards of Embankments, Cuttings, &c., of Heights or Depths up to an average of 80 feet. By JOSEPH BROADBENT, C.E., and FRANCIS CAMPIN, C.E. Crown 8vo, 5s. cloth.

"The way in which accuracy is attained, by a simple division of each cross section into three elements, two of which are constant and one variable, is ingenious."—*Athenæum.*

Heat, Expansion by.

EXPANSION OF STRUCTURES BY HEAT. By JOHN KEILY, C.E., late of the Indian Public Works and Victorian Railway Departments. Crown 8vo, 3s. 6d. cloth.

SUMMARY OF CONTENTS.

Section I. FORMULAS AND DATA.
Section II. METAL BARS.
Section III. SIMPLE FRAMES.
Section IV. COMPLEX FRAMES AND PLATES.
Section V. THERMAL CONDUCTIVITY.
Section VI. MECHANICAL FORCE OF HEAT.
Section VII. WORK OF EXPANSION AND CONTRACTION.
Section VIII. SUSPENSION BRIDGES.
Section IX. MASONRY STRUCTURES.

"The aim the author has set before him, viz., to show the effects of heat upon metallic and other structures, is a laudable one, for this is a branch of physics upon which the engineer or architect can find but little reliable and comprehensive data in books."—*Builder*.

"Whoever is concerned to know the effect of changes of temperature on such structures as suspension bridges and the like, could not do better than consult Mr. Keily's valuable and handy exposition of the geometrical principles involved in these changes."—*Scotsman*.

Earthwork, Measurement of.

A MANUAL ON EARTHWORK. By ALEX. J. S. GRAHAM, C.E. With numerous Diagrams. Second Edition. 18mo, 2s. 6d. cloth.

"A great amount of practical information, very admirably arranged, and available for rough estimates, as well as for the more exact calculations required in the engineer's and contractor's offices."—*Artizan*.

Strains in Ironwork.

THE STRAINS ON STRUCTURES OF IRONWORK; with Practical Remarks on Iron Construction. By F. W. SHEILDS, M.Inst.C.E. Second Edition, with 5 Plates. Royal 8vo, 5s. cloth.

The student cannot find a better little book on this subject."—*Engineer*.

Cast Iron and other Metals, Strength of.

A PRACTICAL ESSAY ON THE STRENGTH OF CAST IRON AND OTHER METALS. By THOMAS TREDGOLD, C.E. Fifth Edition, including HODGKINSON'S Experimental Researches. 8vo, 12s. cloth.

Oblique Arches.

A PRACTICAL TREATISE ON THE CONSTRUCTION OF OBLIQUE ARCHES. By JOHN HART. Third Edition, with Plates. Imperial 8vo, 8s. cloth.

Girders, Strength of.

GRAPHIC TABLE FOR FACILITATING THE COMPUTATION OF THE WEIGHTS OF WROUGHT IRON AND STEEL GIRDERS, etc., for Parliamentary and other Estimates. By J. H. WATSON BUCK, M.Inst.C.E. On a Sheet, 2s. 6d.

MARINE ENGINEERING, SHIPBUILDING, NAVIGATION, etc.

Pocket-Book for Naval Architects and Shipbuilders.

THE NAVAL ARCHITECT'S AND SHIPBUILDER'S POCKET-BOOK of Formulæ, Rules, and Tables, and MARINE ENGINEER'S AND SURVEYOR'S Handy Book of Reference. By CLEMENT MACKROW, Member of the Institution of Naval Architects, Naval Draughtsman. Fifth Edition, Revised and Enlarged to 700 pages, with upwards of 300 Illustrations. Fcap., 12s. 6d. strongly bound in leather. [*Just published.*

SUMMARY OF CONTENTS.

SIGNS AND SYMBOLS, DECIMAL FRACTIONS.— TRIGONOMETRY. — PRACTICAL GEOMETRY.— MENSURATION. — CENTRES AND MOMENTS OF FIGURES.— MOMENTS OF INERTIA AND RADII OF GYRATION. — ALGEBRAICAL EXPRESSIONS FOR SIMPSON'S RULES.—MECHANICAL PRINCIPLES. — CENTRE OF GRAVITY.—LAWS OF MOTION.—DISPLACEMENT, CENTRE OF BUOYANCY.—CENTRE OF GRAVITY OF SHIP'S HULL.—STABILITY CURVES AND METACENTRES.—SEA AND SHALLOW-WATER WAVES.—ROLLING OF SHIPS.—PROPULSION AND RESISTANCE OF VESSELS.—SPEED TRIALS.—SAILING, CENTRE OF EFFORT.--DISTANCES DOWN RIVERS, COAST LINES.—STEERING AND RUDDERS OF VESSELS.—LAUNCHING CALCULATIONS AND VELOCITIES.—WEIGHT OF MATERIAL AND GEAR.—GUN PARTICULARS AND WEIGHT.—STANDARD GAUGES.—RIVETED JOINTS AND RIVETING.—STRENGTH AND TESTS OF MATERIALS. — BINDING AND SHEARING STRESSES, ETC.—STRENGTH OF SHAFTING, PILLARS, WHEELS, ETC. — HYDRAULIC DATA, ETC.—CONIC SECTIONS, CATENARIAN CURVES.—MECHANICAL POWERS, WORK. — BOARD OF TRADE REGULATIONS FOR BOILERS AND ENGINES. — BOARD OF TRADE REGULATIONS FOR SHIPS.—LLOYD'S RULES FOR BOILERS.—LLOYD'S WEIGHT OF CHAINS.—LLOYDS SCANTLINGS FOR SHIPS.—DATA OF ENGINES AND VESSELS. - SHIPS' FITTINGS AND TESTS.—SEASONING PRESERVING TIMBER.—MEASUREMENT OF TIMBER.—ALLOYS, PAINTS, VARNISHES. — DATA FOR STOWAGE. — ADMIRALTY TRANSPORT REGULATIONS. — RULES FOR HORSE-POWER, SCREW PROPELLERS, ETC.—PERCENTAGES FOR BUTT STRAPS, ETC. —PARTICULARS OF YACHTS.—MASTING AND RIGGING VESSELS.—DISTANCES OF FOREIGN PORTS. — TONNAGE TABLES. — VOCABULARY OF FRENCH AND ENGLISH TERMS. — ENGLISH WEIGHTS AND MEASURES.—FOREIGN WEIGHTS AND MEASURES.—DECIMAL EQUIVALENTS. — FOREIGN MONEY.—DISCOUNT AND WAGE TABLES. —USEFUL NUMBERS AND READY RECKONERS —TABLES OF CIRCULAR MEASURES.—TABLES OF AREAS OF AND CIRCUMFERENCES OF CIRCLES.—TABLES OF AREAS OF SEGMENTS OF CIRCLES.—TABLES OF SQUARES AND CUBES AND ROOTS OF NUMBERS. — TABLES OF LOGARITHMS OF NUMBERS.—TABLES OF HYPERBOLIC LOGARITHMS.—TABLES OF NATURAL SINES, TANGENTS, ETC.—TABLES OF LOGARITHMIC SINES, TANGENTS, ETC.

" In these days of advanced knowledge a work like this is of the greatest value. It contains a vast amount of information. We unhesitatingly say that it is the most valuable compilation for its specific purpose that has ever been printed. No naval architect, engineer, surveyor, or seaman, wood or iron shipbuilder, can afford to be without this work."—*Nautical Magazine.*

"Should be used by all who are engaged in the construction or designs of vessels. . . . Will be found to contain the most useful tables and formulæ required by shipbuilders, carefully collected from the best authorities, and put together in a popular and simple form."—*Engineer.*

"The professional shipbuilder has now, in a convenient and accessible form, reliable data for solving many of the numerous problems that present themselves in the course of his work."—*Iron.*

Marine Engineering.

MARINE ENGINES AND STEAM VESSELS (A Treatise on). By ROBERT MURRAY, C.E. Eighth Edition, thoroughly Revised, with considerable Additions by the Author and by GEORGE CARLISLE, C.E., Senior Surveyor to the Board of Trade at Liverpool. 12mo, 5s. cloth boards.

"Well adapted to give the young steamship engineer or marine engine and boiler maker a general introduction into his practical work."—*Mechanical World.*

"We feel sure that this thoroughly revised edition will continue to be as popular in the future as it has been in the past, as, for its size, it contains more useful information than any similar treatise."—*Industries.*

Electric Lighting of Ships.

ELECTRIC SHIP-LIGHTING. By J. W. URQUHART, C.E. Crown 8vo, 7s. 6d. cloth. For full description, see p 24.

c

Pocket-Book for Marine Engineers.

A POCKET-BOOK OF USEFUL TABLES AND FORMULÆ FOR MARINE ENGINEERS. By FRANK PROCTOR, A.I.N.A. Third Edition. Royal 32mo, leather, gilt edges, with strap, 4s.

"We recommend it to our readers as going far to supply a long-felt want."—*Naval Science.*
"A most useful companion to all marine engineers."—*United Service Gazette.*

Introduction to Marine Engineering.

ELEMENTARY ENGINEERING: A Manual for Young Marine Engineers and Apprentices. In the Form of Questions and Answers on Metals, Alloys, Strength of Materials, Construction and Management of Marine Engines and Boilers, Geometry, &c. &c. With an Appendix of Useful Tables. By JOHN SHERREN BREWER, Government Marine Surveyor, Hongkong. Second Edition, Revised. Small crown 8vo, 2s. cloth.

"Contains much valuable information for the class for whom it is intended, especially in the chapters on the management of boilers and engines."—*Nautical Magazine.*
"A useful introduction to the more elaborate text-books."—*Scotsman.*
"To a student who has the requisite desire and resolve to attain a thorough knowledge, Mr. Brewer offers decidedly useful help."—*Athenæum.*

Navigation.

PRACTICAL NAVIGATION. Consisting of THE SAILOR'S SEA-BOOK, by JAMES GREENWOOD and W. H. ROSSER: together with the requisite Mathematical and Nautical Tables for the Working of the Problems, by HENRY LAW, C.E., and Professor J. R. YOUNG. Illustrated. 12mo, 7s. strongly half-bound.

Drawing for Marine Engineers.

LOCKIE'S MARINE ENGINEER'S DRAWING-BOOK. Adapted to the Requirements of the Board of Trade Examinations. By JOHN LOCKIE, C.E. With 22 Plates, Drawn to Scale. Royal 8vo, 3s. 6d. cloth. [*Just published.*

"The student who learns from these drawings will have nothing to unlearn."—*Engineer.*
"The examples chosen are essentially practical, and are such as should prove of service to engineers generally, while admirably fulfilling their specific purpose."—*Mechanical World.*

Sailmaking.

THE ART AND SCIENCE OF SAILMAKING. By SAMUEL B. SADLER, Practical Sailmaker, late in the employment of Messrs. Ratsey and Lapthorne, of Cowes and Gosport. With Plates and other Illustrations. Small 4to, 12s. 6d. cloth. [*Just published.*

SUMMARY OF CONTENTS.

CHAP. I. THE MATERIALS USED AND THEIR RELATION TO SAILS.—II. ON THE CENTRE OF EFFORT.—III. ON MEASURING.—IV. ON DRAWING.—V. ON THE NUMBER OF CLOTHS REQUIRED.—VI. ON ALLOWANCES.—VII. CALCULATION OF GORES.—VIII. ON CUTTING OUT.—IX. ON ROPING.—X. ON DIAGONAL-CUT SAILS.—XI. CONCLUDING REMARKS.

"This work is very ably written, and is illustrated by diagrams and carefully-worked calculations. The work should be in the hands of every sailmaker, whether employer or employed, as it cannot fail to assist them in the pursuit of their important avocations."—*Isle of Wight Herald.*
"This extremely practical work gives a complete education in all the branches of the manufacture, cutting out, roping, seaming, and goring. It is copiously illustrated, and will form a firstrate text-book and guide."—*Portsmouth Times.*
"The author of this work has rendered a distinct service to all interested in the art of sailmaking. The subject of which he treats is a congenial one. Mr. Sadler is a practical sailmaker, and his devoted years of careful observation and study to the subject; and the results of the experience thus gained he has set forth in the volume before us."—*Steamship.*

Chain Cables.

CHAIN CABLES AND CHAINS. Comprising Sizes and Curves of Links, Studs, &c., Iron for Cables and Chains, Chain Cable and Chain Making, Forming and Welding Links, Strength of Cables and Chains, Certificates for Cables, Marking Cables, Prices of Chain Cables and Chains, Historical Notes, Acts of Parliament, Statutory Tests, Charges for Testing, List of Manufacturers of Cables, &c. &c. By THOMAS W. TRAILL, F.E.R.N., M. Inst. C.E., Engineer Surveyor in Chief, Board of Trade, Inspector of Chain Cable and Anchor Proving Establishments, and General Superintendent, Lloyd's Committee on Proving Establishments. With numerous Tables, Illustrations and Lithographic Drawings. Folio, £2 2s. cloth, bevelled boards.

"It contains a vast amount of valuable information. Nothing seems to be wanting to make it a complete and standard work of reference on the subject."—*Nautical Magazine.*

MINING AND METALLURGY.

Metalliferous Mining in the United Kingdom.

BRITISH MINING: A Treatise on the History, Discovery, Practical Development, and Future Prospects of Metalliferous Mines in the United Kingdom. By ROBERT HUNT, F.R.S., Keeper of Mining Records; Editor of "Ure's Dictionary of Arts, Manufactures, and Mines," &c. Upwards of 950 pp., with 230 Illustrations. Second Edition, Revised. Super-royal 8vo, £2 2s. cloth.

"One of the most valuable works of reference of modern times. Mr. Hunt, as keeper of mining records of the United Kingdom, has had opportunities for such a task not enjoyed by anyone else, and has evidently made the most of them. . . . The language and style adopted are good, and the treatment of the various subjects laborious, conscientious, and scientific."—*Engineering*.

"The book is, in fact, a treasure-house of statistical information on mining subjects, and we know of no other work embodying so great a mass of matter of this kind. Were this the only merit of Mr. Hunt's volume, it would be sufficient to render it indispensable to the Library of everyone interested in the development of the mining and metallurgical industries of this country."—*Athenæum*.

"A mass of information not elsewhere available, and of the greatest value to those who may be interested in our great mineral industries."—*Engineer*.

Metalliferous Minerals and Mining.

A TREATISE ON METALLIFEROUS MINERALS AND MINING. By D. C. DAVIES, F.G.S., Mining Engineer, &c., Author of "A Treatise on Slate and Slate Quarrying." Fifth Edition, thoroughly Revised and much Enlarged, by his Son, E. HENRY DAVIES, M.E., F.G.S. With about 150 Illustrations. Crown 8vo, 12s. 6d. cloth. [*Just published.*

"Neither the practical miner nor the general reader interested in mines can have a better book for his companion and his guide."—*Mining Journal.* [*Mining World.*

"We are doing our readers a service in calling their attention to this valuable work."—

"A book that will not only be useful to the geologist, the practical miner, and the metallurgist but also very interesting to the general public."—*Iron*.

"As a history of the present state of mining throughout the world this book has a real value, and it supplies an actual want."—*Athenæum*.

Earthy Minerals and Mining.

A TREATISE ON EARTHY & OTHER MINERALS AND MINING. By D. C. DAVIES, F.G.S., Author of "Metalliferous Minerals," &c. Third Edition. Revised and Enlarged, by his Son, E. HENRY DAVIES, M.E., F.G.S. With about 100 Illusts. Cr. 8vo, 12s. 6d. cl. [*Just published.*

"We do not remember to have met with any English work on mining matters that contains the same amount of information packed in equally convenient form."—*Academy*.

"We should be inclined to rank it as among the very best of the handy technical and trades manuals which have recently appeared."—*British Quarterly Review.*

Mining Machinery.

MACHINERY FOR METALLIFEROUS MINES, including Motive Power, Haulage, Transport, and Electricity as applied to Mining. By E. HENRY DAVIES, M.E., F.G.S., &c. &c. [*In preparation.*

Underground Pumping Machinery.

MINE DRAINAGE. Being a Complete and Practical Treatise on Direct-Acting Underground Steam Pumping Machinery, with a Description of a large number of the best known Engines, their General Utility and the Special Sphere of their Action, the Mode of their Application, and their merits compared with other forms of Pumping Machinery. By STEPHEN MICHELL. 8vo, 15s. cloth.

"Will be highly esteemed by colliery owners and lessees, mining engineers, and students generally who require to be acquainted with the best means of securing the drainage of mines. It is a most valuable work, and stands almost alone in the literature of steam pumping machinery."—*Colliery Guardian.*

"Much valuable information is given, so that the book is thoroughly worthy of an extensive circulation amongst practical men and purchasers of machinery."—*Mining Journal.*

Mining Tools.

A MANUAL OF MINING TOOLS. For the Use of Mine Managers, Agents, Students, &c. By WILLIAM MORGANS, Lecturer on Practical Mining at the Bristol School of Mines. 12mo, 2s. 6d. cloth limp.

ATLAS OF ENGRAVINGS to Illustrate the above, containing 235 Illustrations of Mining Tools, drawn to scale. 4to, 4s. 6d. cloth.

"Students in the science of mining, and overmen, captains, managers, and viewers may gain practical knowledge and useful hints by the study of Mr. Morgans' manual."—*Colliery Guardian.*

"A valuable work, which will tend materially to improve our mining literature."—*Mining Journal.*

Prospecting for Gold and other Metals.

THE PROSPECTOR'S HANDBOOK: A Guide for the Prospector and Traveller in Search of Metal-Bearing or other Valuable Minerals. By J. W. ANDERSON, M.A. (Camb.), F.R.G.S., Author of "Fiji and New Caledonia." Fifth Edition, thoroughly Revised and Enlarged. Small crown 8vo, 3s. 6d. cloth.

"Will supply a much felt want, especially among Colonists, in whose way are so often thrown many mineralogical specimens the value of which it is difficult to determine."—*Engineer.*
"How to find commercial minerals, and how to identify them when they are found, are the leading points to which attention is directed. The author has managed to pack as much practical detail into his pages as would supply material for a book three times its size."—*Mining Journal.*

Mining Notes and Formulæ.

NOTES AND FORMULÆ FOR MINING STUDENTS. By JOHN HERMAN MERIVALE, M.A., Certificated Colliery Manager, Professor of Mining in the Durham College of Science, Newcastle-upon-Tyne. Third Edition, Revised and Enlarged. Small crown 8vo, 2s. 6d. cloth.

"Invaluable to anyone who is working up for an examination on mining subjects."—*Iron and Coal Trades Review.*
"The author has done his work in an exceedingly creditable manner, and has produced a book that will be of service to students, and those who are practically engaged in mining operations."—*Engineer.*
"A vast amount of technical matter of the utmost value to mining engineers, and of considerable interest to students."—*Schoolmaster.*

Miners' and Metallurgists' Pocket-Book.

A POCKET-BOOK FOR MINERS AND METALLURGISTS. Comprising Rules, Formulæ, Tables, and Notes, for Use in Field and Office Work. By F. DANVERS POWER, F.G.S., M.E. Fcap. 8vo, 9s. leather, gilt edges. [*Just published.*

"The book seems to contain an immense amount of useful information in a small space, and no doubt will prove to be a valuable and handy book for mining engineers."—C.LE NEVE FOSTER, Esq.
"Miners and metallurgists will find in this work a useful *vade-mecum* containing a mass of rules, formulæ, tables, and various other information, the necessity for reference to which occurs in their daily duties."—*Iron.*
"A marvellous compendium which every miner who desires to do work rapidly and well should hasten to buy."—*Redruth Times.*
"Mr. Power has succeeded in producing a pocket-book which certainly deserves to become the engineer's *vade-mecum.*"—*Mechanical World.*

Mineral Surveying and Valuing.

THE MINERAL SURVEYOR AND VALUER'S COMPLETE GUIDE, comprising a Treatise on Improved Mining Surveying and the Valuation of Mining Properties, with New Traverse Tables. By WM. LINTERN, Mining and Civil Engineer. Third Edition, with an Appendix on "Magnetic and Angular Surveying," with Records of the Peculiarities of Needle Disturbances. With Four Plates of Diagrams, Plans, &c. 12mo, 4s. cloth.

"Mr. Lintern's book forms a valuable and thoroughly trustworthy guide."—*Iron and Coal Trades Review.*
"This new edition must be of the highest value to colliery surveyors, proprietors, and managers."—*Colliery Guardian.*

Asbestos and its Uses.

ASBESTOS: Its Properties, Occurrence, and Uses. With some Account of the Mines of Italy and Canada. By ROBERT H. JONES. With Eight Collotype Plates and other Illustrations. Crown 8vo, 12s. 6d. cloth.

"An interesting and invaluable work."—*Colliery Guardian.*
"A valuable addition to the architect's and engineer's library."—*Building News.*

Explosives.

A HANDBOOK ON MODERN EXPLOSIVES. Being a Practical Treatise on the Manufacture and Application of Dynamite, Gun-Cotton, Nitro-Glycerine, and other Explosive Compounds. Including the Manufacture of Collodion-Cotton. By M. EISSLER, Mining Engineer and Metallurgical Chemist, Author of "The Metallurgy of Gold," "The Metallurgy of Silver," &c. With about 100 Illusts. Crown 8vo, 10s. 6d. cloth.

"Useful not only to the miner, but also to officers of both services to whom blasting and the use of explosives generally may at any time become a necessary auxiliary."—*Nature.*
"A veritable mine of information on the subject of explosives employed for military, mining, and blasting purposes."—*Army and Navy Gazette.*
"The book is clearly written. Taken as a whole, we consider it an excellent little book and one that should be found of great service to miners and others who are engaged in work requiring the use of explosives."—*Athenæum.*

Colliery Management.

THE COLLIERY MANAGER'S HANDBOOK: A Comprehensive Treatise on the Laying-out and Working of Collieries, Designed as a Book of Reference for Colliery Managers, and for the Use of Coal-Mining Students preparing for First-class Certificates. By CALEB PAMELY, Mining Engineer and Surveyor; Member of the North of England Institute of Mining and Mechanical Engineers; and Member of the South Wales Institute of Mining Engineers. With nearly 500 Plans, Diagrams, and other Illustrations. Medium 8vo, about 600 pages. Price £1 5s. strongly bound.

SUMMARY OF CONTENTS.

GEOLOGY. — SEARCH FOR COAL. — SHAFT SINKING. — FITTING UP THE SHAFT AND SURFACE ARRANGEMENTS. — STEAM BOILERS AND THEIR FITTINGS. — TIMBERING AND WALLING. — NARROW WORK AND METHODS OF WORKING. — UNDERGROUND CONVEYANCE. — DRAINAGE. — THE GASES MET WITH IN MINES; VENTILATION. — ON THE FRICTION OF AIR IN MINES. — SURVEYING AND PLANNING. — SAFETY LAMPS AND FIRE-DAMP DETECTORS. — SUNDRY AND INCIDENTAL OPERATIONS AND APPLIANCES. — MISCELLANEOUS QUESTIONS AND ANSWERS.

Appendix: SUMMARY OF REPORT OF H.M. COMMISSIONERS ON ACCIDENTS IN MINES.

*** OPINIONS OF THE PRESS.

"Mr. Pamely has not only given us a comprehensive reference book of a very high order, suitable to the requirements of mining engineers and colliery managers, but at the same time has provided mining students with a class-book that is as interesting as it is instructive."—*Colliery Manager.*

"Mr. Pamely's work is eminently suited to the purpose for which it is intended—being clear, interesting, exhaustive, rich in detail, and up to date, giving descriptions of the very latest machines in every department. . . . A mining engineer could scarcely go wrong who followed this work."—*Colliery Guardian.*

"This is the most complete 'all round' work on coal-mining published in the English language. . . . No library of coal-mining books is complete without it."—*Colliery Engineer* (Scranton, Pa., U.S.A.).

"Mr. Pamely's work is in all respects worthy of our admiration. No person in any responsible position connected with mines should be without a copy."—*Westminster Review.*

Coal and Iron.

THE COAL AND IRON INDUSTRIES OF THE UNITED KINGDOM. Comprising a Description of the Coal Fields, and of the Principal Seams of Coal, with Returns of their Produce and its Distribution, and Analyses of Special Varieties. Also an Account of the occurrence of Iron Ores in Veins or Seams; Analyses of each Variety; and a History of the Rise and Progress of Pig Iron Manufacture. By RICHARD MEADE, Assistant Keeper of Mining Records. With Maps. 8vo, £1 8s. cloth.

"The book is one which must find a place on the shelves of all interested in coal and iron production, and in the iron, steel, and other metallurgical industries."—*Engineer.*

"Of this book we may unreservedly say that it is the best of its class which we have ever met. . . . A book of reference which no one engaged in the iron or coal trades should omit from his library."—*Iron and Coal Trades Review.*

Coal Mining.

COAL AND COAL MINING: A Rudimentary Treatise on. By the late Sir WARINGTON W. SMYTH, M.A., F.R.S., &c., Chief Inspector of the Mines of the Crown. Seventh Edition, Revised and Enlarged. With numerous Illustrations. 12mo, 4s. cloth boards.

"As an outline is given of every known coal-field in this and other countries, as well as of the principal methods of working, the book will doubtless interest a very large number of readers."—*Mining Journal.*

Subterraneous Surveying.

SUBTERRANEOUS SURVEYING, Elementary and Practical Treatise on, with and without the Magnetic Needle. By THOMAS FENWICK, Surveyor of Mines, and THOMAS BAKER, C.E. Illust. 12mo, 3s. cloth boards.

Granite Quarrying.

GRANITES AND OUR GRANITE INDUSTRIES. By GEORGE F. HARRIS, F.G.S., Membre de la Société Belge de Géologie, Lecturer on Economic Geology at the Birkbeck Institution, &c. With Illustrations. Crown 8vo, 2s. 6d. cloth.

"A clearly and well-written manual for persons engaged or interested in the granite industry."—*Scotsman.*

"An interesting work, which will be deservedly esteemed."—*Colliery Guardian.*

"An exceedingly interesting and valuable monograph on a subject which has hitherto received unaccountably little attention in the shape of systematic literary treatment."—*Scottish Leader.*

Gold, Metallurgy of.

THE METALLURGY OF GOLD: A Practical Treatise on the Metallurgical Treatment of Gold-bearing Ores. Including the Processes of Concentration and Chlorination, and the Assaying, Melting, and Refining of Gold. By M. EISSLER, Mining Engineer and Metallurgical Chemist, formerly Assistant Assayer of the U.S. Mint, San Francisco. Third Edition, Revised and greatly Enlarged. With 187 Illustrations. Crown 8vo, 12s. 6d. cloth.

"This book thoroughly deserves its title of a 'Practical Treatise.' The whole process of gold milling, from the breaking of the quartz to the assay of the bullion, is described in clear and orderly narrative and with much, but not too much, fulness of detail."—*Saturday Review.*
"The work is a storehouse of information and valuable data, and we strongly recommend it to all professional men engaged in the gold-mining industry."—*Mining Journal.*

Silver, Metallurgy of.

THE METALLURGY OF SILVER: A Practical Treatise on the Amalgamation, Roasting, and Lixiviation of Silver Ores. Including the Assaying, Melting and Refining, of Silver Bullion. By M. EISSLER, Author of "The Metallurgy of Gold," &c. Second Edition, Enlarged. With 150 Illustrations. Crown 8vo, 10s. 6d. cloth.

"A practical treatise, and a technical work which we are convinced will supply a long-felt want amongst practical men, and at the same time be of value to students and others indirectly connected with the industries."—*Mining Journal.*
"From first to last the book is thoroughly sound and reliable."—*Colliery Guardian.*
"For chemists, practical miners, assayers, and investors alike, we do not know of any work on the subject so handy and yet so comprehensive."—*Glasgow Herald.*

Lead, Metallurgy of.

THE METALLURGY OF ARGENTIFEROUS LEAD: A Practical Treatise on the Smelting of Silver-Lead Ores and the Refining of Lead Bullion. Including Reports on various Smelting Establishments and Descriptions of Modern Smelting Furnaces and Plants in Europe and America. By M. EISSLER, M.E., Author of "The Metallurgy of Gold," &c. Crown 8vo, 400 pp., with 183 Illustrations, 12s. 6d. cloth.

"This is a very good book."—*Colliery Guardian.*
"The numerous metallurgical processes, which are fully and extensively treated of, embrace all the stages experienced in the passage of the lead from the various natural states to its issue from the refinery as an article of commerce."—*Practical Engineer.*
"The present volume fully maintains the reputation of the author. Those who wish to obtain a thorough insight into the present state of this industry cannot do better than read this volume, and all mining engineers cannot fail to find many useful hints and suggestions in it."—*Industries.*
"It is most carefully written and illustrated with capital drawings and diagrams. In fact, it is the work of an expert for experts, by whom it will be prized as an indispensable text-book."—*Bristol Mercury.*

Iron, Metallurgy of.

METALLURGY OF IRON. Containing History of Iron Manufacture, Methods of Assay, and Analyses of Iron Ores, Processes of Manufacture of Iron and Steel, &c. By H. BAUERMAN, F.G.S. A.R.S.M. With numerous Illustrations. Sixth Edition, Revised and Enlarged. 12mo, 5s. 6d. cloth.

"Carefully written, it has the merit of brevity and conciseness, as to less important points; while all material matters are very fully and thoroughly entered into."—*Standard.*

Iron Mining.

THE IRON ORES OF GREAT BRITAIN AND IRELAND: Their Mode of Occurrence, Age, and Origin, and the Methods of Searching for and Working them, with a Notice of some of the Iron Ores of Spain. By J. D. KENDALL, F.G.S., M.E. Crown 8vo, with Illusts., 16s. cloth.
[*Nearly ready.*

SUMMARY OF CONTENTS.

THE EARLY WORKING OF IRON ORE.—THE HAEMATITE DEPOSITS OF WEST CUMBERLAND AND FURNESS.—THE IRON ORES OF CORNWALL, DEVON, AND WEST SOMERSET.—THE LIMONITE OF THE FOREST OF DEAN AND SOUTH WALES.—THE SIDERITE AND LIMONITE OF ALSTON AND WEARDALE. THE ARGILLACEOUS IRONSTONES OF THE CARBONIFEROUS ROCKS.—THE IRON ORES OF THE SECONDARY ROCKS.—THE IRON ORES OF ANTRIM.—SOME OF THE IRON ORES OF SPAIN.—THE AGE AND ORIGIN OF IRON ORE DEPOSITS.—SEARCHING FOR AND WORKING IRON ORES.—WORKING COSTS AND SELLING PRICES.—RENTS, ROYALTIES, WAY-LEAVES, &c. — EPITOMES OF LEASES, &c. &c.

ELECTRICITY, ELECTRICAL ENGINEERING, etc.

Electrical Engineering.

THE ELECTRICAL ENGINEER'S POCKET-BOOK OF MODERN RULES, FORMULÆ, TABLES, AND DATA. By H. R. KEMPE, M.Inst.E.E., A.M.Inst.C.E., Technical Officer, Postal Telegraphs, Author of "A Handbook of Electrical Testing," &c. Second Edition, thoroughly Revised, with Additions. With numerous Illustrations. Royal 32mo, oblong, 5s. leather. [*Just published.*

"There is very little in the shape of formulæ or data which the electrician is likely to want in a hurry which cannot be found in its pages."—*Practical Engineer.*

"A very useful book of reference for daily use in practical electrical engineering and its various applications to the industries of the present day."—*Iron.*

"It is the best book of its kind."—*Electrical Engineer.*

"Well arranged and compact. The 'Electrical Engineer's Pocket-Book' is a good one."—*Electrician.*

"Strongly recommended to those engaged in the various electrical industries."—*Electrical Review.*

Electric Lighting.

ELECTRIC LIGHT FITTING: A Handbook for Working Electrical Engineers, embodying Practical Notes on Installation Management. By JOHN W. URQUHART, Electrician, Author of "Electric Light," &c. With numerous Illustrations. Crown 8vo, 5s. cloth.

"This volume deals with what may be termed the mechanics of electric lighting, and is addressed to men who are already engaged in the work or are training for it. The work traverses a great deal of ground, and may be read as a sequel to the same author's useful work on 'Electric Light.'"—*Electrician.*

"This is an attempt to state in the simplest language the precautions which should be adopted in installing the electric light, and to give information, for the guidance of those who have to run the plant when installed. The book is well worth the perusal of the workmen for whom it is written."—*Electrical Review.*

"We have read this book with a good deal of pleasure. We believe that the book will be of use to practical workmen, who will not be alarmed by finding mathematical formulæ which they are unable to understand."—*Electrical Plant.*

"Eminently practical and useful. . . . Ought to be in the hands of everyone in charge of an electric light plant."—*Electrical Engineer.*

"Altogether Mr. Urquhart has succeeded in producing a really capital book, which we have no hesitation in recommending to the notice of working electricians and electrical engineers."—*Mechanical World.*

Electric Light.

ELECTRIC LIGHT: Its Production and Use. Embodying Plain Directions for the Treatment of Dynamo-Electric Machines, Batteries, Accumulators, and Electric Lamps. By J. W. URQUHART, C.E., Author of "Electric Light Fitting," "Electroplating," &c. Fifth Edition, carefully Revised, with Large Additions and 145 Illustrations. Crown 8vo, 7s. 6d. cloth. [*Just published.*

"The whole ground of electric lighting is more or less covered and explained in a very clear and concise manner."—*Electrical Review.*

"Contains a good deal of very interesting information, especially in the parts where the author gives dimensions and working costs."—*Electrical Engineer.*

"A miniature *vade-mecum* of the salient facts connected with the science of electric lighting."—*Electrician.*

"You cannot for your purpose have a better book than 'Electric Light,' by Urquhart."—*Engineer.*

"The book is by far the best that we have yet met with on the subject."—*Athenæum.*

Construction of Dynamos.

DYNAMO CONSTRUCTION: A Practical Handbook for the Use of Engineer Constructors and Electricians-in-Charge. Embracing Framework Building, Field Magnet and Armature Winding and Grouping, Compounding, &c. With Examples of leading English, American, and Continental Dynamos and Motors. By J. W. URQUHART, Author of "Electric Light," "Electric Light Fitting," &c. With upwards of 100 Illustrations. Crown 8vo, 7s. 6d. cloth. [*Just published.*

"Mr. Urquhart's book is the first one which deals with these matters in such a way that the engineering student can understand them. The book is very readable, and the author leads his readers up to difficult subjects by reasonably simple tests."—*Engineering Review.*

"The author deals with his subject in a style so popular as to make his volume a handbook of great practical value to engineer contractors and electricians in charge of lighting installations."—*Scotsman.*

"'Dynamo Construction' more than sustains the high character of the author's previous publications. It is sure to be widely read by the large and rapidly increasing number of practical electricians."—*Glasgow Herald.*

"A book for which a demand has long existed."—*Mechanical World.*

Electric Lighting of Ships.
ELECTRIC SHIP-LIGHTING : A Handbook on the Practical Fitting and Running of Ship's Electrical Plant. For the Use of Shipowners and Builders, Marine Electricians, and Sea-going Engineers in Charge. By J. W. URQUHART, C.E., Author of "Electric Light," &c. With numerous Illustrations. Crown 8vo, 7s 6d. cloth. [*Just published.*

"The subject of ship electric lighting is one of vast importance in these days, and Mr Urquhart is to be highly complimented for placing such a valuable work at the service of the practical marine electrician."—*The Steamship.*

"Distinctly a book which of its kind stands almost alone, and for which there should be a demand."—*Electrical Review.*

Electric Lighting.
THE ELEMENTARY PRINCIPLES OF ELECTRIC LIGHTING. By ALAN A. CAMPBELL SWINTON, Associate I.E.E. Second Edition, Enlarged and Revised. With 16 Illustrations. Crown 8vo, 1s. 6d. cloth.

"Anyone who desires a short and thoroughly clear exposition of the elementary principles of electric-lighting cannot do better than read this little work."—*Bradford Observer.*

Dynamic Electricity.
THE ELEMENTS OF DYNAMIC ELECTRICITY AND MAGNETISM. By PHILIP ATKINSON, A.M., Ph.D., Author of "Elements of Static Electricity," "The Elements of Electric Lighting," &c. &c. Crown 8vo, 417 pp., with 120 Illustrations, 10s. 6d. cloth.

Dynamo Construction.
HOW TO MAKE A DYNAMO : *A Practical Treatise for Amateurs.* Containing numerous Illustrations and Detailed Instructions for Constructing a Small Dynamo, to Produce the Electric Light. By ALFRED CROFTS. Fourth Edition, Revised and Enlarged. Crown 8vo, 2s. cloth. [*Just published.*

"The instructions given in this unpretentious little book are sufficiently clear and explicit to enable any amateur mechanic possessed of average skill and the usual tools to be found in an amateur's workshop, to build a practical dynamo machine."—*Electrician.*

Text Book of Electricity.
THE STUDENT'S TEXT-BOOK OF ELECTRICITY. By HENRY M. NOAD, Ph.D., F.R.S. New Edition, carefully Revised. With Introduction and Additional Chapters, by W. H. PREECE, M.I.C.E., Vice-President of Society of Telegraph Engineers, &c. With 470 Illustrations. Crown 8vo, 12s. 6d. cloth.

"We can recommend Dr. Noad's book for clear style, great range of subject, a good Index, and a plethora of woodcuts. Such collections as the present are indispensable."—*Athenæum.*

"An admirable text-book for every student — beginner or advanced — of electricity."—*Engineering.*

Electricity.
A MANUAL OF ELECTRICITY : *Including Galvanism, Magnetism, Dia-Magnetism, Electro-Dynamics, Magno-Electricity, and the Electric Telegraph.* By HENRY M. NOAD, Ph.D., F.R.S., F.C.S. Fourth Edition. With 500 Woodcuts. 8vo, £1 4s. cloth.

*** This is the original work of Dr. Noad (published in 1859) *upon which the* STUDENT'S TEXT-BOOK (*see above*) *may be said to be founded. Very few copies of it are left.*

A New Dictionary of Electricity.
THE STANDARD ELECTRICAL DICTIONARY. A Popular Dictionary of Words and Terms Used in the Practice of Electric Engineering. By T. O'CONNOR SLOANE, A.M., Ph.D., Author of "The Arithmetic of Electricity," &c. Cr. 8vo, 630 pp., 350 Illusts., 12s. 6d. cl. [*Just published.*

NOTE.—*The purpose of this work is to present the public with a concise and practical book of reference. . . . Each title or subject is defined once in the text, and where a title is synonymous with one or more others the definition is given under one title, and the others appear at the foot of the article as synonyms. The work comprises upwards of 3,000 definitions, and will be found indispensable by all who are interested in electrical science and desire to keep abreast with the progress of the times.*

"An encyclopædia of electrical science in the compass of a dictionary. The information given is sound and clear. The book is well printed, well illustrated, and well up to date, and may be confidently recommended."—*Builder.*

ARCHITECTURE, BUILDING, etc.

Sir Wm. Chambers's Treatise on Civil Architecture.
THE DECORATIVE PART OF CIVIL ARCHITECTURE.
By Sir WILLIAM CHAMBERS, F.R.S. With Portrait, Illustrations, Notes, and an Examination of Grecian Architecture, by JOSEPH GWILT, F.S.A. Revised and Edited by W. H. LEEDS, with a Memoir of the Author. 66 Plates, 4to, 21s. cloth.

Mechanics for Architects.
THE MECHANICS OF ARCHITECTURE: A Treatise on Applied Mechanics, especially Adapted to the Use of Architects. By E. W. TARN, M.A., Author of "The Science of Building," &c. Illustrated with 125 Diagrams. Crown 8vo, 7s. 6d. cloth. [*Just published.*

SUMMARY OF CONTENTS.

CHAP. I. FORCES IN EQUILIBRIUM.—II. MOMENTS OF FORCES.—III. CENTRE OF GRAVITY.—IV. RESISTANCE OF MATERIALS TO STRESS.—V. DEFLECTION OF BEAMS.—VI. STRENGTH OF PILLARS.—VII. ROOFS, TRUSSES.—VIII. ARCHES.—IX. DOMES, SPIRES.—X. BUTTRESSES, SHORING, RETAINING WALLS, FOUNDATIONS.—XI. EFFECT OF WIND ON BUILDINGS.—XII. MISCELLANEOUS EXAMPLES AND SOLUTIONS.

Villa Architecture.
A HANDY BOOK OF VILLA ARCHITECTURE: Being a Series of Designs for Villa Residences in various Styles. With Outline Specifications and Estimates. By C. WICKES, Architect, Author of "The Spires and Towers of England," &c. 61 Plates, 4to, £1 11s. 6d. half-morocco, gilt edges.
"The whole of the designs bear evidence of their being the work of an artistic architect, and they will prove very valuable and suggestive."—*Building News.*

Text-Book for Architects.
THE ARCHITECT'S GUIDE: Being a Text-Book of Useful Information for Architects, Engineers, Surveyors, Contractors, Clerks of Works, &c. &c. By FREDERICK ROGERS, Architect, Author of "Specifications for Practical Architecture," &c. Second Edition, Revised and Enlarged. With numerous Illustrations. Crown 8vo, 6s. cloth.
"As a text-book of useful information for architects, engineers, surveyors, &c., it would be hard to find a handier or more complete little volume."—*Standard.*

Taylor and Cresy's Rome.
THE ARCHITECTURAL ANTIQUITIES OF ROME. By the late G. L. TAYLOR, Esq., F.R.I.B.A., and EDWARD CRESY, Esq. New Edition, thoroughly Revised by the Rev. ALEXANDER TAYLOR, M.A. (son of the late G. L. Taylor, Esq.), Fellow of Queen's College, Oxford, and Chaplain of Gray's Inn. Large folio, with 130 Plates, £3 3s. half-bound.
"Taylor and Cresy's work has from its first publication been ranked among those professional books which cannot be bettered. . . . It would be difficult to find examples of drawings, even among those of the most painstaking students of Gothic, more thoroughly worked out than are the one hundred and thirty plates in this volume."—*Architect.*

Linear Perspective.
ARCHITECTURAL PERSPECTIVE: The whole Course and Operations of the Draughtsman in Drawing a Large House in Linear Perspective. Illustrated by 39 Folding Plates. By F. O. FERGUSON. Demy 8vo, 3s. 6d. boards. [*Just published.*
"In a series of graphic illustrations of the actual processes the author shows the practical part of the art. It is all so easy and so clear that a child could follow him, and generations of students yet unborn will bless the name of Ferguson. . . . It is the most intelligible of the treatises on this ill-treated subject that I have met with."—E. INGRESS BELL, Esq., in the *R.I.B.A. Journal.*

Architectural Drawing.
PRACTICAL RULES ON DRAWING, for the Operative Builder and Young Student in Architecture. By GEORGE PYNE. With 14 Plates, 4to, 7s. 6d. boards.

Vitruvius' Architecture.
THE ARCHITECTURE of MARCUS VITRUVIUS POLLIO. Translated by JOSEPH GWILT, F.S.A., F.R.A.S. New Edition, Revised by the Translator. With 23 Plates. Fcap. 8vo, 5s. cloth.

The New Builder's Price Book, 1893.

LOCKWOOD'S BUILDER'S PRICE BOOK FOR 1893. A Comprehensive Handbook of the Latest Prices and Data for Builders, Architects, Engineers, and Contractors. Re-constructed, Re-written, and Greatly Enlarged. By FRANCIS T. W. MILLER, 640 closely-printed pages, crown 8vo, 4s. cloth.

⁎ OPINIONS OF THE PRESS.

"This book is a very useful one, and should find a place in every English office connected with the building and engineering professions."—*Industries.*

"This Price Book has been set up in new type. . . . Advantage has been taken of the transformation to add much additional information, and the volume is now an excellent book of reference."—*Architect.*

"In its new and revised form this Price Book is what a work of this kind should be—comprehensive, reliable, well arranged, legible, and well bound.'—*British Architect.*

"A work of established reputation."—*Athenæum.*

Designing, Measuring, and Valuing.

THE STUDENT'S GUIDE to the PRACTICE of MEASURING AND VALUING ARTIFICERS' WORK. Containing Directions for taking Dimensions, Abstracting the same, and bringing the Quantities into Bill, with Tables of Constants for Valuation of Labour, and for the Calculation of Areas and Solidities. Originally edited by EDWARD DOBSON, Architect. With Additions on Mensuration and Construction, and a New Chapter on Dilapidations, Repairs, and Contracts, by E. WYNDHAM TARN, M.A. Sixth Edition, including a Complete Form of a Bill of Quantities. With 8 Plates and 63 Woodcuts. Crown 8vo, 7s. 6d. cloth.

"Well fulfils the promise of its title-page, and we can thoroughly recommend it to the class for whose use it has been compiled. Mr. Tarn's additions and revisions have much increased the usefulness of the work, and have especially augmented its value to students."—*Engineering.*

"This edition will be found the most complete treatise on the principles of measuring and valuing artificers' work that has yet been published."—*Building News.*

Pocket Estimator and Technical Guide.

THE POCKET TECHNICAL GUIDE, MEASURER, AND ESTIMATOR FOR BUILDERS AND SURVEYORS. Containing Technical Directions for Measuring Work in all the Building Trades, Complete Specifications for Houses, Roads, and Drains, and an easy Method of Estimating the parts of a Building collectively. By A. C. BEATON, Author of "Quantities and Measurements." Sixth Edition. With 53 Woodcuts. Waistcoat-pocket size, 1s. 6d. gilt edges.

"No builder, architect, surveyor, or valuer should be without his 'Beaton.'"—*Building News.*

"Contains an extraordinary amount of information in daily requisition in measuring and estimating. Its presence in the pocket will save valuable time and trouble."—*Building World.*

Donaldson on Specifications.

THE HANDBOOK OF SPECIFICATIONS; or, Practical Guide to the Architect, Engineer, Surveyor, and Builder, in drawing up Specifications and Contracts for Works and Constructions. Illustrated by Precedents of Buildings actually executed by eminent Architects and Engineers. By Professor T. L. DONALDSON, P.R.I.B.A., &c. New Edition. In One large Vol., 8vo, with upwards of 1,000 pages of Text, and 33 Plates, £1 11s. 6d. cloth.

"In this work forty-four specifications of executed works are given, including the specifications for parts of the new Houses of Parliament, by Sir Charles Barry, and for the new Royal Exchange, by Mr. Tite, M.P. The latter, in particular, is a very complete and remarkable document. It embodies, to a great extent, as Mr. Donaldson mentions, 'the bill of quantities with the description of the works.' . . . It is valuable as a record, and more valuable still as a book of precedents. . . . Suffice it to say that Donaldson's 'Handbook of Specifications' must be bought by all architects."—*Builder.*

Bartholomew and Rogers' Specifications.

SPECIFICATIONS FOR PRACTICAL ARCHITECTURE. A Guide to the Architect, Engineer, Surveyor, and Builder. With an Essay on the Structure and Science of Modern Buildings. Upon the Basis of the Work by ALFRED BARTHOLOMEW, thoroughly Revised, Corrected, and greatly added to by FREDERICK ROGERS, Architect. Third Edition, Revised, with Additions. With numerous Illustrations. Medium 8vo, 15s. cloth.

"The collection of specifications prepared by Mr. Rogers on the basis of Bartholomew's work is too well known to need any recommendation from us. It is one of the books with which every young architect must be equipped ; for time has shown that the specifications cannot be set aside through any defect in them."—*Architect.*

Construction.

THE SCIENCE OF BUILDING: *An Elementary Treatise on the Principles of Construction.* By E. WYNDHAM TARN, M.A., Architect. Third Edition, Revised and Enlarged. With 59 Engravings. Fcap. 8vo, 4s. cl.

"A very valuable book, which we strongly recommend to all students."—*Builder.*
"No architectural student should be without this handbook of constructional knowledge."—*Architect.*

House Building and Repairing.

THE HOUSE-OWNER'S ESTIMATOR; or, What will it Cost to Build, Alter, or Repair? A Price Book adapted to the Use of Unprofessional People, as well as for the Architectural Surveyor and Builder. By JAMES D. SIMON, A.R.I.B.A. Edited and Revised by FRANCIS T. W. MILLER, A.R.I.B.A. With numerous Illustrations. Fourth Edition, Revised. Crown 8vo, 3s. 6d. cloth.

"In two years it will repay its cost a hundred times over."—*Field.*
"A very handy book."—*English Mechanic.*

Cottages and Villas.

COUNTRY AND SUBURBAN COTTAGES AND VILLAS: How to Plan and Build Them. Containing 33 Plates, with Introduction, General Explanations, and Description of each Plate. By JAMES W. BOGUE, Architect, Author of "Domestic Architecture," &c. 4to, 10s. 6d. cloth.

Building; Civil and Ecclesiastical.

A BOOK ON BUILDING, Civil and Ecclesiastical, including Church Restoration; with the Theory of Domes and the Great Pyramid, &c. By Sir EDMUND BECKETT, Bart., LL.D., F.R.A.S., Author of "Clocks and Watches, and Bells," &c. Second Edition, Enlarged. Fcap. 8vo, 5s. cloth.

"A book which is always amusing and nearly always instructive. The style throughout is in the highest degree condensed and epigrammatic."—*Times.*

Ventilation of Buildings.

VENTILATION. *A Text Book to the Practice of the Art of Ventilating Buildings.* With a Chapter upon Air Testing. By W. P. BUCHAN, R.P., Sanitary and Ventilating Engineer, Author of "Plumbing," &c. With 170 Illustrations. 12mo, 4s. cloth boards.

"Contains a great amount of useful practical information, as thoroughly interesting as it is technically reliable, and 'Ventilation' forms a worthy companion volume to the author's excellent treatise on 'Plumbing.'"—*British Architect.*
"It is invaluable alike for the architect and builder, and should be in the hands of everyone who has to deal in any way with the subject of ventilation."—*Metropolitan.*

The Art of Plumbing.

PLUMBING. *A Text Book to the Practice of the Art or Craft of the Plumber, with Supplementary Chapters on House Drainage, embodying the latest Improvements.* By WILLIAM PATON BUCHAN, R.P., Sanitary Engineer and Practical Plumber. Sixth Edition, Enlarged to 370 pages, and 380 Illustrations. 12mo, 4s. cloth boards.

"A text-book which may be safely put in the hands of every young plumber, and which will also be found useful by architects and medical professors."—*Builder.*
"A valuable text-book, and the only treatise which can be regarded as a really reliable manual of the plumber's art."—*Building News.*

Geometry for the Architect, Engineer, etc.

PRACTICAL GEOMETRY, for the Architect, Engineer, and Mechanic. Giving Rules for the Delineation and Application of various Geometrical Lines, Figures and Curves. By E. W. TARN, M.A., Architect, Author of "The Science of Building," &c. Second Edition. With 172 Illustrations. Demy 8vo, 9s. cloth.

"No book with the same objects in view has ever been published in which the clearness of the rules laid down and the illustrative diagrams have been so satisfactory."—*Scotsman.*

The Science of Geometry.

THE GEOMETRY OF COMPASSES; or, *Problems Resolved by the mere Description of Circles, and the use of Coloured Diagrams and Symbols.* By OLIVER BYRNE. Coloured Plates. Crown 8vo, 3s. 6d. cloth.

"The treatise is a good one, and remarkable—like all Mr. Byrne's contributions to the science of geometry—for the lucid character of its teaching."—*Building News.*

CARPENTRY, TIMBER, etc.

Tredgold's Carpentry, Revised & Enlarged by Tarn.
THE ELEMENTARY PRINCIPLES OF CARPENTRY.
A Treatise on the Pressure and Equilibrium of Timber Framing, the Resistance of Timber, and the Construction of Floors, Arches, Bridges, Roofs, Uniting Iron and Stone with Timber, &c. To which is added an Essay on the Nature and Properties of Timber, &c., with Descriptions of the kinds of Wood used in Building; also numerous Tables of the Scantlings of Timber for different purposes, the Specific Gravities of Materials, &c. By THOMAS TREDGOLD, C.E. With an Appendix of Specimens of Various Roofs of Iron and Stone, Illustrated. Seventh Edition, thoroughly revised and considerably enlarged by E. WYNDHAM TARN, M.A., Author of "The Science of Building," &c. With 61 Plates, Portrait of the Author, and several Woodcuts. In One large Vol., 4to, price £1 5s. cloth.
"Ought to be in every architect's and every builder's library."—*Builder*.
"A work whose monumental excellence must commend it wherever skilful carpentry is concerned. The author's principles are rather confirmed than impaired by time. The additional plates are of great intrinsic value."—*Building News*.

Woodworking Machinery.
WOODWORKING MACHINERY: Its Rise, Progress, and Construction. With Hints on the Management of Saw Mills and the Economical Conversion of Timber. Illustrated with Examples of Recent Designs by leading English, French, and American Engineers. By M. POWIS BALE, A.M.Inst.C.E., M.I.M.E. Large crown 8vo, 12s. 6d. cloth.
"Mr. Bale is evidently an expert on the subject and he has collected so much information that his book is all-sufficient for builders and others engaged in the conversion of timber."—*Architect*.
"The most comprehensive compendium of wood-working machinery we have seen. The author is a thorough master of his subject."—*Building News*.
"It should be in the office of every wood-working factory."—*English Mechanic*.

Saw Mills.
SAW MILLS: Their Arrangement and Management, and the Economical Conversion of Timber. (A Companion Volume to "Woodworking Machinery.") By M. POWIS BALE. With numerous Illustrations. Crown 8vo, 10s. 6d. cloth.
"The *administration* of a large sawing establishment is discussed, and the subject examined from a financial standpoint. Hence the size, shape, order, and disposition of saw-mills and the like are gone into in detail, and the course of the timber is traced from its reception to its delivery in its converted state. We could not desire a more complete or practical treatise."—*Builder*.
"We highly recommend Mr. Bale's work to the attention and perusal of all those who are engaged in the art of wood conversion, or who are about building or remodelling saw-mills on improved principles."—*Building News*.

Nicholson's Carpentry.
THE CARPENTER'S NEW GUIDE; or, Book of Lines for Carpenters; comprising all the Elementary Principles essential for acquiring knowledge of Carpentry. Founded on the late PETER NICHOLSON's Standard Work. A New Edition, Revised by ARTHUR ASHPITEL, F.S.A. Together with Practical Rules on Drawing, by GEORGE PYNE. With 74 Plates, 4to, £1 1s. cloth.

Handrailing and Stairbuilding.
A PRACTICAL TREATISE ON HANDRAILING: Showing New and Simple Methods for Finding the Pitch of the Plank, Drawing the Moulds, Bevelling, Jointing-up, and Squaring the Wreath. By GEORGE COLLINGS. Second Edition, Revised and Enlarged, to which is added A TREATISE ON STAIRBUILDING. With Plates and Diagrams. 12mo, 2s. 6d. cloth limp.
"Will be found of practical utility in the execution of this difficult branch of joinery."—*Builder*.
"Almost every difficult phase of this somewhat intricate branch of joinery is elucidated by the aid of plates and explanatory letterpress."—*Furniture Gazette*.

Circular Work.
CIRCULAR WORK IN CARPENTRY AND JOINERY: A Practical Treatise on Circular Work of Single and Double Curvature. By GEORGE COLLINGS, Author of "A Practical Treatise on Handrailing." Illustrated with numerous Diagrams. Second Edition. 12mo, 2s. 6d. cloth limp.
"An excellent example of what a book of this kind should be. Cheap in price, clear in definition and practical in the examples selected."—*Builder*.

Timber Merchant's Companion.

THE TIMBER MERCHANT'S AND BUILDER'S COMPANION. Containing New and Copious Tables of the Reduced Weight and Measurement of Deals and Battens, of all sizes, from One to a Thousand Pieces, and the relative Price that each size bears per Lineal Foot to any given Price per Petersburg Standard Hundred; the Price per Cube Foot of Square Timber to any given Price per Load of 50 Feet; the proportionate Value of Deals and Battens by the Standard, to Square Timber by the Load of 50 Feet; the readiest mode of ascertaining the Price of Scantling per Lineal Foot of any size, to any given Figure per Cube Foot, &c. &c. By WILLIAM DOWSING. Fourth Edition, Revised and Corrected. Cr. 8vo, 3s. cl.

"Everything is as concise and clear as it can possibly be made. There can be no doubt that every timber merchant and builder ought to possess it."—*Hull Advertiser.*

"We are glad to see a fourth edition of these admirable tables, which for correctness and simplicity of arrangement leave nothing to be desired."—*Timber Trades Journal.*

Practical Timber Merchant.

THE PRACTICAL TIMBER MERCHANT.' Being a Guide for the use of Building Contractors, Surveyors, Builders, &c., comprising useful Tables for all purposes connected with the Timber Trade, Marks of Wood, Essay on the Strength of Timber, Remarks on the Growth of Timber, &c. By W. RICHARDSON. Fcap. 8vo, 3s. 6d. cloth.

"This handy manual contains much valuable information for the use of timber merchants, builders, foresters, and all others connected with the growth, sale, and manufacture of timber."—*Journal of Forestry.*

Timber Freight Book.

THE TIMBER MERCHANT'S, SAW MILLER'S, AND IMPORTER'S FREIGHT BOOK AND ASSISTANT. Comprising Rules, Tables, and Memoranda relating to the Timber Trade. By WILLIAM RICHARDSON, Timber Broker; together with a Chapter on "SPEEDS OF SAW MILL MACHINERY," by M. POWIS BALE, M.I.M.E., &c., 12mo, 3s. 6d. cl. boards.

"A very useful manual of rules, tables, and memoranda relating to the timber trade. We recommend it as a compendium of calculation to all timber measurers and merchants, and as supplying a real want in the trade."—*Building News.*

Packing-Case Makers, Tables for.

PACKING-CASE TABLES; showing the number of Superficial Feet in Boxes or Packing-Cases, from six inches square and upwards. By W. RICHARDSON, Timber Broker. Third Edition. Oblong 4to, 3s. 6d. cl.

"Invaluable labour-saving tables."—*Ironmonger.*
"Will save much labour and calculation."—*Grocer.*

Superficial Measurement.

THE TRADESMAN'S GUIDE TO SUPERFICIAL MEASUREMENT. Tables calculated from 1 to 200 inches in length, by 1 to 108 inches in breadth. For the use of Architects, Surveyors, Engineers, Timber Merchants, Builders, &c. By. JAMES HAWKINGS. Fourth Edition. Fcap., 3s. 6d. cloth.

"A useful collection of tables to facilitate rapid calculation of surfaces. The exact area of any surface of which the limits have been ascertained can be instantly determined. The book will be found of the greatest utility to all engaged in building operations."—*Scotsman.*

"These tables will be found of great assistance to all who require to make calculations in superficial measurement."—*English Mechanic.*

Forestry.

THE ELEMENTS OF FORESTRY. Designed to afford Information concerning the Planting and Care of Forest Trees for Ornament or Profit, with Suggestions upon the Creation and Care of Woodlands. By F. B. HOUGH. Large crown 8vo, 10s. cloth.

Timber Importer's Guide.

THE TIMBER IMPORTER'S, TIMBER MERCHANT'S, AND BUILDER'S STANDARD GUIDE. By RICHARD E. GRANDY. Comprising an Analysis of Deal Standards, Home and Foreign, with Comparative Values and Tabular Arrangements for fixing Net Landed Cost on Baltic and North American Deals, including all intermediate Expenses, Freight, Insurance, &c. &c. Together with copious Information for the Retailer and Builder. Third Edition, Revised. 12mo, 2s. cloth limp.

"Everything it pretends to be: built up gradually, it leads one from a forest to a treenail, and throws in, as a makeweight, a host of material concerning bricks, columns, cisterns, &c."—*English Mechanic.*

DECORATIVE ARTS, etc.

Woods and Marbles (Imitation of).

SCHOOL OF PAINTING FOR THE IMITATION OF WOODS AND MARBLES, as Taught and Practised by A. R. VAN DER BURG and P. VAN DER BURG, Directors of the Rotterdam Painting Institution. Royal folio, 18¾ by 12¼ in., Illustrated with 24 full-size Coloured Plates; also 12 plain Plates, comprising 154 Figures. Second and Cheaper Edition. Price £1 11s. 6d.

List of Plates.

1. Various Tools required for Wood Painting—2, 3. Walnut: Preliminary Stages of Graining and Finished Specimen—4. Tools used for Marble Painting and Method of Manipulation—5, 6. St. Remi Marble: Earlier Operations and Finished Specimen—7. Methods of Sketching different Grains, Knots, &c.—8, 9. Ash: Preliminary Stages and Finished Specimen—10. Methods of Sketching Marble Grains—11, 12. Breche Marble: Preliminary Stages of Working and Finished Specimen—13. Maple: Methods of Producing the different Grains—14, 15. Bird's-eye Maple: Preliminary Stages and Finished Specimen—16. Methods of Sketching the different Species of White Marble—17, 18. White Marble: Preliminary Stages of Process and Finished Specimen—19. Mahogany: Specimens of various Grains and Methods of Manipulation—20, 21. Mahogany: Earlier Stages and Finished Specimen—22, 23, 24. Sienna Marble: Varieties of Grain, Preliminary Stages and Finished Specimen—25, 26, 27. Juniper Wood: Methods of producing Grain, &c.; Preliminary Stages and Finished Specimen—28, 29, 30. Vert de Mer Marble: Varieties of Grain and Methods of Working Unfinished and Finished Specimens—31, 32, 33. Oak: Varieties of Grain, Tools Employed, and Methods of Manipulation, Preliminary Stages and Finished Specimen—34, 35, 36. Waulsort Marble: Varieties of Grain, Unfinished and Finished Specimens.

*** OPINIONS OF THE PRESS.

"Those who desire to attain skill in the art of painting woods and marbles will find advantage in consulting this book. . . . Some of the Working Men's Clubs should give their young men the opportunity to study it."—*Builder.*

"A comprehensive guide to the art. The explanations of the processes, the manipulation and management of the colours, and the beautifully executed plates will not be the least valuable to the student who aims at making his work a faithful transcript of nature."—*Building News.*

House Decoration.

ELEMENTARY DECORATION. A Guide to the Simpler Forms of Everyday Art, as applied to the Interior and Exterior Decoration of Dwelling Houses, &c. Together with PRACTICAL HOUSE DECORATION: A Guide to the Art of Ornamental Painting, the Arrangement of Colours in Apartments, and the principles of Decorative Design. By JAMES W. FACEY. With numerous Illustrations. In One Vol., 5s. strongly half-bound.

House Painting, Graining, etc.

HOUSE PAINTING, GRAINING, MARBLING, AND SIGN WRITING, A Practical Manual of. By ELLIS A. DAVIDSON. Sixth Edition. With Coloured Plates and Wood Engravings. 12mo, 6s. cloth boards.

"A mass of information, of use to the amateur and of value to the practical man."—*English Mechanic.*

"Simply Invaluable to the youngster entering upon this particular calling, and highly serviceable to the man who is practising it."—*Furniture Gazette.*

Decorators, Receipts for.

THE DECORATOR'S ASSISTANT: A Modern Guide to Decorative Artists and Amateurs, Painters, Writers, Gilders, &c. Containing upwards of 600 Receipts, Rules and Instructions; with a variety of Information for General Work connected with every Class of Interior and Exterior Decorations, &c. Fifth Edition, Revised. 152 pp., crown 8vo, 1s. in wrapper.

"Full of receipts of value to decorators, painters, gilders, &c. The book contains the gist of larger treatises on colour and technical processes. It would be difficult to meet with a work so full of varied information on the painter's art."—*Building News.*

Moyr Smith on Interior Decoration.

ORNAMENTAL INTERIORS, ANCIENT AND MODERN. By J. MOYR SMITH. Super-royal 8vo, with 32 full-page Plates and numerous smaller Illustrations, handsomely bound in cloth, gilt top, price 18s.

"The book is well illustrated and handsomely got up, and contains some true criticism and a good many good examples of decorative treatment."—*The Builder.*

"To all who take an interest in elaborate domestic ornament this handsome volume will be welcome."—*Graphic.*

British and Foreign Marbles.

MARBLE DECORATION and the Terminology of British and Foreign Marbles. A Handbook for Students. By GEORGE H. BLAGROVE, Author of "Shoring and its Application," &c. With 23 Illustrations. Crown 8vo, 3s. 6d. cloth.

"This most useful and much wanted handbook should be in the hands of every architect and builder."—*Building World.*
"A carefully and usefully written treatise; the work is essentially practical."—*Scotsman.*

Marble Working, etc.

MARBLE AND MARBLE WORKERS: A Handbook for Architects, Artists, Masons, and Students. By ARTHUR LEE, Author of "A Visit to Carrara," "The Working of Marble," &c. Small crown 8vo, 2s. cloth.

"A really valuable addition to the technical literature of architects and masons."—*Building News.*

DELAMOTTE'S WORKS ON ILLUMINATION AND ALPHABETS.

A PRIMER OF THE ART OF ILLUMINATION, for the Use of Beginners: with a Rudimentary Treatise on the Art, Practical Directions for its Exercise, and Examples taken from Illuminated MSS., printed in Gold and Colours. By F. DELAMOTTE. New and Cheaper Edition. Small 4to, 6s. ornamental boards.

"The examples of ancient MSS. recommended to the student, which, with much good sense, the author chooses from collections accessible to all, are selected with judgment and knowledge, as well as taste."—*Athenæum.*

ORNAMENTAL ALPHABETS, Ancient and Mediæval, from the Eighth Century, with Numerals; including Gothic, Church-Text, large and small, German, Italian, Arabesque, Initials for Illumination, Monograms, Crosses, &c. &c., for the use of Architectural and Engineering Draughtsmen, Missal Painters, Masons, Decorative Painters, Lithographers, Engravers, Carvers, &c. &c. Collected and Engraved by F. DELAMOTTE, and printed in Colours. New and Cheaper Edition. Royal 8vo, oblong, 2s. 6d. ornamental boards.

"For those who insert enamelled sentences round gilded chalices, who blazon shop legends over shop-doors, who letter church walls with pithy sentences from the Decalogue, this book will be useful."—*Athenæum.*

EXAMPLES OF MODERN ALPHABETS, Plain and Ornamental; including German, Old English, Saxon, Italic, Perspective, Greek, Hebrew, Court Hand, Engrossing, Tuscan, Riband, Gothic, Rustic, and Arabesque; with several Original Designs, and an Analysis of the Roman and Old English Alphabets, large and small, and Numerals, for the use of Draughtsmen, Surveyors, Masons, Decorative Painters, Lithographers, Engravers, Carvers, &c. Collected and Engraved by F. DELAMOTTE, and printed in Colours. New and Cheaper Edition. Royal 8vo, oblong, 2s. 6d. ornamental boards.

"There is comprised in it every possible shape into which the letters of the alphabet and numerals can be formed, and the talent which has been expended in the conception of the various plain and ornamental letters is wonderful."—*Standard.*

MEDIÆVAL ALPHABETS AND INITIALS FOR ILLUMINATORS. By F. G. DELAMOTTE. Containing 21 Plates and Illuminated Title, printed in Gold and Colours. With an Introduction by J. WILLIS BROOKS. Fourth and Cheaper Edition. Small 4to, 4s. ornamental boards.

"A volume in which the letters of the alphabet come forth glorified in gilding and all the colours of the prism interwoven and intertwined and intermingled."—*Sun.*

THE EMBROIDERER'S BOOK OF DESIGN. Containing Initials, Emblems, Cyphers, Monograms, Ornamental Borders, Ecclesiastical Devices, Mediæval and Modern Alphabets, and National Emblems. Collected by F. DELAMOTTE, and printed in Colours. Oblong royal 8vo, 1s. 6d. ornamental wrapper.

"The book will be of great assistance to ladies and young children who are endowed with the art of plying the needle in this most ornamental and useful pretty work."—*East Anglian Times.*

Wood Carving.

INSTRUCTIONS IN WOOD-CARVING, for Amateurs; with Hints on Design. By A LADY. With Ten Plates. New and Cheaper Edition. Crown 8vo, 2s. in emblematic wrapper.

"The handicraft of the wood-carver, so well as a book can impart it, may be learnt from 'A Lady's' publication."—*Athenæum.*

NATURAL SCIENCE, etc.

The Heavens and their Origin.

THE VISIBLE UNIVERSE: Chapters on the Origin and Construction of the Heavens. By J. E. GORE, F.R.A.S., Author of "Star Groups," &c. Illustrated by 6 Stellar Photographs and 12 Lithographic Plates. Demy 8vo, 16s. cloth, gilt top. [*Just published*.

"A valuable and lucid summary of recent astronomical theory, rendered more valuable and attractive by a series of stellar photographs and other illustrations."—*The Times.*

"In presenting a clear and concise account of the present state of our knowledge, Mr. Gore has made a valuable addition to the literature of the subject."—*Nature.*

"Mr. Gore's 'Visible Universe' is one of the finest works on astronomical science that has recently appeared in our language. In spirit and in method it is scientific from cover to cover, but the style is so clear and attractive that it will be as acceptable and as readable to those who make no scientific pretensions as to those who devote themselves specially to matters astronomical."—*Leeds Mercury.*

"We are glad to bear witness to the fulness, the accuracy, and the entire honesty of the latest and the best compilation of the kind which has appeared of late years. . . . The illustrations are also admirable."—*Daily Chronicle.*

"As interesting as a novel, and instructive withal; the text being made still more luminous by stellar photographs and other illustrations. . . . A most valuable book."—*Manchester Examiner.*

The Constellations.

STAR GROUPS: A Student's Guide to the Constellations. By J. ELLARD GORE, F.R.A.S., M.R.I.A., &c., Author of "The Visible Universe," "The Scenery of the Heavens." With 30 Maps. Small 4to, 5s. cloth, silvered.

"A knowledge of the principal constellations visible in our latitudes may be easily acquired by the thirty maps and accompanying text contained in this work."—*Nature.*

"The volume contains thirty maps showing stars of the sixth magnitude—the usual naked-eye limit—and each is accompanied by a brief commentary, adapted to facilitate recognition and bring to notice objects of special interest. For the purpose of a preliminary survey of the 'midnight pomp' of the heavens, nothing could be better than a set of delineations averaging scarcely twenty square inches in area, and including nothing that cannot at once be identified."—*Saturday Review.*

"A very compact and handy guide to the constellations."—*Athenæum.*

The Microscope.

THE MICROSCOPE: Its Construction and Management, including Technique, Photo-micrography, and the Past and Future of the Microscope. By Dr. HENRI VAN HEURCK, Director of the Antwerp Botanical Gardens. English Edition, Re-Edited and Augmented by the Author from the Fourth French Edition, and Translated by WYNNE E. BAXTER, F.R.M.S., F.G.S., &c. About 400 pages, with Three Plates and upwards of 250 Woodcuts. Imp. 8vo, 18s. cloth gilt. [*Just published.*

"This is a translation of a well-known work, at once popular and comprehensive, on the structure, mechanism, and use of the microscope. Of adequate English manuals on the use of the microscope there is certainly no lack; but, as the translator very truly says, such a book as Professor van Heurck's must necessarily be of interest to all who devote serious attention to microscopic work as a means of comparing the continental views and modes of thought with those of their own and other countries.'—*Times.*

Astronomy.

ASTRONOMY. By the late Rev. ROBERT MAIN, M.A., F.R.S., formerly Radcliffe Observer at Oxford. Third Edition, Revised and Corrected to the present time, by WILLIAM THYNNE LYNN, B.A., F.R.A.S., formerly of the Royal Observatory, Greenwich. 12mo, 2s. cloth limp.

"A sound and simple treatise, very carefully edited, and a capital book for beginners."—*Knowledge.*

"Accurately brought down to the requirements of the present time by Mr. Lynn."—*Educational Times.*

Recent and Fossil Shells.

A MANUAL OF THE MOLLUSCA: Being a Treatise on Recent and Fossil Shells. By S. P. WOODWARD, A.L.S., F.G.S., late Assistant Palæontologist in the British Museum. With an Appendix on *Recent and Fossil Conchological Discoveries*, by RALPH TATE, A.L.S., F.G.S. Illustrated by A. N. WATERHOUSE and JOSEPH WILSON LOWRY. With 23 Plates and upwards of 300 Woodcuts. Reprint of Fourth Ed., 1880. Cr. 8vo, 7s. 6d. cl.

"A most valuable storehouse of conchological and geological information."—*Science Gossip.*

Geology and Genesis.

THE TWIN RECORDS OF CREATION; or, Geology and Genesis: their Perfect Harmony and Wonderful Concord. By GEORGE W. VICTOR LE VAUX. Numerous Illustrations. Fcap. 8vo, 5s. cloth.

"A valuable contribution to the evidences of Revelation, and disposes very conclusively of the arguments of those who would set God's Works against God's Word. No real difficulty is shirked and no sophistry is left unexposed."—*The Rock.*

DR. LARDNER'S COURSE OF NATURAL PHILOSOPHY.

THE HANDBOOK OF MECHANICS. Enlarged and almost rewritten by BENJAMIN LOEWY, F.R.A.S. With 378 Illustrations. Post 8vo, 6s. cloth.

"The perspicuity of the original has been retained, and chapters which had become obsolete have been replaced by others of more modern character. The explanations throughout are studiously popular, and care has been taken to show the application of the various branches of physics to the industrial arts, and to the practical business of life."—*Mining Journal.*

"Mr. Loewy has carefully revised the book, and brought it up to modern requirements."—*Nature.*

"Natural philosophy has had few exponents more able or better skilled in the art of popularising the subject than Dr. Lardner; and Mr. Loewy is doing good service in fitting this treatise, and the others of the series, for use at the present time."—*Scotsman.*

THE HANDBOOK OF HYDROSTATICS AND PNEUMATICS. New Edition, Revised and Enlarged, by BENJAMIN LOEWY, F.R.A.S. With 236 Illustrations. Post 8vo, 5s. cloth.

"For those 'who desire to attain an accurate knowledge of physical science without the profound methods of mathematical investigation,' this work is not merely intended, but well adapted."—*Chemical News.*

"The volume before us has been carefully edited, augmented to nearly twice the bulk of the former edition, and all the most recent matter has been added. . . . It is a valuable text-book."—*Nature.*

"Candidates for pass examinations will find it, we think, specially suited to their requirements."—*English Mechanic.*

THE HANDBOOK OF HEAT. Edited and almost entirely Re-written by BENJAMIN LOEWY, F.R.A.S., &c. 117 Illusts. Post 8vo, 6s. cloth.

"The style is always clear and precise, and conveys instruction without leaving any cloudiness or lurking doubts behind."—*Engineering.*

"A most exhaustive book on the subject on which it treats, and is so arranged that it can be understood by all who desire to attain an accurate knowledge of physical science. . . . Mr. Loewy has included all the latest discoveries in the varied laws and effects of heat."—*Standard.*

"A complete and handy text-book for the use of students and general readers."—*English Mechanic.*

THE HANDBOOK OF OPTICS. By DIONYSIUS LARDNER, D.C.L., formerly Professor of Natural Philosophy and Astronomy in University College, London. New Edition. Edited by T. OLVER HARDING, B.A. Lond., of University College, London. With 298 Illustrations. Small 8vo, 448 pages, 5s. cloth.

"Written by one of the ablest English scientific writers, beautifully and elaborately illustrated."—*Mechanic's Magazine.*

THE HANDBOOK OF ELECTRICITY, MAGNETISM, AND ACOUSTICS. By Dr. LARDNER. Ninth Thousand. Edit. by GEORGE CAREY FOSTER, B.A., F.C.S. With 400 Illustrations. Small 8vo, 5s. cloth.

"The book could not have been entrusted to anyone better calculated to preserve the terse and lucid style of Lardner, while correcting his errors and bringing up his work to the present state of scientific knowledge."—*Popular Science Review.*

THE HANDBOOK OF ASTRONOMY. Forming a Companion to the "Handbook of Natural Philosophy." By DIONYSIUS LARDNER, D.C.L., formerly Professor of Natural Philosophy and Astronomy in University College, London. Fourth Edition, Revised and Edited by EDWIN DUNKIN, F.R.A.S., Royal Observatory, Greenwich. With 38 Plates and upwards of 100 Woodcuts. In One Vol., small 8vo, 550 pages, 9s. 6d. cloth.

"Probably no other book contains the same amount of information in so compendious and well-arranged a form—certainly none at the price at which this is offered to the public."—*Athenæum.*

"We can do no other than pronounce this work a most valuable manual of astronomy, and we strongly recommend it to all who wish to acquire a general—but at the same time correct—acquaintance with this sublime science."—*Quarterly Journal of Science.*

"One of the most deservedly popular books on the subject . . . We would recommend not only the student of the elementary principles of the science, but he who aims at mastering the higher and mathematical branches of astronomy, not to be without this work beside him."—*Practical Magazine.*

Geology.

RUDIMENTARY TREATISE ON GEOLOGY, PHYSICAL AND HISTORICAL. Consisting of "Physical Geology," which sets forth the leading Principles of the Science; and "Historical Geology," which treats of the Mineral and Organic Conditions of the Earth at each successive epoch, especial reference being made to the British Series of Rocks. By RALPH TATE, A.L.S., F.G.S., &c. With 250 Illustrations. 12mo, 5s. cl. bds.

"The fulness of the matter has elevated the book into a manual. Its information is exhaustive and well arranged."—*School Board Chronicle.*

DR. LARDNER'S MUSEUM OF SCIENCE AND ART.

THE MUSEUM OF SCIENCE AND ART. Edited by DIONYSIUS LARDNER, D.C.L., formerly Professor of Natural Philosophy and Astronomy in University College, London. With upwards of 1,200 Engravings on Wood. In 6 Double Volumes, £1 1s. in a new and elegant cloth binding; or handsomely bound in half-morocco, 31s. 6d.

*** OPINIONS OF THE PRESS.

"This series, besides affording popular but sound instruction on scientific subjects, with which the humblest man in the country ought to be acquainted, also undertakes that teaching of 'Common Things' which every well-wisher of his kind is anxious to promote. Many thousand copies of this serviceable publication have been printed, in the belief and hope that the desire for instruction and improvement widely prevails; and we have no fear that such enlightened faith will meet with disappointment."—*Times.*

"A cheap and interesting publication, alike informing and attractive. The papers combine subjects of importance and great scientific knowledge, considerable inductive powers, and a popular style of treatment."—*Spectator.*

"The 'Museum of Science and Art' is the most valuable contribution that has ever been made to the Scientific Instruction of every class of society."—Sir DAVID BREWSTER, in the *North British Review.*

"Whether we consider the liberality and beauty of the illustrations, the charm of the writing, or the durable interest of the matter, we must express our belief that there is hardly to be found among the new books one that would be welcomed by people of so many ages and classes as a valuable present."—*Examiner.*

*** *Separate books formed from the above, suitable for Workmen's Libraries, Science Classes, etc.*

Common Things Explained. Containing Air, Earth, Fire, Water, Time, Man, the Eye, Locomotion, Colour, Clocks and Watches, &c. 233 Illustrations, cloth gilt, 5s.

The Microscope. Containing Optical Images, Magnifying Glasses, Origin and Description of the Microscope, Microscopic Objects, the Solar Microscope, Microscopic Drawing and Engraving, &c. 147 Illustrations, cloth gilt, 2s.

Popular Geology. Containing Earthquakes and Volcanoes, the Crust of the Earth, &c. 201 Illustrations, cloth gilt, 2s. 6d.

Popular Physics. Containing Magnitude and Minuteness, the Atmosphere, Meteoric Stones, Popular Fallacies, Weather Prognostics, the Thermometer, the Barometer, Sound, &c. 85 Illustrations, cloth gilt, 2s. 6d.

Steam and its Uses. Including the Steam Engine, the Locomotive, and Steam Navigation. 89 Illustrations, cloth gilt, 2s.

Popular Astronomy. Containing How to observe the Heavens—The Earth, Sun, Moon, Planets, Light, Comets, Eclipses, Astronomical Influences, &c. 182 Illustrations, cloth gilt, 4s. 6d.

The Bee and White Ants: Their Manners and Habits. With Illustrations of Animal Instinct and Intelligence. 135 Illustrations, cloth gilt, 2s.

The Electric Telegraph Popularized. To render intelligible to all who can Read, irrespective of any previous Scientific Acquirements, the various forms of Telegraphy in Actual Operation. 100 Illustrations, cloth gilt, 1s. 6d.

Dr. Lardner's School Handbooks.

NATURAL PHILOSOPHY FOR SCHOOLS. By Dr. LARDNER. 328 Illustrations. Sixth Edition. One Vol., 3s. 6d. cloth.

"A very convenient class-book for junior students in private schools. It is intended to convey in clear and precise terms, general notions of all the principal divisions of Physical Science."—*British Quarterly Review.*

ANIMAL PHYSIOLOGY FOR SCHOOLS. By Dr. LARDNER. With 190 Illustrations. Second Edition. One Vol., 3s. 6d. cloth.

"Clearly written, well arranged, and excellently illustrated."—*Gardener's Chronicle.*

Lardner and Bright on the Electric Telegraph.

THE ELECTRIC TELEGRAPH. By Dr. LARDNER. Revised and Re-written by E. B. BRIGHT, F.R.A.S. 140 Illustrations. Small 8vo, 2s. 6d. cloth.

"One of the most readable books extant on the Electric Telegraph."—*English Mechanic.*

CHEMICAL MANUFACTURES, CHEMISTRY.

Alkali Trade, Manufacture of Sulphuric Acid, etc.
A MANUAL OF THE ALKALI TRADE, including the Manufacture of Sulphuric Acid, Sulphate of Soda, and Bleaching Powder. By JOHN LOMAS, Alkali Manufacturer, Newcastle-upon-Tyne and London. With 232 Illustrations and Working Drawings, and containing 390 pages of Text. Second Edition, with Additions. Super-royal 8vo, £1 10s. cloth.

"This book is written by a manufacturer for manufacturers. The working details of the most approved forms of apparatus are given, and these are accompanied by no less than 232 wood engravings, all of which may be used for the purposes of construction. Every step in the manufacture is very fully described in this manual, and each improvement explained."—*Athenæum.*

"We find not merely a sound and luminous explanation of the chemical principles of the trade, but a notice of numerous matters which have a most important bearing on the successful conduct of alkali works, but which are generally overlooked by even experienced technological authors."—*Chemical Review.*

The Blowpipe.
THE BLOWPIPE IN CHEMISTRY, MINERALOGY, AND GEOLOGY. Containing all known Methods of Anhydrous Analysis, many Working Examples, and Instructions for Making Apparatus. By Lieut.-Colonel W. A. Ross, R.A., F.G.S. With 120 Illustrations. Second Edition, Revised and Enlarged. Crown 8vo, 5s. cloth.

"The student who goes conscientiously through the course of experimentation here laid down will gain a better insight into inorganic chemistry and mineralogy than if he had 'got up' any of the best text-books of the day, and passed any number of examinations in their contents."—*Chemical News.*

Commercial Chemical Analysis.
THE COMMERCIAL HANDBOOK OF CHEMICAL ANALYSIS; or, Practical Instructions for the determination of the Intrinsic or Commercial Value of Substances used in Manufactures, in Trades, and in the Arts. By A. NORMANDY, Editor of Rose's "Treatise on Chemical Analysis." New Edition, to a great extent re-written by HENRY M. NOAD, Ph.D., F.R.S. With numerous Illustrations. Crown 8vo, 12s. 6d. cloth.

"We strongly recommend this book to our readers as a guide, alike indispensable to the housewife as to the pharmaceutical practitioner."—*Medical Times.*

"Essential to the analysts appointed under the new Act. The most recent results are given, and the work is well edited and carefully written."—*Nature.*

Chemistry for Engineers, etc.
ENGINEERING CHEMISTRY: A Practical Treatise for the Use of Analytical Chemists, Engineers, Iron Masters, Iron Founders, Students, and others. Comprising Methods of Analysis and Valuation of the Principal Materials used in Engineering Work, with numerous Analyses, Examples, and Suggestions. By H. JOSHUA PHILLIPS, F.I.C., F.C.S. Analytical and Consulting Chemist to the Great Eastern Railway. Crown 8vo, 320 pp., with Illustrations, 10s. 6d. cloth. [*Just published.*

"In this work the author has rendered no small service to a numerous body of practical men. . . . The analytical methods may be pronounced most satisfactory, being as accurate as the despatch required of engineering chemists permits."—*Chemical News.*

"Those in search of a handy treatise on the subject of analytical chemistry as applied to the every-day requirements of workshop practice will find this volume of great assistance."—*Iron.*

"The first attempt to bring forward a Chemistry specially written for the use of engineers, and we have no hesitation whatever in saying that it should at once be in the possession of every railway engineer."—*The Railway Engineer.*

"The book will be very useful to those who require a handy and concise *resumé* of approved methods of analysing and valuing metals, oils, fuels, &c. It is, in fact, a work for chemists, a guide to the routine of the engineering laboratory. . . . The book is full of good things. As a handbook of technical analysis, it is very welcome."—*Builder.*

Dye-Wares and Colours.
THE MANUAL OF COLOURS AND DYE-WARES: Their Properties, Applications, Valuations, Impurities, and Sophistications. For the use of Dyers, Printers, Drysalters, Brokers, &c. By J. W. SLATER. Second Edition, Revised and greatly Enlarged. Crown 8vo, 7s. 6d. cloth.

"A complete encyclopædia of the *materia tinctoria*. The information given respecting each article is full and precise, and the methods of determining the value of articles such as these, so liable to sophistication, are given with clearness, and are practical as well as valuable."—*Chemist and Druggist.*

"There is no other work which covers precisely the same ground. To students preparing for examinations in dyeing and printing it will prove exceedingly useful."—*Chemical News.*

Modern Brewing and Malting.

A HANDYBOOK FOR BREWERS: Being a Practical Guide to the Art of Brewing and Malting. Embracing the Conclusions of Modern Research which bear upon the Practice of Brewing. By HERBERT EDWARDS WRIGHT, M.A., Author of "A Handbook for Young Brewers." Crown 8vo, 530 pp., 12s. 6d. cloth. [*Just published.*

"May be consulted with advantage by the student who is preparing himself for examinational tests, while the scientific brewer will find in it a *resume* of all the most important discoveries of modern times. The work is written throughout in a clear and concise manner, and the author takes great care to discriminate between vague theories and well-established facts."—*Brewers' Journal.*

"We have very great pleasure in recommending this handybook, and have no hesitation in saying that it is one of the best—if not the best—which has yet been written on the subject of beer-brewing in this country, and it should have a place on the shelves of every brewer's library."—*The Brewer's Guardian.*

"Well arranged, under special headings which separate each paragraph, and furnished with a good index, every facility for speedy reference is afforded. . . . On every debatable subject we have presented in an unbiased fashion the opinions which have been advanced in explanation of these points, making the work exactly what it purports to be, a comprehensive review of the conclusions of modern research in regard to brewing."—*Chemical Trade Journal.*

Analysis and Valuation of Fuels.

FUELS: SOLID, LIQUID, AND GASEOUS, Their Analysis and Valuation. For the Use of Chemists and Engineers. By H. J. PHILLIPS, F.C.S., Analytical and Consulting Chemist to the Great Eastern Railway, Author of "Engineering Chemistry," &c. Second Edition, Revised and Enlarged. Crown 8vo, 5s. cloth. [*Just published.*

"Ought to have its place in the laboratory of every metallurgical establishment, and wherever fuel is used on a large scale."—*Chemical News.*

"Mr. Phillips' new book cannot fail to be of wide interest, especially at the present time."—*Railway News.*

Pigments.

THE ARTIST'S MANUAL OF PIGMENTS. Showing their Composition, Conditions of Permanency, Non-Permanency, and Adulterations; Effects in Combination with Each Other and with Vehicles; and the most Reliable Tests of Purity. Together with the Science and Arts Department's Examination Questions on Painting. By H. C. STANDAGE. Second Edition, crown 8vo, 2s. 6d. cloth.

"This work is indeed *multum-in-parvo,* and we can, with good conscience, recommend it to all who come in contact with pigments, whether as makers, dealers or users."—*Chemical Review.*

Gauging. Tables and Rules for Revenue Officers, Brewers, etc.

A POCKET BOOK OF MENSURATION AND GAUGING: Containing Tables, Rules and Memoranda for Revenue Officers, Brewers, Spirit Merchants, &c. By J. B. MANT (Inland Revenue). Second Edition, Revised. Oblong 18mo, 4s. leather, with elastic band.

"This handy and useful book is adapted to the requirements of the Inland Revenue Department, and will be a favourite book of reference. The range of subjects is comprehensive, and the arrangement simple and clear."—*Civilian.*

"Should be in the hands of every practical brewer."—*Brewers' Journal.*

INDUSTRIAL ARTS, TRADES, AND MANUFACTURES.

Flour Manufacture, Milling, etc.

FLOUR MANUFACTURE: A Treatise on Milling Science and Practice. By FRIEDRICH KICK, Imperial Regierungsrath, Professor of Mechanical Technology in the Imperial German Polytechnic Institute, Prague. Translated from the Second Enlarged and Revised Edition with Supplement. By H. H. P. POWLES, Assoc. Memb. Institution of Civil Engineers. Nearly 400 pp. Illustrated with 28 Folding Plates, and 167 Woodcuts. Royal 8vo, 25s. cloth.

"This valuable work is, and will remain, the standard authority on the science of milling. . . The miller who has read and digested this work will have laid the foundation, so to speak, of a successful career; he will have acquired a number of general principles which he can proceed to apply. In this handsome volume we at last have the accepted text-book of modern milling in good, sound English, which has little, if any, trace of the German idiom."—*The Miller.*

"The appearance of this celebrated work in English is very opportune, and British millers will, we are sure, not be slow in availing themselves of its pages."—*Millers' Gazette.*

Soap-making.

THE ART OF SOAP-MAKING: A Practical Handbook of the *Manufacture of Hard and Soft Soaps, Toilet Soaps, etc.* Including many New Processes, and a Chapter on the Recovery of Glycerine from Waste Leys. By ALEXANDER WATT, Author of " Electro-Metallurgy Practically Treated,' &c. With numerous Illustrations. Fourth Edition, Revised and Enlarged. Crown 8vo, 7s. 6d. cloth.

"The work will prove very useful, not merely to the technological student, but to the practical soap-boiler who wishes to understand the theory of his art."—*Chemical News.*

" Really an excellent example of a technical manual, entering, as it does, thoroughly and exhaustively, both into the theory and practice of soap manufacture. The book is well and honestly done, and deserves the considerable circulation with which it will doubtless meet."—*Knowledge.*

"Mr. Watt's book is a thoroughly practical treatise on an art which has almost no literature in our language. We congratulate the author on the success of his endeavours to fill a void in English technical literature."—*Nature.*

Paper Making.

THE ART OF PAPER MAKING: A Practical Handbook of the *Manufacture of Paper from Rags, Esparto, Straw, and other Fibrous Materials,* Including the Manufacture of Pulp from Wood Fibre, with a Description of the Machinery and Appliances used. To which are added Details of Processes for Recovering Soda from Waste Liquors. By ALEXANDER WATT, Author of " The Art of Soap-Making," " The Art of Leather Manufacture," &c. With Illustrations. Crown 8vo, 7s. 6d. cloth.

" This book is succinct, lucid, thoroughly practical, and includes everything of interest to the modern paper-maker. The book, besides being all the student of paper-making will require in his apprenticeship, will be found of interest to the paper-maker himself. It is the latest, most practical, and most complete work on the paper-making art before the British public."—*Paper Record.*

" It may be regarded as the standard work on the subject. The book is full of valuable information. The 'Art of Paper-making,' is in every respect a model of a text-book, either for a technical class or for the private student."—*Paper and Printing Trades Journal.*

Leather Manufacture.

THE ART OF LEATHER MANUFACTURE. Being a Practical Handbook, in which the Operations of Tanning, Currying, and Leather Dressing are fully Described, and the Principles of Tanning Explained, and many Recent Processes Introduced; as also the Methods for the Estimation of Tannin, and a Description of the Arts of Glue Boiling, Gut Dressing, &c. By ALEXANDER WATT, Author of " Soap-Making," " Electro-Metallurgy," &c. With numerous Illustrations. Second Edition. Crown 8vo, 9s. cloth.

"A sound, comprehensive treatise on tanning and its accessories. The book is an eminently valuable production, which redounds to the credit of both author and publishers."—*Chemical Review.*

" This volume is technical without being tedious, comprehensive and complete without being prosy, and it bears on every page the impress of a master hand. We have never come across a better trade treatise, nor one that so thoroughly supplied an absolute want."—*Shoe and Leather Trades' Chronicle.*

Boot and Shoe Making.

THE ART OF BOOT AND SHOE-MAKING. A Practical Handbook, including Measurement, Last-Fitting, Cutting-Out, Closing, and Making, with a Description of the most approved Machinery employed. By JOHN B. LENO, late Editor of *St. Crispin,* and *The Boot and Shoe-Maker.* With numerous Illustrations. Third Edition. 12mo, 2s. cloth limp.

" This excellent treatise is by far the best work ever written on the subject. The chapter on clicking, which shows how waste may be prevented, will save fifty times the price of the book."—*Scottish Leather Trader.*

Dentistry Construction.

MECHANICAL DENTISTRY: A Practical Treatise on the *Construction of the various kinds of Artificial Dentures.* Comprising also Useful Formulæ, Tables, and Receipts for Gold Plate, Clasps, Solders, &c. &c. By CHARLES HUNTER. Third Edition, Revised. With upwards of 100 Wood Engravings. Crown 8vo, 3s. 6d. cloth.

"The work is very practical."—*Monthly Review of Dental Surgery.*

"We can strongly recommend Mr. Hunter's treatise to all students preparing for the profession of dentistry, as well as to every mechanical dentist."—*Dublin Journal of Medical Science.*

Wood Engraving.

WOOD ENGRAVING: A Practical and Easy Introduction to the *Study of the Art.* By WILLIAM NORMAN BROWN. Second Edition. With numerous Illustrations. 12mo, 1s. 6d. cloth limp.

"The book is clear and complete, and will be useful to anyone wanting to understand the first elements of the beautiful art of wood engraving."—*Graphic.*

Horology.

A TREATISE ON MODERN HOROLOGY, in Theory and Practice. Translated from the French of CLAUDIUS SAUNIER, ex-Director of the School of Horology at Maçon, by JULIEN TRIPPLIN, F.R.A.S., Besançon Watch Manufacturer, and EDWARD RIGG, M.A., Assayer in the Royal Mint. With 78 Woodcuts and 22 Coloured Copper Plates. Second Edition. Super-royal 8vo, £2 2s. cloth; £2 10s. half-calf.

"There is no horological work in the English language at all to be compared to this production of M. Saunier's for clearness and completeness. It is alike good as a guide for the student and as a reference for the experienced horologist and skilled workman."—*Horological Journal.*

"The latest, the most complete, and the most reliable of those literary productions to which continental watchmakers are indebted for the mechanical superiority over their English brethren—in fact, the Book of Books, is M. Saunier's 'Treatise.'"—*Watchmaker, Jeweller and Silversmith.*

Watchmaking.

THE WATCHMAKER'S HANDBOOK. Intended as a Workshop Companion for those engaged in Watchmaking and the Allied Mechanical Arts. Translated from the French of CLAUDIUS SAUNIER, and considerably enlarged by JULIEN TRIPPLIN, F.R.A.S., Vice-President of the Horological Institute, and EDWARD RIGG, M.A., Assayer in the Royal Mint. With numerous Woodcuts and 14 Copper Plates. Third Edition. Crown 8vo, 9s. cloth.

"Each part is truly a treatise in itself. The arrangement is good and the language is clear and concise. It is an admirable guide for the young watchmaker."—*Engineering.*

"It is impossible to speak too highly of its excellence. It fulfils every requirement in a handbook intended for the use of a workman. Should be found in every workshop."—*Watch and Clockmaker.*

"This book contains an immense number of practical details bearing on the daily occupation of a watchmaker."—*Watchmaker and Metalworker* (Chicago).

Watches and Timekeepers.

A HISTORY OF WATCHES AND OTHER TIMEKEEPERS. By JAMES F. KENDAL, M.B.H.Inst. 250 pp., with 88 Illustrations, 1s. 6d. boards; or 2s. 6d. cloth gilt. [*Just published.*

"Mr. Kendal's book, for its size, is the best which has yet appeared on this subject in the English language."—*Industries.*

"Open the book where you may, there is interesting matter in it concerning the ingenious devices of the ancient or modern horologer. The subject is treated in a liberal and entertaining spirit, as might be expected of a historian who is a master of the craft."—*Saturday Review.*

Electrolysis of Gold, Silver, Copper, etc.

ELECTRO-DEPOSITION: A Practical Treatise on the Electrolysis of Gold, Silver, Copper, Nickel, and other Metals and Alloys. With descriptions of Voltaic Batteries, Magneto and Dynamo-Electric Machines, Thermopiles, and of the Materials and Processes used in every Department of the Art, and several Chapters on Electro-Metallurgy. By ALEXANDER WATT, Author of "Electro-Metallurgy," &c. With numerous Illustrations. Third Edition, Revised and Corrected. Crown 8vo, 9s. cloth.

"Eminently a book for the practical worker in electro-deposition. It contains practical descriptions of methods, processes and materials as actually pursued and used in the workshop."—*Engineer.*

Electro-Metallurgy.

ELECTRO-METALLURGY; Practically Treated. By ALEXANDER WATT, Author of "Electro-Deposition," &c. Ninth Edition, Enlarged and Revised, with Additional Illustrations, and including the most recent Processes. 12mo, 4s. cloth boards.

"From this book both amateur and artisan may learn everything necessary for the successful prosecution of electroplating."—*Iron.*

Working in Gold.

THE JEWELLER'S ASSISTANT IN THE ART OF WORKING IN GOLD: A Practical Treatise for Masters and Workmen, Compiled from the Experience of Thirty Years' Workshop Practice. By GEORGE E. GEE, Goldsmith and Silversmith, Author of "The Goldsmith's Handbook," &c. Crown 8vo, 7s. 6d. cloth. [*Just published.*

"This manual of technical education is apparently destined to be a valuable auxiliary to a handicraft which is certainly capable of great improvement."—*The Times.*

"This volume will be very useful in the workshop, as the knowledge is practical, having been acquired by long experience, and all the recipes and directions are guaranteed to be successful if properly worked out."—*Jeweller and Metalworker.*

Electroplating.

ELECTROPLATING: A Practical Handbook on the Deposition of Copper, Silver, Nickel, Gold, Aluminium, Brass, Platinum, &c. &c. With Descriptions of the Chemicals, Materials, Batteries, and Dynamo Machines used in the Art. By J. W. URQUHART, C.E., Author of "Electric Light," &c. Second Edition, Revised, with Additions. Numerous Illustrations. Crown 8vo, 5s. cloth.

"An excellent practical manual."—*Engineering.*
"An excellent work, giving the newest information."—*Horological Journal.*

Electrotyping.

ELECTROTYPING: The Reproduction and Multiplication of Printing Surfaces and Works of Art by the Electro-deposition of Metals. By J. W. URQUHART, C.E. Crown 8vo, 5s. cloth.

"The book is thoroughly practical. The reader is, therefore, conducted through the leading laws of electricity, then through the metals used by electrotypers, the apparatus, and the depositing processes, up to the final preparation of the work."—*Art Journal.*

Goldsmiths' Work.

THE GOLDSMITH'S HANDBOOK. By GEORGE E. GEE, Jeweller, &c. Third Edition, considerably Enlarged. 12mo, 3s. 6d. cl. bds.

"A good, sound educator, and will be generally accepted as an authority."—*Horological Journal.*

Silversmiths' Work.

THE SILVERSMITH'S HANDBOOK. By GEORGE E. GEE, Jeweller, &c. Second Edition, Revised, with numerous Illustrations. 12mo, 3s. 6d. cloth boards.

"The chief merit of the work is its practical character. . . . The workers in the trade will speedily discover its merits when they sit down to study it."—*English Mechanic.*

⁎ *The above two works together, strongly half-bound, price 7s.*

Bread and Biscuit Baking.

THE BREAD AND BISCUIT BAKER'S AND SUGAR-BOILER'S ASSISTANT. Including a large variety of Modern Recipes. With Remarks on the Art of Bread-making. By ROBERT WELLS, Practical Baker. Second Edition, with Additional Recipes. Crown 8vo, 2s. cloth.

"A large number of wrinkles for the ordinary cook, as well as the baker."—*Saturday Review.*

Confectionery for Hotels and Restaurants.

THE PASTRYCOOK AND CONFECTIONER'S GUIDE. For Hotels, Restaurants and the Trade in general, adapted also for Family Use. By ROBERT WELLS, Author of "The Bread and Biscuit Baker's and Sugar-Boiler's Assistant." Crown 8vo, 2s. cloth.

"We cannot speak too highly of this really excellent work. In these days of keen competition our readers cannot do better than purchase this book."—*Bakers' Times.*

Ornamental Confectionery.

ORNAMENTAL CONFECTIONERY: A Guide for Bakers, Confectioners and Pastrycooks; including a variety of Modern Recipes, and Remarks on Decorative and Coloured Work. With 129 Original Designs. By ROBERT WELLS, Practical Baker, Author of "The Bread and Biscuit Baker's and Sugar-Boiler's Assistant," &c. Crown 8vo, cloth gilt, 5s.

"A valuable work, practical, and should be in the hands of every baker and confectioner. The illustrative designs are alone worth treble the amount charged for the whole work."—*Bakers' Times.*

Flour Confectionery.

THE MODERN FLOUR CONFECTIONER. Wholesale and Retail. Containing a large Collection of Recipes for Cheap Cakes, Biscuits, &c. With Remarks on the Ingredients used in their Manufacture. To which are added Recipes for Dainties for the Working Man's Table. By R. WELLS, Author of "The Bread and Biscuit Baker," &c. Crown 8vo, 2s. cl.

"The work is of a decidedly practical character, and in every recipe regard is had to economical working."—*North British Daily Mail.*

Laundry Work.

LAUNDRY MANAGEMENT. A Handbook for Use in Private and Public Laundries, Including Descriptive Accounts of Modern Machinery and Appliances for Laundry Work. By the EDITOR of "The Laundry Journal." With numerous Illustrations. Crown 8vo, 2s. 6d. cloth.

"This book should certainly occupy an honoured place on the shelves of all housekeepers who wish to keep themselves *au courant* of the newest appliances and methods."—*The Queen.*

HANDYBOOKS FOR HANDICRAFTS.
By PAUL N. HASLUCK,
EDITOR OF "WORK" (NEW SERIES); AUTHOR OF "LATHEWORK," "MILLING MACHINES AND PROCESSES," etc.

Crown 8vo, 144 pages, cloth, price 1s. each.

☞ *These* HANDYBOOKS *have been written to supply information for* WORKMEN, STUDENTS, *and* AMATEURS *in the several Handicrafts, on the actual* PRACTICE *of the* WORKSHOP, *and are intended to convey in plain language* TECHNICAL KNOWLEDGE *of the several* CRAFTS. *In describing the processes employed, and the manipulation of material, workshop terms are used; workshop practice is fully explained; and the text is freely illustrated with drawings of modern tools, appliances, and processes. The information given will thus be found useful, not only by the young beginner, but by the veteran whose range of experience has been narrowed under a system of divided labour; while the amateur will find himself introduced to the very atmosphere and surroundings of the workshop.*

In view of the wide circulation which the HANDYBOOKS *have already attained, and the yet wider circulation which must accrue from the facilities for* MANUAL INSTRUCTION *now provided by* LOCAL AUTHORITIES *in pursuance of recent legislation, it has been decided to issue them at the price of* One Shilling each.

*** *The following Volumes are now ready:*

THE METAL TURNER'S HANDBOOK. *A Practical Manual for Workers at the Foot-Lathe.* With over 100 Illustrations. Price 1s.

"The book will be of service alike to the amateur and the artisan turner. It displays thorough knowledge of the subject."—*Scotsman.*

THE WOOD TURNER'S HANDBOOK. *A Practical Manual for Workers at the Lathe.* With over 100 Illustrations. Price 1s.

"We recommend the book to young turners and amateurs. A multitude of workmen have hitherto sought in vain for a manual of this special industry."—*Mechanical World.*

THE WATCH JOBBER'S HANDBOOK. *A Practical Manual on Cleaning, Repairing, and Adjusting.* With upwards of 100 Illustrations. Price 1s.

"We strongly advise all young persons connected with the watch trade to acquire and study this inexpensive work."—*Clerkenwell Chronicle.*

THE PATTERN MAKER'S HANDBOOK. A Practical Manual on the Construction of Patterns for Founders. With upwards of 100 Illustrations. Price 1s.

"A most valuable, if not indispensable, manual for the pattern maker."—*Knowledge.*

THE MECHANIC'S WORKSHOP HANDBOOK. *A Practical Manual on Mechanical Manipulation.* Embracing Information on various Handicraft Processes, with Useful Notes and Miscellaneous Memoranda. Comprising about 200 Subjects. Price 1s.

"A very clever and useful book, which should be found in every workshop; and it should certainly find a place in all technical schools."—*Saturday Review.*

THE MODEL ENGINEER'S HANDBOOK. *A Practical Manual on the Construction of Model Steam Engines.* With upwards of 100 Illustrations. Price 1s.

"Mr. Hasluck has produced a very good little book."—*Builder.*

THE CLOCK JOBBER'S HANDBOOK. *A Practical Manual on Cleaning, Repairing, and Adjusting.* With upwards of 100 Illustrations. Price 1s.

"It is of inestimable service to those commencing the trade."—*Coventry Standard.*

THE CABINET WORKER'S HANDBOOK: A Practical Manual on the Tools, Materials, Appliances, and Processes employed in Cabinet Work. With upwards of 100 Illustrations. Price 1s.

"Mr. Hasluck's thoroughgoing little Handybook is amongst the most practical guides we have seen for beginners in cabinet-work."—*Saturday Review.*

* *The following are in preparation:*

THE WOODWORKER'S HANDBOOK.
THE METALWORKER'S HANDBOOK.

COMMERCE, COUNTING-HOUSE WORK, TABLES, etc.

Commercial Education.
LESSONS IN COMMERCE. By Professor R. GAMBARO, of the Royal High Commercial School at Genoa. Edited and Revised by JAMES GAULT, Professor of Commerce and Commercial Law in King's College, London. Crown 8vo, 3s. 6d. cloth.

"The publishers of this work have rendered considerable service to the cause of commercial education by the opportune production of this volume. . . . The work is peculiarly acceptable to English readers and an admirable addition to existing class-books. In a phrase, we think the work attains its object in furnishing a brief account of those laws and customs of British trade with which the commercial man interested therein should be familiar."—*Chamber of Commerce Journal.*

Foreign Commercial Correspondence.
THE FOREIGN COMMERCIAL CORRESPONDENT: Being Aids to Commercial Correspondence in Five Languages—English, French, German, Italian, and Spanish. By CONRAD E. BAKER. Second Edition. Crown 8vo, 3s. 6d. cloth.

"Whoever wishes to correspond in all the languages mentioned by Mr. Baker cannot do better than study this work, the materials of which are excellent and conveniently arranged."—*Athenæum.*
"A careful examination has convinced us that it is unusually complete, well arranged, and reliable. The book is a thoroughly good one."—*Schoolmaster.*

Accounts for Manufacturers.
FACTORY ACCOUNTS: Their Principles and Practice. A Handbook for Accountants and Manufacturers, with Appendices on the Nomenclature of Machine Details; the Income Tax Acts; the Rating of Factories; Fire and Boiler Insurance; the Factory and Workshop Acts, &c., including also a Glossary of Terms and a large number of Specimen Rulings. By EMILE GARCKE and J. M. FELLS. Third Edition. Demy 8vo, 250 pages, price 6s. strongly bound.

"A very interesting description of the requirements of Factory Accounts. . . . the principle of assimilating the Factory Accounts to the general commercial books is one which we thoroughly agree with."—*Accountants' Journal.*
"Characterised by extreme thoroughness. There are few owners of factories who would not derive great benefit from the perusal of this most admirable work."—*Local Government Chronicle*

Intuitive Calculations.
THE COMPENDIOUS CALCULATOR; or, Easy and Concise Methods of Performing the various Arithmetical Operations required in Commercial and Business Transactions, together with Useful Tables. By DANIEL O'GORMAN. Corrected and Extended by Professor J. R. YOUNG. Twenty-seventh Edition, Revised by C. NORRIS. Fcap. 8vo, 2s. 6d. cloth limp; or, 3s. 6d. strongly half-bound in leather.

"It would be difficult to exaggerate the usefulness of a book like this to everyone engaged in commerce or manufacturing industry. It is crammed full of rules and formulæ for shortening and employing calculations."—*Knowledge.*

Modern Metrical Units and Systems.
MODERN METROLOGY: A Manual of the Metrical Units and Systems of the Present Century. With an Appendix containing a proposed English System. By LOWIS D'A. JACKSON, A.M.Inst.C.E., Author of "Aid to Survey Practice," &c. Large crown 8vo, 12s. 6d. cloth.

"We recommend the work to all interested in the practical reform of our weights and measures."—*Nature.*

The Metric System and the British Standards.
A SERIES OF METRIC TABLES, in which the British Standard Measures and Weights are compared with those of the Metric System at present in Use on the Continent. By C. H. DOWLING, C.E. 8vo, 10s. 6d. strongly bound.

"Mr. Dowling's Tables are well put together as a ready-reckoner for the conversion of one system into the other."—*Athenæum.*

Iron and Metal Trades' Calculator.
THE IRON AND METAL TRADES' COMPANION. For expeditiously ascertaining the Value of any Goods bought or sold by Weight, from 1s. per cwt. to 112s. per cwt., and from one farthing per pound to one shilling per pound. By THOMAS DOWNIE. Strongly bound in leather, 396 pp., 9s.

"A most useful set of tables; nothing like them before existed."—*Building News.*
"Although specially adapted to the iron and metal trades, the tables will be found useful in every other business in which merchandise is bought and sold by weight."—*Railway News.*

Calculator for Numbers and Weights Combined.

THE NUMBER, WEIGHT, AND FRACTIONAL CALCULATOR. Containing upwards of 250,000 Separate Calculations, showing at a glance the value at 422 different rates, ranging from $\frac{1}{32}$th of a Penny to 20s. each, or per cwt., and £20 per ton, of any number of articles consecutively, from 1 to 470.—Any number of cwts., qrs., and lbs., from 1 cwt. to 470 cwts.—Any number of tons, cwts., qrs., and lbs., from 1 to 1,000 tons. By WILLIAM CHADWICK, Public Accountant. Third Edition, Revised and Improved. 8vo, price 18s., strongly bound for Office wear and tear.

☞ *Is adapted for the use of Accountants and Auditors, Railway Companies, Canal Companies, Shippers, Shipping Agents, General Carriers, etc. Ironfounders, Brassfounders, Metal Merchants, Iron Manufacturers, Ironmongers, Engineers, Machinists, Boiler Makers, Millwrights, Roofing, Bridge and Girder Makers, Colliery Proprietors, etc. Timber Merchants, Builders, Contractors, Architects, Surveyors, Auctioneers, Valuers, Brokers, Mill Owners and Manufacturers, Mill Furnishers, Merchants, and General Wholesale Tradesmen. Also for the Apportionment of Mileage Charges for Railway Traffic.*

*** OPINIONS OF THE PRESS.

"It is easy of reference for any answer or any number of answers as a dictionary, and the references are even more quickly made. For making up accounts or estimates the book must prove invaluable to all who have any considerable quantity of calculations involving price and measure in any combination to do."—*Engineer.*

"The most complete and practical ready reckoner which it has been our fortune yet to see. It is difficult to imagine a trade or occupation in which it could not be of the greatest use, either in saving human labour or in checking work. The publishers have placed within the reach of every commercial man an invaluable and unfailing assistant."—*The Miller.*

Harben's Comprehensive Weight Calculator.

THE WEIGHT CALCULATOR. Being a Series of Tables upon a New and Comprehensive Plan, exhibiting at One Reference the exact Value of any Weight from 1 lb. to 15 tons, at 300 Progressive Rates, from 1d. to 168s. per cwt., and containing 186,000 Direct Answers, which, with their Combinations, consisting of a single addition (mostly to be performed at sight), will afford an aggregate of 10,266,000 Answers; the whole being calculated and designed to ensure correctness and promote despatch. By HENRY HARBEN, Accountant. Fourth Edition, carefully Corrected. Royal 8vo, £1 5s., strongly half-bound.

"A practical and useful work of reference for men of business generally; it is the best of the kind we have seen."—*Ironmonger.*

"Of priceless value to business men. It is a necessary book in all mercantile offices."—*Sheffield Independent.*

Harben's Comprehensive Discount Guide.

THE DISCOUNT GUIDE. Comprising several Series of Tables for the use of Merchants, Manufacturers, Ironmongers, and others, by which may be ascertained the exact Profit arising from any mode of using Discounts, either in the Purchase or Sale of Goods, and the method of either Altering a Rate of Discount or Advancing a Price, so as to produce, by one operation, a sum that will realise any required profit after allowing one or more Discounts: to which are added Tables of Profit or Advance from 1¼ to 90 per cent., Tables of Discount from 1¼ to 98¾ per cent., and Tables of Commission, &c., from ⅛ to 10 per cent. By HENRY HARBEN, Accountant, Author of "The Weight Calculator." New Edition, carefully Revised and Corrected. Demy 8vo, 544 pp., £1 5s. half-bound.

"A book such as this can only be appreciated by business men, to whom the saving of time means saving of money. We have the high authority of Professor J. R. Young that the tables throughout the work are constructed upon strictly accurate principles. The work is a model of typographical clearness, and must prove of great value to merchants, manufacturers, and general traders."—*British Trade Journal.*

Iron Shipbuilders' and Merchants' Weight Tables.

IRON-PLATE WEIGHT TABLES: For Iron Shipbuilders, Engineers, and Iron Merchants. Containing the Calculated Weights of upwards of 150,000 different sizes of Iron Plates, from 1 foot by 6 in. by ¼ in. to 10 feet by 5 feet by 1 in. Worked out on the basis of 40 lbs. to the square foot of Iron of 1 inch in thickness. Carefully compiled and thoroughly Revised by H. BURLINSON and W. H. SIMPSON. Oblong 4to, 25s. half-bound.

"This work will be found of great utility. The authors have had much practical experience of what is wanting in making estimates; and the use of the book will save much time in making elaborate calculations."—*English Mechanic.*

AGRICULTURE, FARMING, GARDENING, etc.

"*The Standard Treatise on Agriculture.*"

THE COMPLETE GRAZIER, and FARMER'S and CATTLE-BREEDER'S ASSISTANT: A Compendium of Husbandry. Originally Written by WILLIAM YOUATT. Thirteenth Edition, entirely Re-written, considerably Enlarged, and brought up to the Present Requirements of Agricultural Practice, by WILLIAM FREAM, LL.D., Steven Lecturer in the University of Edinburgh, Author of "The Elements of Agriculture," &c. Royal 8vo, 1,100 pp., with over 450 Illustrations. Price £1 11s. 6d. strongly and handsomely bound. [*Just published.*

EXTRACT FROM PUBLISHERS' ADVERTISEMENT.

" A treatise that made its original appearance in the first decade of the century, and that enters upon its Thirteenth Edition before the century has run its course, has undoubtedly established its position as a work of permanent value. It has been deemed expedient, therefore, to retain, as far as possible, in the present edition those features of Youatt's Work which must have commended themselves to general approval.

" The phenomenal progress of the last dozen years in the Practice and Science of Farming has rendered it necessary, however, that the volume should be re-written, and the publishers were fortunate enough to secure for the revision the services of Dr. FREAM, whose high attainments in all matters pertaining to agriculture have been so emphatically recognised by the highest professional and official authorities. In carrying out his editorial duties, Dr. FREAM has been favoured with valuable contributions by Prof. J. WORTLEY AXE, Mr. E. BROWN, Dr. BERNARD DYER, Mr. W. J. MALDEN, Mr. R. H. REW, Prof. SHELDON, Mr. J. SINCLAIR, Mr. SANDERS SPENCER, and others.

" No pains have been spared to make the illustrations as representative and characteristic as possible; those of Live Stock (with one or two exceptions) being new to the work; and amongst them will be found portraits of prize-winning animals of the leading breeds."

" On the whole, it may be safely said that no effort has been lacking on the part of either Editor or Publishers to make this New Edition of THE COMPLETE GRAZIER a faithful mirror of agricultural progress and a reliable record of modern practice in farming, and, as such, deserving of the reputation gained by the work (*vide Mark Lane Express*) as '*a treatise which will remain a standard work on the subject as long as British agriculture endures.*' "

SUMMARY OF CONTENTS.

BOOK I. ON THE VARIETIES, BREEDING, REARING, FATTENING, AND GENERAL MANAGEMENT OF CATTLE.

BOOK II. ON THE ECONOMY AND MANAGEMENT OF THE DAIRY.

BOOK III. ON THE BREEDING, REARING, AND MANAGEMENT OF HORSES.

BOOK IV. ON THE BREEDING, REARING, AND FATTENING OF SHEEP.

BOOK V. ON THE BREEDING, REARING, AND FATTENING OF SWINE.

BOOK VI. ON THE DISEASES OF LIVE STOCK.

BOOK VII. ON THE BREEDING, REARING, AND MANAGEMENT OF POULTRY.

BOOK VIII. ON FARM OFFICES AND IMPLEMENTS OF HUSBANDRY.

BOOK IX. ON THE CULTURE AND MANAGEMENT OF GRASS LANDS.

BOOK X. ON THE CULTIVATION AND APPLICATION OF GRASSES, PULSE, AND ROOTS.

BOOK XI. ON MANURES AND THEIR APPLICATION.

BOOK XII. MONTHLY CALENDARS OF FARMWORK THROUGHOUT THE YEAR.

** OPINIONS OF THE PRESS ON PREVIOUS EDITIONS.

" The standard text-book with the farmer and grazier."—*Farmer's Magazine.*

" A treatise which will remain a standard work on the subject as long as British agriculture endures."—*Mark Lane Express* (first notice).

" The book deals with all departments of agriculture, and contains an immense amount of valuable information. It is, in fact, an encyclopædia of agriculture put into readable form, and it is the only work, equally comprehensive brought down to present date. It deserves a place in the library of every agriculturist."—*Mark Lane Express* (second notice).

British Farm Live Stock.

FARM LIVE STOCK OF GREAT BRITAIN. By ROBERT WALLACE, F.L.S., F.R.S.E., &c., Professor of Agriculture and Rural Economy in the University of Edinburgh. Third Edition, thoroughly Revised and considerably Enlarged. With over 120 Phototypes of Stock. Demy 8vo, 384 pp., with 79 Plates and Maps, price 12s. 6d. cloth. [*Just published.*

" A valuable, if not an indispensable, addition to every agricultural library worthy of the name, and an excellent gift-book to all who are, or are likely to become, concerned in the care and management of live stock in any position."—*Agricultural Economist.*

" Few country gentlemen who take up this book will care to put it down again until they have looked at its hundred phototypes of prize cattle, sheep, pigs, and horses—the very best collection we have ever seen."—*Saturday Review.*

Dairy Farming.

BRITISH DAIRYING. A Handy Volume on the Work of the Dairy-Farm. For the Use of Technical Instruction Classes, Students in Agricultural Colleges, and the Working Dairy-Farmer. By Prof. J. P. SHELDON, late Special Commissioner of the Canadian Government, Author of "Dairy Farming," "The Farm and the Dairy," &c. With numerous Illustrations. Crown 8vo, 2s. 6d. cloth. [*Just published*.

Agricultural Facts and Figures.

NOTE-BOOK OF AGRICULTURAL FACTS AND FIGURES FOR FARMERS AND FARM STUDENTS. By PRIMROSE MCCONNELL, B.Sc., Fellow of the Highland and Agricultural Society. Fourth Edition. Royal 32mo, roan, gilt edges, with band, 4s.

"Literally teems with information, and we can cordially recommend it to all connected with agriculture."—*North British Agriculturist.*

Small Farming.

SYSTEMATIC SMALL FARMING; or, *The Lessons of my Farm.* Being an Introduction to Modern Farm Practice for Small Farmers. By ROBERT SCOTT BURN, Author of "Outlines of Modern Farming," &c. With numerous Illustrations, crown 8vo, 6s. cloth.

"This is the completest book of its class we have seen, and one which every amateur farmer will read with pleasure and accept as a guide."—*Field.*

"The volume contains a vast amount of useful information. No branch of farming is left untouched, from the labour to be done to the results achieved. It may be safely recommended to all who think they will be in paradise when they buy or rent a three-acre farm."—*Glasgow Herald.*

Modern Farming.

OUTLINES OF MODERN FARMING. By R. SCOTT BURN. Soils, Manures, and Crops—Farming and Farming Economy—Cattle, Sheep, and Horses — Management of Dairy, Pigs, and Poultry — Utilisation of Town-Sewage, Irrigation, &c. Sixth Edition. In One Vol., 1,250 pp., half-bound, profusely Illustrated, 12s.

"The aim of the author has been to make his work at once comprehensive and trustworthy, and in this aim he has succeeded to a degree which entitles him to much credit."—*Morning Advertiser.* "No farmer should be without this book."—*Banbury Guardian.*

Agricultural Engineering.

FARM ENGINEERING, THE COMPLETE TEXT-BOOK OF. Comprising Draining and Embanking; Irrigation and Water Supply; Farm Roads, Fences, and Gates; Farm Buildings, their Arrangement and Construction, with Plans and Estimates; Barn Implements and Machines; Field Implements and Machines; Agricultural Surveying, Levelling, &c. By Prof. JOHN SCOTT, late Professor of Agriculture and Rural Economy at the Royal Agricultural College, Cirencester, &c. &c. In One Vol., 1,150 pages, half-bound, with over 600 Illustrations, 12s.

"Written with great care, as well as with knowledge and ability. The author has done his work well; we have found him a very trustworthy guide wherever we have tested his statements. The volume will be of great value to agricultural students."—*Mark Lane Express.*

"For a young agriculturist we know of no handy volume likely to be more usefully studied."—*Bell's Weekly Messenger.*

Agricultural Text-Book.

THE FIELDS OF GREAT BRITAIN: A Text-Book of Agriculture, adapted to the Syllabus of the Science and Art Department. For Elementary and Advanced Students. By HUGH CLEMENTS (Board of Trade). Second Edition, Revised, with Additions. 18mo, 2s. 6d. cloth.

"A most comprehensive volume, giving a mass of information."—*Agricultural Economist.*

"It is a long time since we have seen a book which has pleased us more, or which contains such a vast and useful fund of knowledge."—*Educational Times.*

Tables for Farmers, etc.

TABLES, MEMORANDA, AND CALCULATED RESULTS *for Farmers, Graziers, Agricultural Students, Surveyors, Land Agents, Auctioneers, etc.* Selected and Arranged by SIDNEY FRANCIS. Second Edition, Revised. 272 pp., waistcoat-pocket size, 1s. 6d. limp leather.

"Weighing less than 1 oz., and occupying no more space than a match box, it contains a mass of facts and calculations which has never before, in such handy form, been obtainable. We cordially recommend it."—*Bell's Weekly Messenger.*

AGRICULTURE, FARMING, GARDENING, etc. 45

The Management of Bees.
BEES FOR PLEASURE AND PROFIT: A Guide to the Manipulation of Bees, the Production of Honey, and the General Management of the Apiary. By G. GORDON SAMSON. With numerous Illustrations. Crown 8vo, 1s. cloth.

"The intending bee-keeper will find exactly the kind of information required to enable him to make a successful start with his hives. The author is a thoroughly competent teacher, and his book may be commended."—*Morning Post.*

Farm and Estate Book-keeping.
BOOK-KEEPING FOR FARMERS & ESTATE OWNERS. A Practical Treatise, presenting, in Three Plans, a System adapted for all Classes of Farms. By JOHNSON M. WOODMAN, Chartered Accountant. Second Edition, Revised. Cr. 8vo, 3s. 6d. cl. bds.; or 2s. 6d. cl. limp.

"The volume is a capital study of a most important subject."—*Agricultural Gazette.*

Farm Account Book.
WOODMAN'S YEARLY FARM ACCOUNT BOOK. Giving a Weekly Labour Account and Diary, and showing the Income and Expenditure under each Department of Crops, Live Stock, Dairy, &c. &c. With Valuation, Profit and Loss Account, and Balance Sheet at the end of the Year. By JOHNSON M. WOODMAN, Chartered Accountant, Author of Book-keeping for Farmers." Folio, 7s. 6d. half-bound.

"Contains every requisite form for keeping farm accounts readily and accurately."—*Agriculture.*

Early Fruits, Flowers, and Vegetables.
THE FORCING GARDEN; or, How to Grow Early Fruits, Flowers, and Vegetables. With Plans and Estimates for Building Glasshouses, Pits, and Frames. With Illustrations. By SAMUEL WOOD. Crown 8vo, 3s. 6d. cloth.

"A good book, and fairly fills a place that was in some degree vacant." The book is written with great care, and contains a great deal of valuable teaching."—*Gardeners' Magazine.*

Good Gardening.
A PLAIN GUIDE TO GOOD GARDENING; or, How to Grow Vegetables, Fruits, and Flowers. By S. WOOD. Fourth Edition, with considerable Additions, &c., and numerous Illustrations. Crown 8vo, 3s. 6d. cl.

"May be recommended to young gardeners, cottagers, and specially to amateurs, for the plain, simple, and trustworthy information it gives on common matters too often neglected."—*Gardeners' Chronicle.*

Gainful Gardening.
MULTUM-IN-PARVO GARDENING; or, How to make One Acre of Land produce £620 a-year by the Cultivation of Fruits and Vegetables; also, How to Grow Flowers in Three Glass Houses, so as to realise £176 per annum clear Profit. By SAMUEL WOOD, Author of "Good Gardening," &c. Fifth and Cheaper Edition, Revised, with Additions. Crown 8vo, 1s. sewed.

"We are bound to recommend it as not only suited to the case of the amateur and gentleman's gardener, but to the market grower."—*Gardeners' Magazine.*

Gardening for Ladies.
THE LADIES' MULTUM-IN-PARVO FLOWER GARDEN, and Amateurs' Complete Guide. With Illusts. By S. WOOD. Cr. 8vo, 3s. 6d. cl.

"This volume contains a good deal of sound, common sense instruction."—*Florist.*
"Full of shrewd hints and useful instructions, based on a lifetime of experience."—*Scotsman.*

Receipts for Gardeners.
GARDEN RECEIPTS. Edited by CHARLES W. QUIN. 12mo, 1s. 6d. cloth limp.

"A useful and handy book, containing a good deal of valuable information."—*Athenæum.*

Market Gardening.
MARKET AND KITCHEN GARDENING. By Contributors to "The Garden." Compiled by C. W. SHAW, late Editor of "Gardening Illustrated." 12mo, 3s. 6d. cloth boards.

"The most valuable compendium of kitchen and market-garden work published."—*Farmer.*

Cottage Gardening.
COTTAGE GARDENING; or, Flowers, Fruits, and Vegetables for Small Gardens. By E. HOBDAY. 12mo, 1s. 6d. cloth limp.

"Contains much useful information at a small charge."—*Glasgow Herald.*

AUCTIONEERING, VALUING, LAND SURVEYING ESTATE AGENCY, etc.

Auctioneer's Assistant.

THE APPRAISER, AUCTIONEER, BROKER, HOUSE AND ESTATE AGENT AND VALUER'S POCKET ASSISTANT, for the Valuation for Purchase, Sale, or Renewal of Leases, Annuities and Reversions, and of property generally; with Prices for Inventories, &c. By JOHN WHEELER, Valuer, &c. Sixth Edition, Re-written and greatly extended by C. NORRIS, Surveyor, Valuer, &c. Royal 32mo, 5s. cloth.

"A neat and concise book of reference, containing an admirable and clearly-arranged list of prices for inventories, and a very practical guide to determine the value of furniture, &c."—*Standard*.
"Contains a large quantity of varied and useful information as to the valuation for purchase, sale, or renewal of leases, annuities and reversions, and of property generally, with prices for inventories, and a guide to determine the value of interior fittings and other effects."—*Builder*.

Auctioneering.

AUCTIONEERS: THEIR DUTIES AND LIABILITIES. A Manual of Instruction and Counsel for the Young Auctioneer. By ROBERT SQUIBBS, Auctioneer. Second Edition, Revised and partly Re-written. Demy 8vo, 12s. 6d. cloth.

"The standard text-book on the topics of which it treats."—*Athenæum*.
"The work is one of general excellent character, and gives much information in a compendious and satisfactory form."—*Builder*.
"May be recommended as giving a great deal of information on the law relating to auctioneers, in a very readable form."—*Law Journal*.
"Auctioneers may be congratulated on having so pleasing a writer to minister to their special needs."—*Solicitors' Journal*.

Inwood's Estate Tables.

TABLES FOR THE PURCHASING OF ESTATES, Freehold, Copyhold, or Leasehold; Annuities, Advowsons, etc., and for the Renewing of Leases held under Cathedral Churches, Colleges, or other Corporate bodies, for Terms of Years certain, and for Lives; also for Valuing Reversionary Estates, Deferred Annuities, Next Presentations, &c.; together with SMART'S Five Tables of Compound Interest, and an Extension of the same to Lower and Intermediate Rates. By W. INWOOD. 23rd Edition, with considerable Additions, and new and valuable Tables of Logarithms for the more Difficult Computations of the Interest of Money, Discount, Annuities, &c., by M. FEDOR THOMAN, of the Société Crédit Mobilier of Paris. Crown 8vo, 8s. cloth.

"Those interested in the purchase and sale of estates, and in the adjustment of compensation cases, as well as in transactions in annuities, life insurances, &c., will find the present edition of eminent service."—*Engineering*.

Agricultural Valuer's Assistant.

THE AGRICULTURAL VALUER'S ASSISTANT. A Practical Handbook on the Valuation of Landed Estates; including Rules and Data for Measuring and Estimating the Contents, Weights, and Values of Agricultural Produce and Timber, and the Values of Feeding Stuffs, Manures, and Labour; with Forms of Tenant-Right-Valuations, Lists of Local Agricultural Customs, Scales of Compensation under the Agricultural Holdings Act, &c. &c. By TOM BRIGHT, Agricultural Surveyor. Second Edition, much Enlarged. Crown 8vo, 5s. cloth. [*Just published.*

"Full of tables and examples in connection with the valuation of tenant-right, estates, labour, contents, and weights of timber, and farm produce of all kinds."—*Agricultural Gazette*.
"An eminently practical handbook, full of practical tables and data of undoubted interest and value to surveyors and auctioneers in preparing valuations of all kinds."—*Farmer*.

Plantations and Underwoods.

POLE PLANTATIONS AND UNDERWOODS: A Practical Handbook on Estimating the Cost of Forming, Renovating, Improving, and Grubbing Plantations and Underwoods, their Valuation for Purposes of Transfer, Rental, Sale, or Assessment. By TOM BRIGHT, Author of "The Agricultural Valuer's Assistant," &c. Crown 8vo, 3s. 6d. cloth.

"To valuers, foresters and agents it will be a welcome aid."—*North British Agriculturist*.
"Well calculated to assist the valuer in the discharge of his duties, and of undoubted interest and use both to surveyors and auctioneers in preparing valuations of all kinds."—*Kent Herald*.

AUCTIONEERING, VALUING, LAND SURVEYING, etc. 47

Hudson's Land Valuer's Pocket-Book.
THE LAND VALUER'S BEST ASSISTANT: Being Tables on a very much Improved Plan, for Calculating the Value of Estates. With Tables for reducing Scotch, Irish, and Provincial Customary Acres to Statute Measure, &c. By R. HUDSON, C.E. New Edition. Royal 32mo, leather, elastic band, 4s.

Ewart's Land Improver's Pocket-Book.
THE LAND IMPROVER'S POCKET-BOOK OF FORMULÆ, TABLES, and MEMORANDA required in any Computation relating to the Permanent Improvement of Landed Property. By JOHN EWART, Land Surveyor and Agricultural Engineer. Second Edition, Revised. Royal 32mo, oblong, leather, gilt edges, with elastic band, 4s.
"A compendious and handy little volume."—*Spectator.*

Complete Agricultural Surveyor's Pocket-Book.
THE LAND VALUER'S AND LAND IMPROVER'S COMPLETE POCKET-BOOK. Consisting of the above Two Works bound together. Leather, gilt edges, with strap, 7s. 6d.

House Property.
HANDBOOK OF HOUSE PROPERTY. A Popular and Practical Guide to the Purchase, Mortgage, Tenancy, and Compulsory Sale of Houses and Land, including the Law of Dilapidations and Fixtures; with Examples of all kinds of Valuations, Useful Information on Building, and Suggestive Elucidations of Fine Art. By E. L. TARBUCK, Architect and Surveyor. Fifth Edition, Enlarged. 12mo, 5s. cloth.
"The advice is thoroughly practical."—*Law Journal.*
"For all who have dealings with house property, this is an indispensable guide."—*Decoration.*
"Carefully brought up to date, and much improved by the addition of a division on fine art. . . . A well-written and thoughtful work."—*Land Agent's Record.*

LAW AND MISCELLANEOUS.

Private Bill Legislation and Provisional Orders.
HANDBOOK FOR THE USE OF SOLICITORS AND ENGINEERS Engaged in Promoting Private Acts of Parliament and Provisional Orders, for the Authorization of Railways, Tramways, Works for the Supply of Gas and Water, and other undertakings of a like character. By L. LIVINGSTON MACASSEY, of the Middle Temple, Barrister-at-Law, M.Inst.C.E.; Author of "Hints on Water Supply." Demy 8vo, 950 pp., 25s. cl.
"The author's double experience as an engineer and barrister has enabled him to approach the subject alike from an engineering and legal point of view."—*Local Government Chronicle.*

Law of Patents.
PATENTS FOR INVENTIONS, AND HOW TO PROCURE THEM. Compiled for the Use of Inventors, Patentees and others. By G. G. M. HARDINGHAM, Assoc.Mem,Inst.C.E., &c. Demy 8vo, 2s. 6d. cloth.

Metropolitan Rating Appeals.
REPORTS OF APPEALS HEARD BEFORE THE COURT OF GENERAL ASSESSMENT SESSIONS, from the Year 1871 to 1885. By EDWARD RYDE and ARTHUR LYON RYDE. Fourth Edition, with Introduction and Appendix by WALTER C. RYDE, of the Inner Temple, Barrister-at-Law. 8vo, 16s. cloth.

Pocket-Book for Sanitary Officials.
THE HEALTH OFFICER'S POCKET-BOOK: A Guide to Sanitary Practice and Law. For Medical Officers of Health, Sanitary Inspectors, Members of Sanitary Authorities, &c. By EDWARD F. WILLOUGHBY, M.D. (Lond.), &c., Author of "Hygiene and Public Health." Fcap. 8vo, 7s. 6d. cloth, red edges, rounded corners. [*Just published.*
"A mine of condensed information of a pertinent and useful kind on the various subjects of which it treats. The matter seems to have been carefully compiled and arranged for facility of reference, and it is well illustrated by diagrams and woodcuts. The different subjects are succinctly but fully and scientifically dealt with."—*The Lancet.*

A Complete Epitome of the Laws of this Country.

EVERY MAN'S OWN LAWYER: A Handy-Book of the Principles of Law and Equity. By A BARRISTER. Thirtieth Edition, carefully Revised, and including the Legislation of 1892. Comprising (amongst other Acts) the *Betting and Loans (Infants) Act*, 1892; the *Small Holdings Act*, 1892; the *Clergy Discipline Act*, 1892; the *Conveyancing and Law of Property Act*, 1892, &c.; as well as the *Forged Transfers Act*, 1891; the *Custody of Children Act*, 1891; the *Slander of Women Act*, 1891; the *Bankruptcy Act*, 1890; the *Directors' Liability Act*, 1890; the *Partnership Act*, 1890; the *Intestates' Estates Act*, 1890, and many other new Acts. Crown 8vo, 700 pp., price 6s. 8d. (saved at every consultation!), strongly bound in cloth.

[*Just published.*

*** The Book will be found to comprise (amongst other matter)—
THE RIGHTS AND WRONGS OF INDIVIDUALS—LANDLORD AND TENANT—VENDORS AND PURCHASERS—PARTNERS AND AGENTS—COMPANIES AND ASSOCIATIONS—MASTERS, SERVANTS, AND WORKMEN—LEASES AND MORTGAGES—CHURCH AND CLERGY, RITUAL—LIBEL AND SLANDER—CONTRACTS AND AGREEMENTS—BONDS AND BILLS OF SALE—CHEQUES, BILLS, AND NOTES—RAILWAY AND SHIPPING LAW—BANKRUPTCY AND INSURANCE—BORROWERS, LENDERS, AND SURETIES—CRIMINAL LAW—PARLIAMENTARY ELECTIONS—COUNTY COUNCILS—MUNICIPAL CORPORATIONS—PARISH LAW, CHURCHWARDENS, ETC.—INSANITARY DWELLINGS AND AREAS—PUBLIC HEALTH AND NUISANCES—FRIENDLY AND BUILDING SOCIETIES—COPYRIGHT AND PATENTS—TRADE MARKS AND DESIGNS—HUSBAND AND WIFE, DIVORCE, ETC.—TRUSTEES AND EXECUTORS—GUARDIAN AND WARD, INFANTS, ETC.—GAME LAWS AND SPORTING—HORSES, HORSE-DEALING, AND DOGS—INNKEEPERS, LICENSING, ETC.—FORMS OF WILLS, AGREEMENTS, ETC. ETC

☞ *The object of this work is to enable those who consult it to help themselves to the law; and thereby to dispense, as far as possible, with professional assistance and advice. There are many wrongs and grievances which persons submit to from time to time through not knowing how or where to apply for redress; and many persons have as great a dread of a lawyer's office as of a lion's den. With this book at hand it is believed that many a* SIX-AND-EIGHTPENCE *may be saved; many a wrong redressed; many a right reclaimed; many a law suit avoided; and many an evil abated. The work has established itself as the standard legal adviser of all classes, and has also made a reputation for itself as a useful book of reference for lawyers residing at a distance from law libraries, who are glad to have at hand a work embodying recent decisions and enactments.*

*** OPINIONS OF THE PRESS.

" It is a complete code of English Law, written in plain language, which all can understand. . . Should be in the hands of every business man, and all who wish to abolish lawyers' bills.'—*Weekly Times*.

" A useful and concise epitome of the law, compiled with considerable care."—*Law Magazine*.

" A complete digest of the most useful facts which constitute English law."—*Globe*.

" This excellent handbook. . . . Admirably done, admirably arranged, and admirably cheap."—*Leeds Mercury*.

" A concise, cheap and complete epitome of the English law. So plainly written that he who runs may read, and he who reads may understand."—*Figaro*.

" A dictionary of legal facts well put together. The book is a very useful one."—*Spectator*.

" A work which has long been wanted, which is thoroughly well done, and which we most cordially recommend."—*Sunday Times*.

" The latest edition of this popular book ought to be in every business establishment, and on every library table."—*Sheffield Post*.

" A complete epitome of the law; thoroughly intelligible to non-professional readers."
Bell's Life.

Legal Guide for Pawnbrokers.

THE PAWNBROKERS', FACTORS' AND MERCHANTS' GUIDE TO THE LAW OF LOANS AND PLEDGES. With the Statutes and a Digest of Cases on Rights and Liabilities, Civil and Criminal, as to Loans and Pledges of Goods, Debentures, Mercantile and other Securities. By H. C. FOLKARD, Esq., Barrister-at-Law, Author of " The Law of Slander and Libel," &c. With Additions and Corrections. Fcap. 8vo, 3s. 6d. cloth.

" This work contains simply everything that requires to be known concerning the department of the law of which it treats. We can safely commend the book as unique and very nearly perfect."—*Iron*.

The Law of Contracts.

LABOUR CONTRACTS: A Popular Handbook on the Law of Contracts for Works and Services. By DAVID GIBBONS. Fourth Edition, with Appendix of Statutes by T. F. UTTLEY, Solicitor. Fcap. 8vo, 3s. 6d. cloth. [*Just published.*

Weale's Rudimentary Series.

LONDON, 1862.
THE PRIZE MEDAL
Was awarded to the Publishers of
"WEALE'S SERIES."

A NEW LIST OF

WEALE'S SERIES
RUDIMENTARY SCIENTIFIC, EDUCATIONAL, AND CLASSICAL.

Comprising nearly Three Hundred and Fifty distinct works in almost every department of Science, Art, and Education, recommended to the notice of Engineers, Architects, Builders, Artisans, and Students generally, as well as to those interested in Workmen's Libraries, Literary and Scientific Institutions, Colleges, Schools, Science Classes, &c., &c.

☞ "WEALE'S SERIES includes Text-Books on almost every branch of Science and Industry, comprising such subjects as Agriculture, Architecture and Building, Civil Engineering, Fine Arts, Mechanics and Mechanical Engineering, Physical and Chemical Science, and many miscellaneous Treatises. The whole are constantly undergoing revision, and new editions, brought up to the latest discoveries in scientific research, are constantly issued. The prices at which they are sold are as low as their excellence is assured."—*American Literary Gazette.*

"Amongst the literature of technical education, WEALE'S SERIES has ever enjoyed a high reputation, and the additions being made by Messrs. CROSBY LOCKWOOD & SON render the series more complete, and bring the information upon the several subjects down to the present time."—*Mining Journal.*

"It is not too much to say that no books have ever proved more popular with, or more useful to, young engineers and others than the excellent treatises comprised in WEALE'S SERIES."—*Engineer.*

"The excellence of WEALE'S SERIES is now so well appreciated, that it would be wasting our space to enlarge upon their general usefulness and value."—*Builder.*

"The volumes of WEALE'S SERIES form one of the best collections of elementary technical books in any language."—*Architect.*

"WEALE'S SERIES has become a standard as well as an unrivalled collection of treatises in all branches of art and science."—*Public Opinion.*

PHILADELPHIA, 1876.
THE PRIZE MEDAL
Was awarded to the Publishers for
Books: Rudimentary, Scientific,
"WEALE'S SERIES," ETC.

CROSBY LOCKWOOD & SON,
7, STATIONERS' HALL COURT, LUDGATE HILL, LONDON, E.C.

WEALE'S RUDIMENTARY SCIENTIFIC SERIES.

※※ The volumes of this Series are freely Illustrated with Woodcuts, or otherwise, where requisite. Throughout the following List it must be understood that the books are bound in limp cloth, unless otherwise stated; *but the volumes marked with a ‡ may also be had strongly bound in cloth boards for 6d. extra.*

N.B.—In ordering from this List it is recommended, as a means of facilitating business and obviating error, to quote the numbers affixed to the volumes, as well as the titles and prices.

CIVIL ENGINEERING, SURVEYING, ETC.

No.
31. *WELLS AND WELL-SINKING.* By JOHN GEO. SWINDELL, A.R.I.B.A., and G. R. BURNELL, C.E. Revised Edition. With a New Appendix on the Qualities of Water. Illustrated. 2s.
35. *THE BLASTING AND QUARRYING OF STONE*, for Building and other Purposes. By Gen. Sir J. BURGOYNE, Bart. 1s. 6d.
43. *TUBULAR, AND OTHER IRON GIRDER BRIDGES*, particularly describing the Britannia and Conway Tubular Bridges. By G. DRYSDALE DEMPSEY, C.E. Fourth Edition. 2s.
44. *FOUNDATIONS AND CONCRETE WORKS*, with Practical Remarks on Footings, Sand, Concrete, Béton, Pile-driving, Caissons, and Cofferdams, &c. By E. DOBSON. Seventh Edition. 1s. 6d.
60. *LAND AND ENGINEERING SURVEYING.* By T. BAKER, C.E. Fifteenth Edition, revised by Professor J. R. YOUNG. 2s.‡
80*. *EMBANKING LANDS FROM THE SEA.* With examples and Particulars of actual Embankments, &c. By J. WIGGINS, F.G.S. 2s.
81. *WATER WORKS*, for the Supply of Cities and Towns. With a Description of the Principal Geological Formations of England as influencing Supplies of Water, &c. By S. HUGHES, C.E. New Edition. 4s.‡
118. *CIVIL ENGINEERING IN NORTH AMERICA*, a Sketch of. By DAVID STEVENSON, F.R.S.E., &c. Plates and Diagrams. 3s.
167. *IRON BRIDGES, GIRDERS, ROOFS, AND OTHER WORKS.* By FRANCIS CAMPIN, C.E. 2s. 6d.‡
197. *ROADS AND STREETS.* By H. LAW, C.E., revised and enlarged by D. K. CLARK, C.E., including pavements of Stone, Wood, Asphalte, &c. 4s. 6d.‡
203. *SANITARY WORK IN THE SMALLER TOWNS AND IN VILLAGES.* By C. SLAGG, A.M.I.C.E. Revised Edition. 3s.‡
212. *GAS-WORKS, THEIR CONSTRUCTION AND ARRANGEMENT*; and the Manufacture and Distribution of Coal Gas. Originally written by SAMUEL HUGHES, C.E. Re-written and enlarged by WILLIAM RICHARDS, C.E. Eighth Edition, with important additions. 5s. 6d.‡
213. *PIONEER ENGINEERING.* A Treatise on the Engineering Operations connected with the Settlement of Waste Lands in New Countries. By EDWARD DOBSON, Assoc. Inst. C.E. 4s. 6d.‡
216. *MATERIALS AND CONSTRUCTION*; A Theoretical and Practical Treatise on the Strains, Designing, and Erection of Works of Construction. By FRANCIS CAMPIN, C.E. Second Edition, revised. 3s.‡
219. *CIVIL ENGINEERING.* By HENRY LAW, M.Inst. C.E. Including HYDRAULIC ENGINEERING by GEO. R. BURNELL, M.Inst. C.E. Seventh Edition, revised, with large additions by D. KINNEAR CLARK, M.Inst. C.E. 6s. 6d., Cloth boards, 7s. 6d.
268. *THE DRAINAGE OF LANDS, TOWNS, & BUILDINGS.* By G. D. DEMPSEY, C.E. Revised, with large Additions on Recent Practice in Drainage Engineering, by D. KINNEAR CLARK, M.I.C.E. Second Edition, Corrected. 4s. 6d.‡

☞ *The ‡ indicates that these vols. may be had strongly bound at 6d. extra.*

LONDON : CROSBY LOCKWOOD AND SON,

MECHANICAL ENGINEERING, ETC.

33. *CRANES*, the Construction of, and other Machinery for Raising Heavy Bodies. By JOSEPH GLYNN, F.R.S. Illustrated. 1s. 6d.
34. *THE STEAM ENGINE*. By Dr. LARDNER. Illustrated. 1s. 6d.
59. *STEAM BOILERS:* their Construction and Management. By R. ARMSTRONG, C.E. Illustrated. 1s. 6d.
82. *THE POWER OF WATER*, as applied to drive Flour Mills, and to give motion to Turbines, &c. By JOSEPH GLYNN, F.R.S. 2s.‡
98. *PRACTICAL MECHANISM*, the Elements of; and Machine Tools. By T. BAKER, C.E. With Additions by J. NASMYTH, C.E. 2s. 6d.‡
139. *THE STEAM ENGINE*, a Treatise on the Mathematical Theory of, with Rules and Examples for Practical Men. By T. BAKER, C.E. 1s. 6d.
164. *MODERN WORKSHOP PRACTICE*, as applied to Steam Engines, Bridges, Ship-building, &c. By J. G. WINTON. New Edition. 3s. 6d.‡
165. *IRON AND HEAT*, exhibiting the Principles concerned in the Construction of Iron Beams, Pillars, and Girders. By J. ARMOUR. 2s. 6d.‡
166. *POWER IN MOTION:* Horse-Power, Toothed-Wheel Gearing, Long and Short Driving Bands, and Angular Forces. By J. ARMOUR, 2s.‡
171. *THE WORKMAN'S MANUAL OF ENGINEERING DRAWING.* By J. MAXTON. 7th Edn. With 7 Plates and 350 Cuts. 3s. 6d.‡
190. *STEAM AND THE STEAM ENGINE*, Stationary and Portable. By J. SEWELL and D. K. CLARK, C.E. 3s. 6d.‡
200. *FUEL*, its Combustion and Economy. By C. W. WILLIAMS. With Recent Practice in the Combustion and Economy of Fuel—Coal, Coke, Wood, Peat, Petroleum, &c.—by D. K. CLARK, M.I.C.E. 3s. 6d.‡
202. *LOCOMOTIVE ENGINES.* By G. D. DEMPSEY, C.E.; with large additions by D. KINNEAR CLARK, M.I.C.E. 3s.‡
211. *THE BOILERMAKER'S ASSISTANT* in Drawing, Templating, and Calculating Boiler and Tank Work. By JOHN COURTNEY, Practical Boiler Maker. Edited by D. K. CLARK, C.E. 100 Illustrations. 2s.
217. *SEWING MACHINERY:* Its Construction, History, &c., with full Technical Directions for Adjusting, &c. By J. W. URQUHART, C.E. 2s.‡
223. *MECHANICAL ENGINEERING.* Comprising Metallurgy, Moulding, Casting, Forging, Tools, Workshop Machinery, Manufacture of the Steam Engine, &c. By FRANCIS CAMPIN, C.E. Second Edition. 2s. 6d.‡
236. *DETAILS OF MACHINERY.* Comprising Instructions for the Execution of various Works in Iron. By FRANCIS CAMPIN, C.E. 3s.‡
237. *THE SMITHY AND FORGE;* including the Farrier's Art and Coach Smithing. By W. J. E. CRANE. Illustrated. 2s. 6d.†
238. *THE SHEET-METAL WORKER'S GUIDE;* a Practical Handbook for Tinsmiths, Coppersmiths, Zincworkers, &c. With 94 Diagrams and Working Patterns. By W. J. E. CRANE. Second Edition, revised. 1s. 6d.
251. *STEAM AND MACHINERY MANAGEMENT:* with Hints on Construction and Selection. By M. POWIS BALE, M.I.M.E. 2s. 6d.‡
254. *THE BOILERMAKER'S READY-RECKONER.* By J. COURTNEY. Edited by D. K. CLARK, C.E. 4s.
. *Nos.* 211 *and* 254 *in One Vol., half-bound, entitled* "THE BOILERMAKER'S READY-RECKONER AND ASSISTANT." By J. COURTNEY and D. K. CLARK. 7s.
255. *LOCOMOTIVE ENGINE-DRIVING.* A Practical Manual for Engineers in charge of Locomotive Engines. By MICHAEL REYNOLDS, M.S.E Eighth Edition. 3s. 6d., limp; 4s. 6d. cloth boards.
256. *STATIONARY ENGINE-DRIVING.* A Practical Manual Engineers in charge of Stationary Engines. By MICHAEL REYNOLDS, M.S.E. Fourth Edition. 3s. 6d. limp; 4s. 6d. cloth boards.
260. *IRON BRIDGES OF MODERATE SPAN:* their Construction and Erection. By HAMILTON W. PENDRED, C.E. 2s.

☞ *The ‡ indicates that these vols. may be had strongly bound at* 6*d. extra.*

7, STATIONERS' HALL COURT, LUDGATE HILL, E.C.

MINING, METALLURGY, ETC.

4. *MINERALOGY*, Rudiments of; a concise View of the General Properties of Minerals. By A. RAMSAY, F.G.S., F.R.G.S., &c. Third Edition, revised and enlarged. Illustrated. 3s. 6d.‡

117. *SUBTERRANEOUS SURVEYING*, with and without the Magnetic Needle. By T. FENWICK and T. BAKER, C.E. Illustrated. 2s. 6d.‡

135. *ELECTRO-METALLURGY;* Practically Treated. By ALEXANDER WATT. Ninth Edition, enlarged and revised, with additional Illustrations, and including the most recent Processes. 3s. 6d.‡

172. *MINING TOOLS*, Manual of. For the Use of Mine Managers, Agents, Students, &c. By WILLIAM MORGANS. 2s. 6d.

172*. *MINING TOOLS, ATLAS* of Engravings to Illustrate the above, containing 235 Illustrations, drawn to Scale. 4to. 4s. 6d.

176. *METALLURGY OF IRON.* Containing History of Iron Manufacture, Methods of Assay, and Analyses of Iron Ores, Processes of Manufacture of Iron and Steel, &c. By H. BAUERMAN, F.G.S. Sixth Edition, revised and enlarged. 5s.‡

180. *COAL AND COAL MINING.* By the late Sir WARINGTON W. SMYTH, M.A., F.R.S. Seventh Edition, revised. 3s. 6d.‡

195. *THE MINERAL SURVEYOR AND VALUER'S COMPLETE GUIDE.* By W. LINTERN, M.E. Third Edition, including Magnetic and Angular Surveying. With Four Plates. 3s. 6d.‡

214. *SLATE AND SLATE QUARRYING*, Scientific, Practical, and Commercial. By D. C. DAVIES, F.G.S., Mining Engineer, &c. 3s.‡

264. *A FIRST BOOK OF MINING AND QUARRYING*, with the Sciences connected therewith, for Primary Schools and Self Instruction. By J. H. COLLINS, F.G.S. Second Edition, with additions. 1s. 6d.

ARCHITECTURE, BUILDING, ETC.

16. *ARCHITECTURE—ORDERS*—The Orders and their Æsthetic Principles. By W. H. LEEDS. Illustrated. 1s. 6d.

17. *ARCHITECTURE—STYLES*—The History and Description of the Styles of Architecture of Various Countries, from the Earliest to the Present Period. By T. TALBOT BURY, F.R.I.B.A., &c. Illustrated. 2s.
*** ORDERS AND STYLES OF ARCHITECTURE, *in One Vol.*, 3s. 6d.

18. *ARCHITECTURE—DESIGN*—The Principles of Design in Architecture, as deducible from Nature and exemplified in the Works of the Greek and Gothic Architects. By E. L. GARBETT, Architect. Illustrated. 2s. 6d.
*** *The three preceding Works, in One handsome Vol., half bound, entitled* "MODERN ARCHITECTURE," *price* 6s.

22. *THE ART OF BUILDING*, Rudiments of. General Principles of Construction, Materials used in Building, Strength and Use of Materials, Working Drawings, Specifications, and Estimates. By E. DOBSON, 2s.‡

25. *MASONRY AND STONECUTTING:* Rudimentary Treatise on the Principles of Masonic Projection and their application to Construction. By EDWARD DOBSON, M.R.I.B.A., &c. 2s. 6d.‡

42. *COTTAGE BUILDING.* By C. BRUCE ALLEN, Architect. Eleventh Edition, revised and enlarged. With a Chapter on Economic Cottages for Allotments, by EDWARD E. ALLEN, C.E. 2s.

45. *LIMES, CEMENTS, MORTARS, CONCRETES, MASTICS,* PLASTERING, &c. By G. R. BURNELL, C.E Fourteenth Edition. 1s. 6d

57. *WARMING AND VENTILATION.* An Exposition of the General Principles as applied to Domestic and Public Buildings, Mines, Lighthouses, Ships, &c. By C. TOMLINSON, F.R.S., &c. Illustrated. 3s.

111. *ARCHES, PIERS, BUTTRESSES, &c.:* Experimental Essays on the Principles of Construction. By W. BLAND. Illustrated. 1s. 6d.

☞ *The* ‡ *indicates that these vols. may be had strongly bound at 6d. extra.*

LONDON : CROSBY LOCKWOOD AND SON,

Architecture, Building, etc., *continued.*

116. **THE ACOUSTICS OF PUBLIC BUILDINGS;** or, The Principles of the Science of Sound applied to the purposes of the Architect and Builder. By T. ROGER SMITH, M.R.I.B.A., Architect. Illustrated. 1s. 6d.

127. **ARCHITECTURAL MODELLING IN PAPER,** the Art of. By T. A. RICHARDSON, Architect. Illustrated. 1s. 6d.

128. **VITRUVIUS—THE ARCHITECTURE OF MARCUS VITRUVIUS POLLO.** In Ten Books. Translated from the Latin by JOSEPH GWILT, F.S.A., F.R.A.S. With 23 Plates. 5s.

130. **GRECIAN ARCHITECTURE,** An Inquiry into the Principles of Beauty in; with an Historical View of the Rise and Progress of the Art in Greece. By the EARL OF ABERDEEN. 1s.

⁎ *The two preceding Works in One handsome Vol., half bound, entitled "ANCIENT ARCHITECTURE," price 6s.*

132. **THE ERECTION OF DWELLING-HOUSES.** Illustrated by a Perspective View, Plans, Elevations, and Sections of a pair of Semi-detached Villas, with the Specification, Quantities, and Estimates, &c. By S. H. BROOKS. New Edition, with Plates. 2s. 6d.

156. **QUANTITIES & MEASUREMENTS** in Bricklayers', Masons', Plasterers', Plumbers', Painters', Paperhangers', Gilders', Smiths', Carpenters' and Joiners' Work. By A. C. BEATON, Surveyor. Ninth Edition. 1s. 6d.

175. **LOCKWOOD'S BUILDER'S PRICE BOOK FOR 1893.** A Comprehensive Handbook of the Latest Prices and Data for Builders, Architects, Engineers, and Contractors. Re-constructed, Re-written, and further Enlarged. By FRANCIS T. W. MILLER, A.R.I.B.A. 700 pages. 3s. 6d.; cloth boards, 4s. [*Just Published.*

182. **CARPENTRY AND JOINERY**—THE ELEMENTARY PRINCIPLES OF CARPENTRY. Chiefly composed from the Standard Work of THOMAS TREDGOLD, C.E. With a TREATISE ON JOINERY by E. WYNDHAM TARN, M.A. Fifth Edition. Revised. 3s. 6d.‡

182*. **CARPENTRY AND JOINERY. ATLAS** of 35 Plates to accompany the above. With Descriptive Letterpress. 4to. 6s.

185. **THE COMPLETE MEASURER;** the Measurement of Boards, Glass, &c.; Unequal-sided, Square-sided, Octagonal-sided, Round Timber and Stone, and Standing Timber, &c. By RICHARD HORTON. Fifth Edition. 4s.; strongly bound in leather, 5s.

187. **HINTS TO YOUNG ARCHITECTS.** By G. WIGHTWICK. New Edition. By G. H. GUILLAUME. Illustrated. 3s. 6d.‡

188. **HOUSE PAINTING, GRAINING, MARBLING, AND SIGN WRITING:** with a Course of Elementary Drawing for House-Painters, Sign-Writers, &c., and a Collection of Useful Receipts. By ELLIS A. DAVIDSON. Sixth Edition. With Coloured Plates. 5s. cloth limp; 6s. cloth boards.

189. **THE RUDIMENTS OF PRACTICAL BRICKLAYING.** In Six Sections: General Principles; Arch Drawing, Cutting, and Setting; Pointing; Paving, Tiling, Materials; Slating and Plastering; Practical Geometry, Mensuration, &c. By ADAM HAMMOND. Seventh Edition. 1s. 6d.

191. **PLUMBING.** A Text-Book to the Practice of the Art or Craft of the Plumber. With Chapters upon House Drainage and Ventilation. Sixth Edition. With 380 Illustrations. By W. P. BUCHAN. 3s. 6d.‡

192. **THE TIMBER IMPORTER'S, TIMBER MERCHANT'S,** and BUILDER'S STANDARD GUIDE. By R. E. GRANDY. 2s.

206. **A BOOK ON BUILDING,** *Civil and Ecclesiastical,* including CHURCH RESTORATION. With the Theory of Domes and the Great Pyramid, &c. By Sir EDMUND BECKETT, Bart., LL.D., Q.C., F.R.A.S. 4s. 6d.‡

226. **THE JOINTS MADE AND USED BY BUILDERS** in the Construction of various kinds of Engineering and Architectural Works. By WYVILL J. CHRISTY, Architect. With upwards of 160 Engravings on Wood. 3s.‡

228. **THE CONSTRUCTION OF ROOFS OF WOOD AND IRON.** By E. WYNDHAM TARN, M.A., Architect. Second Edition, revised. 1s. 6d.

☞ *The ‡ indicates that these vols. may be had strongly bound at 6d. extra.*

Architecture, Building, etc., *continued.*

229. *ELEMENTARY DECORATION:* as applied to the Interior and Exterior Decoration of Dwelling-Houses, &c. By J. W. FACEY. 2s.

257. *PRACTICAL HOUSE DECORATION.* A Guide to the Art of Ornamental Painting. By JAMES W. FACEY. 2s. 6d.

*** *The two preceding Works, in One handsome Vol., half-bound, entitled* "HOUSE DECORATION, ELEMENTARY AND PRACTICAL," *price* 5s.

230. *A PRACTICAL TREATISE ON HANDRAILING.* Showing New and Simple Methods. By G. COLLINGS. Second Edition, Revised, including A TREATISE ON STAIRBUILDING. Plates. 2s. 6d.

247. *BUILDING ESTATES:* a Rudimentary Treatise on the Development, Sale, Purchase, and General Management of Building Land. By FOWLER MAITLAND, Surveyor. Second Edition, revised. 2s.

248. *PORTLAND CEMENT FOR USERS.* By HENRY FAIJA, Assoc. M. Inst. C.E. Third Edition, corrected. Illustrated. 2s.

252. *BRICKWORK:* a Practical Treatise, embodying the General and Higher Principles of Bricklaying, Cutting and Setting, &c. By F. WALKER. Third Edition, Revised and Enlarged. 1s. 6d.

23. *THE PRACTICAL BRICK AND TILE BOOK.* Comprising:
189. BRICK AND TILE MAKING, by E. DOBSON, A.I.C.E.; PRACTICAL BRICKLAY-
265. ING, by A. HAMMOND; BRICKCUTTING AND SETTING, by A. HAMMOND. 534 pp. with 270 Illustrations. 6s. Strongly half-bound.

253. *THE TIMBER MERCHANT'S, SAW-MILLER'S, AND IMPORTER'S FREIGHT-BOOK AND ASSISTANT.* By WM. RICHARDSON. With Additions by M. POWIS BALE, A.M.Inst.C.E. 3s.‡

258. *CIRCULAR WORK IN CARPENTRY AND JOINERY.* A Practical Treatise on Circular Work of Single and Double Curvature. By GEORGE COLLINGS. Second Edition, 2s. 6d.

259. *GAS FITTING:* A Practical Handbook treating of every Description of Gas Laying and Fitting. By JOHN BLACK. 2s. 6d.‡

261. *SHORING AND ITS APPLICATION:* A Handbook for the Use of Students. By GEORGE H. BLAGROVE. 1s. 6d.

265. *THE ART OF PRACTICAL BRICK CUTTING & SETTING.* By ADAM HAMMOND. With 90 Engravings. 1s. 6d.

267. *THE SCIENCE OF BUILDING:* An Elementary Treatise on the Principles of Construction. By E. WYNDHAM TARN, M.A. Lond. Third Edition, Revised and Enlarged. 3s. 6d.‡

271. *VENTILATION:* a Text-book to the Practice of the Art of Ventilating Buildings. By W. P. BUCHAN, R.P., Sanitary Engineer, Author of "Plumbing," &c. 3s. 6d.‡

272. *ROOF CARPENTRY;* Practical Lessons in the Framing of Wood Roofs. For the Use of Working Carpenters. By GEO. COLLINGS, Author of "Handrailing and Stairbuilding," &c. 2s. [*Just published.*

273. *THE PRACTICAL PLASTERER:* A Compendium of Plain and Ornamental Plaster Work. By WILFRED KEMP. 2s. [*Just published.*

SHIPBUILDING, NAVIGATION, ETC.

51. *NAVAL ARCHITECTURE.* An Exposition of the Elementary Principles. By J. PEAKE. Fifth Edition, with Plates. 3s. 6d.‡

53*. *SHIPS FOR OCEAN & RIVER SERVICE,* Elementary and Practical Principles of the Construction of. By H. A. SOMMERFELDT. 1s. 6d.

53**. *AN ATLAS OF ENGRAVINGS* to Illustrate the above. Twelve large folding plates. Royal 4to, cloth. 7s. 6d.

54. *MASTING, MAST-MAKING, AND RIGGING OF SHIPS,* Also Tables of Spars, Rigging, Blocks; Chain, Wire, and Hemp Ropes, &c., relative to every class of vessels. By ROBERT KIPPING, N.A. 2s.

☞ *The* ‡ *indicates that these vols. may be had strongly bound at* 6d. *extra.*

LONDON: CROSBY LOCKWOOD AND SON,

Shipbuilding, Navigation, Marine Engineering, etc., *cont.*

54*. *IRON SHIP-BUILDING.* With Practical Examples and Details. By JOHN GRANTHAM, C.E. Fifth Edition. 4s.

55. *THE SAILOR'S SEA BOOK:* a Rudimentary Treatise on Navigation. By JAMES GREENWOOD, B.A. With numerous Woodcuts and Coloured Plates. New and enlarged edition. By W. H. ROSSER. 2s. 6d.‡

80. *MARINE ENGINES AND STEAM VESSELS.* By ROBERT MURRAY, C.E. Eighth Edition, thoroughly Revised, with Additions by the Author and by GEORGE CARLISLE, C.E. 4s. 6d. limp; 5s. cloth boards.

83*bis*. *THE FORMS OF SHIPS AND BOATS.* By W. BLAND. Eighth Edition, Revised, with numerous Illustrations and Models. 1s. 6d.

99. *NAVIGATION AND NAUTICAL ASTRONOMY,* in Theory and Practice. By Prof. J. R. YOUNG. New Edition. 2s. 6d.

106. *SHIPS' ANCHORS,* a Treatise on. By G. COTSELL, N.A. 1s. 6d.

149. *SAILS AND SAIL-MAKING.* With Draughting, and the Centre of Effort of the Sails; Weights and Sizes of Ropes; Masting, Rigging, and Sails of Steam Vessels, &c. 12th Edition. By R. KIPPING, N.A., 2s. 6d.‡

155. *ENGINEER'S GUIDE TO THE ROYAL & MERCANTILE* NAVIES. By a PRACTICAL ENGINEER. Revised by D. F. M'CARTHY. 3s.

55 & 204. *PRACTICAL NAVIGATION.* Consisting of The Sailor's Sea-Book. By JAMES GREENWOOD and W. H. ROSSER. Together with the requisite Mathematical and Nautical Tables for the Working of the Problems. By H. LAW, C.E., and Prof. J. R. YOUNG. 7s. Half-bound.

AGRICULTURE, GARDENING, ETC.

61*. *A COMPLETE READY RECKONER FOR THE ADMEA-SUREMENT* OF LAND, &c. By A. ARMAN. Third Edition, revised and extended by C. NORRIS, Surveyor, Valuer, &c. 2s.

131. *MILLER'S, CORN MERCHANT'S, AND FARMER'S* READY RECKONER. Second Edition, with a Price List of Modern Flour-Mill Machinery, by W. S. HUTTON, C.E. 2s.

140. *SOILS, MANURES, AND CROPS.* (Vol. 1. OUTLINES OF MODERN FARMING.) By R. SCOTT BURN. Woodcuts. 2s.

141. *FARMING & FARMING ECONOMY,* Notes, Historical and Practical, on. (Vol. 2. OUTLINES OF MODERN FARMING.) By R. SCOTT BURN. 3s.

142. *STOCK; CATTLE, SHEEP, AND HORSES.* (Vol. 3. OUTLINES OF MODERN FARMING.) By R. SCOTT BURN. Woodcuts. 2s. 6d.

145. *DAIRY, PIGS, AND POULTRY,* Management of the. By R. SCOTT BURN. (Vol. 4. OUTLINES OF MODERN FARMING.) 2s.

146. *UTILIZATION OF SEWAGE, IRRIGATION, AND* RECLAMATION OF WASTE LAND. (Vol. 5. OUTLINES OF MODERN FARMING.) By R. SCOTT BURN. Woodcuts. 2s. 6d.

⁎ Nos. 140-1-2-5-6, *in One Vol., handsomely half-bound, entitled "*OUTLINES OF MODERN FARMING." *By* ROBERT SCOTT BURN. *Price* 12s.

177. *FRUIT TREES,* The Scientific and Profitable Culture of. From the French of DU BREUIL. Revised by GEO. GLENNY. 187 Woodcuts. 3s. 6d.‡

198. *SHEEP; THE HISTORY, STRUCTURE, ECONOMY, AND* DISEASES OF. By W. C. SPOONER, M.R.V.C., &c. Fifth Edition, enlarged, including Specimens of New and Improved Breeds. 3s. 6d.‡

201. *KITCHEN GARDENING MADE EASY.* By GEORGE M. F. GLENNY. Illustrated. 1s. 6d.‡

207. *OUTLINES OF FARM MANAGEMENT, and the Organi-zation of Farm Labour.* By R. SCOTT BURN. 2s. 6d.‡

208. *OUTLINES OF LANDED ESTATES MANAGEMENT.* By R. SCOTT BURN. 2s. 6d.

⁎ Nos. 207 & 208 *in One Vol., handsomely half-bound, entitled "*OUTLINES OF LANDED ESTATES AND FARM MANAGEMENT." *By* R. SCOTT BURN. *Price* 6s.

☞ *The ‡ indicates that these vols. may be had strongly bound at 6d. extra.*

7, STATIONERS' HALL COURT, LUDGATE HILL, E.C.

Agriculture, Gardening, etc., *continued.*

209. *THE TREE PLANTER AND PLANT PROPAGATOR.* A Practical Manual on the Propagation of Forest Trees, Fruit Trees, Flowering Shrubs, Flowering Plants, &c. By SAMUEL WOOD. 2s.

210. *THE TREE PRUNER.* A Practical Manual on the Pruning of Fruit Trees, including also their Training and Renovation; also the Pruning of Shrubs, Climbers, and Flowering Plants. By SAMUEL WOOD. 1s. 6d.

*** *Nos.* 209 & 210 *in One Vol.*, *handsomely half-bound, entitled* "THE TREE PLANTER, PROPAGATOR, AND PRUNER." By SAMUEL WOOD. *Price* 3s. 6d.

218. *THE HAY AND STRAW MEASURER:* Being New Tables for the Use of Auctioneers, Valuers, Farmers, Hay and Straw Dealers, &c. By JOHN STEELE. Fifth Edition. 2s.

222. *SUBURBAN FARMING.* The Laying-out and Cultivation of Farms, adapted to the Produce of Milk, Butter, and Cheese, Eggs, Poultry, and Pigs. By Prof. JOHN DONALDSON and R. SCOTT BURN. 3s. 6d.‡

231. *THE ART OF GRAFTING AND BUDDING.* By CHARLES BALTET. With Illustrations. 2s. 6d.‡

232. *COTTAGE GARDENING;* or, Flowers, Fruits, and Vegetables for Small Gardens. By E. HOBDAY. 1s. 6d.

233. *GARDEN RECEIPTS.* Edited by CHARLES W. QUIN. 1s. 6d.

234. *MARKET AND KITCHEN GARDENING.* By C. W. SHAW, late Editor of "Gardening Illustrated." 3s.‡

239. *DRAINING AND EMBANKING.* A Practical Treatise, embodying the most recent experience in the Application of Improved Methods. By JOHN SCOTT, late Professor of Agriculture and Rural Economy at the Royal Agricultural College, Cirencester. With 68 Illustrations. 1s. 6d.

240. *IRRIGATION AND WATER SUPPLY.* A Treatise on Water Meadows, Sewage Irrigation, and Warping; the Construction of Wells, Ponds, and Reservoirs, &c. By Prof. JOHN SCOTT. With 34 Illus. 1s. 6d.

241. *FARM ROADS, FENCES, AND GATES.* A Practical Treatise on the Roads, Tramways, and Waterways of the Farm; the Principles of Enclosures; and the different kinds of Fences, Gates, and Stiles. By Professor JOHN SCOTT. With 75 Illustrations. 1s. 6d.

242. *FARM BUILDINGS.* A Practical Treatise on the Buildings necessary for various kinds of Farms, their Arrangement and Construction, with Plans and Estimates. By Prof. JOHN SCOTT. With 105 Illus. 2s.

243. *BARN IMPLEMENTS AND MACHINES.* A Practical Treatise on the Application of Power to the Operations of Agriculture; and on various Machines used in the Threshing-barn, in the Stock-yard, and in the Dairy, &c. By Prof. J. SCOTT. With 123 Illustrations. 2s.

244. *FIELD IMPLEMENTS AND MACHINES.* A Practical Treatise on the Varieties now in use, with Principles and Details of Construction, their Points of Excellence, and Management. By Professor JOHN SCOTT. With 138 Illustrations. 2s.

245. *AGRICULTURAL SURVEYING.* A Practical Treatise on Land Surveying, Levelling, and Setting-out; and on Measuring and Estimating Quantities, Weights, and Values of Materials, Produce, Stock, &c. By Prof. JOHN SCOTT. With 62 Illustrations. 1s. 6d.

*** *Nos.* 239 *to* 245 *in One Vol.*, *handsomely half-bound, entitled* "THE COMPLETE TEXT-BOOK OF FARM ENGINEERING." By Professor JOHN SCOTT. *Price* 12s.

250. *MEAT PRODUCTION.* A Manual for Producers, Distributors, &c. By JOHN EWART. 2s. 6d.‡

266. *BOOK-KEEPING FOR FARMERS & ESTATE OWNERS.* By J. M. WOODMAN, Chartered Accountant. 2s. 6d. cloth limp; 3s. 6d. cloth boards.

☞ *The ‡ indicates that these vols. may be had strongly bound at 6d. extra.*

LONDON : CROSBY LOCKWOOD AND SON,

MATHEMATICS, ARITHMETIC, ETC.

32. *MATHEMATICAL INSTRUMENTS*, a Treatise on; Their Construction, Adjustment, Testing, and Use concisely Explained. By J. F. HEATHER, M.A. Fourteenth Edition, revised, with additions, by A. T. WALMISLEY, M.I.C.E., Fellow of the Surveyors' Institution. Original Edition, in 1 vol., Illustrated. 2s.‡

⁎ *In ordering the above, be careful to say, "Original Edition" (No. 32), to distinguish it from the Enlarged Edition in 3 vols. (Nos. 168-9-70.)*

76. *DESCRIPTIVE GEOMETRY*, an Elementary Treatise on; with a Theory of Shadows and of Perspective, extracted from the French of G. MONGE. To which is added, a description of the Principles and Practice of Isometrical Projection. By J. F. HEATHER, M.A. With 14 Plates. 2s.

178. *PRACTICAL PLANE GEOMETRY:* giving the Simplest Modes of Constructing Figures contained in one Plane and Geometrical Construction of the Ground. By J. F. HEATHER, M.A. With 215 Woodcuts. 2s.

83. *COMMERCIAL BOOK-KEEPING.* With Commercial Phrases and Forms in English, French, Italian, and German. By JAMES HADDON, M.A., Arithmetical Master of King's College School, London. 1s. 6d.

84. *ARITHMETIC*, a Rudimentary Treatise on: with full Explanations of its Theoretical Principles, and numerous Examples for Practice. By Professor J. R. YOUNG. Eleventh Edition. 1s. 6d.

84*. A KEY to the above, containing Solutions in full to the Exercises, together with Comments, Explanations, and Improved Processes, for the Use of Teachers and Unassisted Learners. By J. R. YOUNG. 1s. 6d.

85. *EQUATIONAL ARITHMETIC*, applied to Questions of Interest, Annuities, Life Assurance, and General Commerce; with various Tables by which all Calculations may be greatly facilitated. By W. HIPSLEY. 2s.

86. *ALGEBRA*, the Elements of. By JAMES HADDON, M.A. With Appendix, containing miscellaneous Investigations, and a Collection of Problems in various parts of Algebra. 2s.

86*. A KEY AND COMPANION to the above Book, forming an extensive repository of Solved Examples and Problems in Illustration of the various Expedients necessary in Algebraical Operations. By J. R. YOUNG. 1s. 6d.

88. *EUCLID*, THE ELEMENTS OF: with many additional Propositions
89. and Explanatory Notes: to which is prefixed, an Introductory Essay o Logic. By HENRY LAW, C.E. 2s. 6d.‡

⁎ *Sold also separately, viz.:—*

88. EUCLID, The First Three Books. By HENRY LAW, C.E. 1s. 6d.
89. EUCLID, Books 4, 5, 6, 11, 12. By HENRY LAW, C.E. 1s. 6d.

90. *ANALYTICAL GEOMETRY AND CONIC SECTIONS*, By JAMES HANN. A New Edition, by Professor J. R. YOUNG. 2s.‡

91. *PLANE TRIGONOMETRY*, the Elements of. By JAMES HANN, formerly Mathematical Master of King's College, London. 1s. 6d.

92. *SPHERICAL TRIGONOMETRY*, the Elements of. By JAMES HANN. Revised by CHARLES H. DOWLING, C.E. 1s.

⁎ *Or with "The Elements of Plane Trigonometry," in One Volume, 2s. 6d.*

93. *MENSURATION AND MEASURING.* With the Mensuration and Levelling of Land for the Purposes of Modern Engineering. By T. BAKER, C.E. New Edition by E. NUGENT, C.E. Illustrated. 1s. 6d.

101. *DIFFERENTIAL CALCULUS*, Elements of the. By W. S. B. WOOLHOUSE, F.R.A.S., &c. 1s. 6d.

102. *INTEGRAL CALCULUS*, Rudimentary Treatise on the. By HOMERSHAM COX, B.A. Illustrated. 1s.

136. *ARITHMETIC*, Rudimentary, for the Use of Schools and Self-Instruction. By JAMES HADDON, M.A. Revised by A. ARMAN. 1s. 6d.

137. A KEY TO HADDON'S RUDIMENTARY ARITHMETIC. By A. ARMAN. 1s. 6d.

☞ *The ‡ indicates that these vols. may be had strongly bound at 6d. extra.*

7, STATIONERS' HALL COURT, LUDGATE HILL, E.C.

Mathematics, Arithmetic, etc., *continued*.

168. *DRAWING AND MEASURING INSTRUMENTS.* Including—I. Instruments employed in Geometrical and Mechanical Drawing, and in the Construction, Copying, and Measurement of Maps and Plans. II. Instruments used for the purposes of Accurate Measurement, and for Arithmetical Computations. By J. F. Heather, M.A. Illustrated. 1s. 6d

169. *OPTICAL INSTRUMENTS.* Including (more especially) Telescopes, Microscopes, and Apparatus for producing copies of Maps and Plans by Photography. By J. F. Heather, M.A. Illustrated. 1s. 6d.

170. *SURVEYING AND ASTRONOMICAL INSTRUMENTS.* Including—I. Instruments Used for Determining the Geometrical Features of a portion of Ground. II. Instruments Employed in Astronomical Observations. By J. F. Heather, M.A. Illustrated. 1s. 6d.

*** *The above three volumes form an enlargement of the Author's original work* "Mathematical Instruments." (See No. 32 in the Series.).

168. } *MATHEMATICAL INSTRUMENTS.* By J. F. Heather,
169. } M.A. Enlarged Edition, for the most part entirely re-written. The 3 Parts as
170. } above, in One thick Volume. With numerous Illustrations. 4s. 6d.‡

158. *THE SLIDE RULE, AND HOW TO USE IT;* containing full, easy, and simple Instructions to perform all Business Calculations with unexampled rapidity and accuracy. By Charles Hoare, C.E. Sixth Edition. With a Slide Rule in tuck of cover. 2s. 6d.‡

196. *THEORY OF COMPOUND INTEREST AND ANNUITIES;* with Tables of Logarithms for the more Difficult Computations of Interest, Discount, Annuities, &c. By Fédor Thoman. Fourth Edition. 4s.‡

199. *THE COMPENDIOUS CALCULATOR;* or, Easy and Concise Methods of Performing the various Arithmetical Operations required in Commercial and Business Transactions; together with Useful Tables. By D. O'Gorman. Twenty-seventh Edition, carefully revised by C. Norris. 2s. 6d., cloth limp; 3s. 6d., strongly half-bound in leather.

204. *MATHEMATICAL TABLES*, for Trigonometrical, Astronomical, and Nautical Calculations; to which is prefixed a Treatise on Logarithms. By Henry Law, C.E. Together with a Series of Tables for Navigation and Nautical Astronomy. By Prof. J. R. Young. New Edition. 4s.

204*. *LOGARITHMS.* With Mathematical Tables for Trigonometrical, Astronomical, and Nautical Calculations. By Henry Law, M.Inst.C.E. New and Revised Edition. (Forming part of the above Work). 3s.

221. *MEASURES, WEIGHTS, AND MONEYS OF ALL NATIONS*, and an Analysis of the Christian, Hebrew, and Mahometan Calendars. By W. S. B. Woolhouse, F.R.A.S., F.S.S. Seventh Edition, 2s. 6d.‡

227. *MATHEMATICS AS APPLIED TO THE CONSTRUCTIVE ARTS.* Illustrating the various processes of Mathematical Investigation, by means of Arithmetical and Simple Algebraical Equations and Practical Examples. By Francis Campin. C.E. Second Edition. 3s.‡

PHYSICAL SCIENCE, NATURAL PHILOSOPHY, ETC.

1. *CHEMISTRY.* By Professor George Fownes, F.R.S. With an Appendix on the Application of Chemistry to Agriculture. 1s.

2. *NATURAL PHILOSOPHY*, Introduction to the Study of. By C. Tomlinson. Woodcuts. 1s. 6d.

6. *MECHANICS*, Rudimentary Treatise on. By Charles Tomlinson. Illustrated. 1s. 6d.

7. *ELECTRICITY;* showing the General Principles of Electrical Science, and the purposes to which it has been applied. By Sir W. Snow Harris, F.R.S., &c. With Additions by R. Sabine, C.E., F.S.A. 1s. 6d.

7*. *GALVANISM.* By Sir W. Snow Harris. New Edition by Robert Sabine, C.E., F.S.A. 1s. 6d.

8. *MAGNETISM;* being a concise Exposition of the General Principles of Magnetical Science. By Sir W. Snow Harris. New Edition, revised by H. M. Noad, Ph.D. With 165 Woodcuts. 3s. 6d.‡

☞ *The ‡ indicates that these vols. may be had strongly bound at 6d. extra.*

LONDON: CROSBY LOCKWOOD AND SON,

Physical Science, Natural Philosophy, etc., *continued.*

11. *THE ELECTRIC TELEGRAPH;* its History and Progress; with Descriptions of some of the Apparatus. By R. SABINE, C.E., F.S.A. 3s.
12. *PNEUMATICS,* including Acoustics and the Phenomena of Wind Currents, for the Use of Beginners By CHARLES TOMLINSON, F.R.S. Fourth Edition, enlarged. Illustrated. 1s. 6d.
72. *MANUAL OF THE MOLLUSCA;* a Treatise on Recent and Fossil Shells. By Dr. S. P. WOODWARD, A.L.S. Fourth Edition. With Plates and 300 Woodcuts. 7s. 6d., cloth.
96. *ASTRONOMY.* By the late Rev. ROBERT MAIN, M.A. Third Edition, by WILLIAM THYNNE LYNN, B.A., F.R.A.S. 2s.
97. *STATICS AND DYNAMICS,* the Principles and Practice of; embracing also a clear development of Hydrostatics, Hydrodynamics, and Central Forces. By T. BAKER, C.E. Fourth Edition. 1s. 6d.
173. *PHYSICAL GEOLOGY,* partly based on Major-General PORTLOCK's "Rudiments of Geology." By RALPH TATE, A.L.S., &c. Woodcuts. 2s.
174. *HISTORICAL GEOLOGY,* partly based on Major-General PORTLOCK's "Rudiments." By RALPH TATE, A.L.S., &c. Woodcuts. 2s. 6d.
173 & 174. *RUDIMENTARY TREATISE ON GEOLOGY,* Physical and Historical. Partly based on Major-General PORTLOCK's "Rudiments of Geology." By RALPH TATE, A.L.S., F.G.S., &c. In One Volume. 4s. 6d.‡
183 & 184. *ANIMAL PHYSICS,* Handbook of. By Dr. LARDNER, D.C.L., formerly Professor of Natural Philosophy and Astronomy in University College, Lond. With 520 Illustrations. In One Vol. 7s. 6d., cloth boards.
*** *Sold also in Two Parts, as follows :—*
183. ANIMAL PHYSICS. By Dr. LARDNER. Part I., Chapters I.—VII. 4s.
184. ANIMAL PHYSICS. By Dr. LARDNER. Part II., Chapters VIII.—XVIII. 3s.
269. *LIGHT:* an Introduction to the Science of Optics, for the Use of Students of Architecture, Engineering, and other Applied Sciences. By E. WYNDHAM TARN, M.A. 1s. 6d. [*Just published.*

FINE ARTS.

20. *PERSPECTIVE FOR BEGINNERS.* Adapted to Young Students and Amateurs in Architecture, Painting, &c. By GEORGE PYNE. 2s.
40 *GLASS STAINING, AND THE ART OF PAINTING ON GLASS.* From the German of Dr. GESSERT and EMANUEL OTTO FROMBERG. With an Appendix on THE ART OF ENAMELLING. 2s. 6d.
69. *MUSIC,* A Rudimentary and Practical Treatise on. With numerous Examples. By CHARLES CHILD SPENCER. 2s. 6d.
71. *PIANOFORTE,* The Art of Playing the. With numerous Exercises & Lessons from the Best Masters. By CHARLES CHILD SPENCER. 1s.6d.
69-71. *MUSIC & THE PIANOFORTE.* In one vol. Half bound, 5s.
181. *PAINTING POPULARLY EXPLAINED,* including Fresco, Oil, Mosaic, Water Colour, Water-Glass, Tempera, Encaustic, Miniature, Painting on Ivory, Vellum, Pottery, Enamel, Glass, &c. With Historical Sketches of the Progress of the Art by THOMAS JOHN GULLICK, assisted JOHN TIMBS, F.S.A. Sixth Edition, revised and enlarged. 5s.‡
186. *A GRAMMAR OF COLOURING,* applied to Decorative Painting and the Arts. By GEORGE FIELD. New Edition, enlarged and adapted to the Use of the Ornamental Painter and Designer. By ELLIS A. DAVIDSON. With two new Coloured Diagrams, &c. 3s.‡
246. *A DICTIONARY OF PAINTERS, AND HANDBOOK FOR PICTURE AMATEURS;* including Methods of Painting, Cleaning, Relining and Restoring, Schools of Painting, &c. With Notes on the Copyists and Imitators of each Master. By PHILIPPE DARYL. 2s. 6d.‡

☞ *The ‡ indicates that these vols. may be had strongly bound at 6d. extra.*

INDUSTRIAL AND USEFUL ARTS.

23. *BRICKS AND TILES*, Rudimentary Treatise on the Manufacture of. By E. Dobson, M.R.I.B.A. Illustrated, 3s.‡
67. *CLOCKS, WATCHES, AND BELLS*, a Rudimentary Treatise on. By Sir Edmund Beckett, LL.D., Q.C. Seventh Edition, revised and enlarged. 4s. 6d. limp; 5s. 6d. cloth boards.
83**. *CONSTRUCTION OF DOOR LOCKS.* Compiled from the Papers of A. C. Hobbs, and Edited by Charles Tomlinson, F.R.S. 2s. 6d.
162. *THE BRASS FOUNDER'S MANUAL;* Instructions for Modelling, Pattern-Making, Moulding, Turning, Filing, Burnishing, Bronzing, &c. With copious Receipts, &c. By Walter Graham. 2s.‡
205. *THE ART OF LETTER PAINTING MADE EASY.* By J. G. Badenoch. Illustrated with 12 full-page Engravings of Examples. 1s. 6d.
215. *THE GOLDSMITH'S HANDBOOK*, containing full Instructions for the Alloying and Working of Gold. By George E. Gee, 3s.‡
225. *THE SILVERSMITH'S HANDBOOK*, containing full Instructions for the Alloying and Working of Silver. By George E. Gee. 3s.‡
⁎ *The two preceding Works, in One handsome Vol., half-bound, entitled "*The Goldsmith's & Silversmith's Complete Handbook*," 7s.*
249. *THE HALL-MARKING OF JEWELLERY PRACTICALLY CONSIDERED.* By George E. Gee. 3s.‡
224. *COACH BUILDING*, A Practical Treatise, Historical and Descriptive. By J. W. Burgess. 2s. 6d.‡
235. *PRACTICAL ORGAN BUILDING.* By W. E. Dickson, M.A., Precentor of Ely Cathedral. Illustrated. 2s. 6d.‡
262. *THE ART OF BOOT AND SHOEMAKING.* By John Bedford Leno. Numerous Illustrations. Third Edition. 2s.
263. *MECHANICAL DENTISTRY:* A Practical Treatise on the Construction of the Various Kinds of Artificial Dentures, with Formulæ, Tables, Receipts, &c. By Charles Hunter. Third Edition. 3s.‡
270. *WOOD ENGRAVING:* A Practical and Easy Introduction to the Study of the Art. By W. N. Brown. 1s. 6d.

MISCELLANEOUS VOLUMES.

36. *A DICTIONARY OF TERMS used in ARCHITECTURE, BUILDING, ENGINEERING, MINING, METALLURGY, ARCHÆOLOGY, the FINE ARTS, &c.* By John Weale. Sixth Edition. Revised by Robert Hunt, F.R.S. Illustrated. 5s. limp; 6s. cloth boards.
50. *LABOUR CONTRACTS.* A Popular Handbook on the Law of Contracts for Works and Services. By David Gibbons. Fourth Edition, Revised, with Appendix of Statutes by T. F. Uttley, Solicitor, 3s. 6d. cloth.
112. *MANUAL OF DOMESTIC MEDICINE.* By R. Gooding, B.A., M.D. A Family Guide in all Cases of Accident and Emergency. 2s.
112*. *MANAGEMENT OF HEALTH.* A Manual of Home and Personal Hygiene. By the Rev. James Baird, B.A. 1s.
150. *LOGIC*, Pure and Applied. By S. H. Emmens. 1s. 6d.
153. *SELECTIONS FROM LOCKE'S ESSAYS ON THE HUMAN UNDERSTANDING.* With Notes by S. H. Emmens. 2s.
154. *GENERAL HINTS TO EMIGRANTS.* 2s.
157. *THE EMIGRANT'S GUIDE TO NATAL.* By R. Mann. 2s.
193. *HANDBOOK OF FIELD FORTIFICATION.* By Major W. W. Knollys, F.R.G.S. With 163 Woodcuts. 3s.‡
194. *THE HOUSE MANAGER:* Being a Guide to Housekeeping, Practical Cookery, Pickling and Preserving, Household Work, Dairy Management, &c. By An Old Housekeeper. 3s. 6d.‡
194, *HOUSE BOOK (The).* Comprising:—I. The House Manager.
112 & By an Old Housekeeper. II. Domestic Medicine. By R. Gooding, M.D.
112*. III. Management of Health. By J. Baird. In One Vol., half-bound, 6s.

☞ *The ‡ indicates that these vols may be had strongly bound at 6d. extra.*

LONDON : CROSBY LOCKWOOD AND SON.

EDUCATIONAL AND CLASSICAL SERIES.

HISTORY.

1. **England, Outlines of the History of;** more especially with reference to the Origin and Progress of the English Constitution. By WILLIAM DOUGLAS HAMILTON, F.S.A., of Her Majesty's Public Record Office. 4th Edition, revised. 5s.; cloth boards, 6s.
5. **Greece, Outlines of the History of;** in connection with the Rise of the Arts and Civilization in Europe. By W. DOUGLAS HAMILTON, of University College, London, and EDWARD LEVIEN, M.A., of Balliol College, Oxford. 2s. 6d.; cloth boards, 3s. 6d.
7. **Rome, Outlines of the History of:** from the Earliest Period to the Christian Era and the Commencement of the Decline of the Empire. By EDWARD LEVIEN, of Balliol College, Oxford. Map, 2s. 6d.; cl. bds. 3s. 6d.
9. **Chronology of History, Art, Literature, and Progress,** from the Creation of the World to the Present Time. The Continuation by W. D. HAMILTON, F.S.A. 3s.; cloth boards, 3s. 6d.
50. **Dates and Events in English History,** for the use of Candidates in Public and Private Examinations. By the Rev. E. RAND. 1s.

ENGLISH LANGUAGE AND MISCELLANEOUS.

11. **Grammar of the English Tongue,** Spoken and Written. With an Introduction to the Study of Comparative Philology. By HYDE CLARKE, D.C.L. Fifth Edition. 1s. 6d.
12. **Dictionary of the English Language,** as Spoken and Written. Containing above 100,000 Words. By HYDE CLARKE, D.C.L. 3s. 6d.; cloth boards, 4s. 6d.; complete with the GRAMMAR, cloth bds., 5s. 6d.
48. **Composition and Punctuation,** familiarly Explained for those who have neglected the Study of Grammar. By JUSTIN BRENAN. 18th Edition. 1s. 6d.
49. **Derivative Spelling-Book:** Giving the Origin of Every Word from the Greek, Latin, Saxon, German, Teutonic, Dutch, French, Spanish, and other Languages; with their present Acceptation and Pronunciation. By J. ROWBOTHAM, F.R.A.S. Improved Edition. 1s. 6d.
51. **The Art of Extempore Speaking:** Hints for the Pulpit, the Senate, and the Bar. By M. BAUTAIN, Vicar-General and Professor at the Sorbonne. Translated from the French. 8th Edition, carefully corrected. 2s. 6d.
54. **Analytical Chemistry,** Qualitative and Quantitative, a Course of. To which is prefixed, a Brief Treatise upon Modern Chemical Nomenclature and Notation. By WM. W. PINK and GEORGE E. WEBSTER. 2s.

THE SCHOOL MANAGERS' SERIES OF READING BOOKS,

Edited by the Rev. A. R. GRANT, Rector of Hitcham, and Honorary Canon of Ely; formerly H.M. Inspector of Schools.

INTRODUCTORY PRIMER, 3d.

	s. d.		s. d.
FIRST STANDARD	0 6	FOURTH STANDARD	1 2
SECOND ,,	0 10	FIFTH ,,	1 6
THIRD ,,	1 0	SIXTH ,,	1 6

LESSONS FROM THE BIBLE. Part I. Old Testament. 1s.
LESSONS FROM THE BIBLE. Part II. New Testament, to which is added THE GEOGRAPHY OF THE BIBLE, for very young Children. By Rev. C. THORNTON FORSTER.' 1s. 2d. *** Or the Two Parts in One Volume. 2s.

7, STATIONERS' HALL COURT, LUDGATE HILL, E.C.

FRENCH.

24. **French Grammar.** With Complete and Concise Rules on the Genders of French Nouns. By G. L. STRAUSS, Ph.D. 1s. 6d.
25. **French-English Dictionary.** Comprising a large number of New Terms used in Engineering, Mining, &c. By ALFRED ELWES. 1s. 6d.
26. **English-French Dictionary.** By ALFRED ELWES. 2s.
25,26. **French Dictionary** (as above). Complete, in One Vol., 3s.; cloth boards, 3s. 6d. *⁎* Or with the GRAMMAR, cloth boards, 4s. 6d.
47. **French and English Phrase Book:** containing Introductory Lessons, with Translations, several Vocabularies of Words, a Collection of suitable Phrases, and Easy Familiar Dialogues. 1s. 6d.

GERMAN.

39. **German Grammar.** Adapted for English Students, from Heyse's Theoretical and Practical Grammar, by Dr. G. L. STRAUSS. 1s. 6d.
40. **German Reader:** A Series of Extracts, carefully culled from the most approved Authors of Germany; with Notes, Philological and Explanatory. By G. L. STRAUSS, Ph.D. 1s.
41-43. **German Triglot Dictionary.** By N. E. S. A. HAMILTON. In Three Parts. Part I. German-French-English. Part II. English-German-French. Part III. French-German-English. 3s., or cloth boards, 4s.
41-43 **German Triglot Dictionary** (as above), together with German
& 39. Grammar (No. 39), in One Volume, cloth boards, 5s.

ITALIAN.

27. **Italian Grammar,** arranged in Twenty Lessons, with a Course of Exercises. By ALFRED ELWES. 1s. 6d.
28. **Italian Triglot Dictionary,** wherein the Genders of all the Italian and French Nouns are carefully noted down. By ALFRED ELWES. Vol. 1. Italian-English-French. 2s. 6d.
30. **Italian Triglot Dictionary.** By A. ELWES. Vol. 2. English-French-Italian. 2s. 6d.
32. **Italian Triglot Dictionary.** By ALFRED ELWES. Vol. 3. French-Italian-English. 2s. 6d.
28,30, **Italian Triglot Dictionary** (as above). In One Vol., 7s. 6d.
32. Cloth boards.

SPANISH AND PORTUGUESE.

34. **Spanish Grammar,** in a Simple and Practical Form. With a Course of Exercises. By ALFRED ELWES. 1s. 6d.
35. **Spanish-English and English-Spanish Dictionary.** Including a large number of Technical Terms used in Mining, Engineering, &c. with the proper Accents and the Gender of every Noun. By ALFRED ELWES 4s.; cloth boards, 5s. *⁎* Or with the GRAMMAR, cloth boards, 6s.
55. **Portuguese Grammar,** in a Simple and Practical Form. With a Course of Exercises. By ALFRED ELWES. 1s. 6d.
56. **Portuguese-English and English-Portuguese Dictionary.** Including a large number of Technical Terms used in Mining, Engineering, &c., with the proper Accents and the Gender of every Noun. By ALFRED ELWES. Second Edition, Revised, 5s.; cloth boards, 6s. *⁎* Or with the GRAMMAR, cloth boards, 7s.

HEBREW.

46*. **Hebrew Grammar.** By Dr. BRESSLAU. 1s. 6d.
44. **Hebrew and English Dictionary,** Biblical and Rabbinical; containing the Hebrew and Chaldee Roots of the Old Testament Post-Rabbinical Writings. By Dr. BRESSLAU. 6s.
46. **English and Hebrew Dictionary.** By Dr. BRESSLAU. 3s.
44,46. **Hebrew Dictionary** (as above), in Two Vols., complete, with
46*. the GRAMMAR, cloth boards, 12s.

LONDON: CROSBY LOCKWOOD AND SON,

LATIN.

19. **Latin Grammar.** Containing the Inflections and Elementary Principles of Translation and Construction. By the Rev. THOMAS GOODWIN, M.A., Head Master of the Greenwich Proprietary School. 1s. 6d.
20. **Latin-English Dictionary.** By the Rev. THOMAS GOODWIN, M.A. 2s.
22. **English-Latin Dictionary;** together with an Appendix of French and Italian Words which have their origin from the Latin. By the Rev. THOMAS GOODWIN, M.A. 1s. 6d.
20,22. **Latin Dictionary** (as above). Complete in One Vol., 3s. 6d. cloth boards, 4s. 6d. *.* Or with the GRAMMAR, cloth boards, 5s. 6d.

LATIN CLASSICS. With Explanatory Notes in English.
1. **Latin Delectus.** Containing Extracts from Classical Authors, with Genealogical Vocabularies and Explanatory Notes, by H. YOUNG. 1s. 6d.
2. **Cæsaris Commentarii de Bello Gallico.** Notes, and a Geographical Register for the Use of Schools, by H. YOUNG. 2s.
3. **Cornelius Nepos.** With Notes. By H. YOUNG. 1s.
4. **Virgilii Maronis Bucolica et Georgica.** With Notes on the Bucolics by W. RUSHTON, M.A., and on the Georgics by H. YOUNG. 1s. 6d.
5. **Virgilii Maronis Æneis.** With Notes, Critical and Explanatory, by H. YOUNG. New Edition, revised and improved With copious Additional Notes by Rev. T. H. L. LEARY, D.C.L., formerly Scholar of Brasenose College, Oxford. 3s.
5*. ——— Part 1. Books i.—vi., 1s. 6d.
5**. ——— Part 2. Books vii.—xii., 2s.
6. **Horace;** Odes, Epode, and Carmen Sæculare. Notes by H. YOUNG. 1s. 6d.
7. **Horace;** Satires, Epistles, and Ars Poetica. Notes by W. BROWNRIGG SMITH, M.A., F.R.G.S. 1s. 6d.
8. **Sallustii Crispi Catalina et Bellum Jugurthinum.** Notes, Critical and Explanatory, by W. M. DONNE, B.A., Trin. Coll., Cam. 1s. 6d.
9. **Terentii Andria et Heautontimorumenos.** With Notes, Critical and Explanatory, by the Rev. JAMES DAVIES, M.A. 1s. 6d.
10. **Terentii Adelphi, Hecyra, Phormio.** Edited, with Notes, Critical and Explanatory, by the Rev. JAMES DAVIES, M.A. 2s.
11. **Terentii Eunuchus, Comœdia.** Notes, by Rev. J. DAVIES, M.A. 1s. 6d.
12. **Ciceronis Oratio pro Sexto Roscio Amerino.** Edited, with an Introduction, Analysis, and Notes, Explanatory and Critical, by the Rev JAMES DAVIES, M.A. 1s. 6d.
13. **Ciceronis Orationes in Catilinam, Verrem, et pro Archia.** With Introduction, Analysis, and Notes, Explanatory and Critical, by Rev. T. H. L. LEARY, D.C.L. formerly Scholar of Brasenose College, Oxford. 1s. 6d.
14. **Ciceronis Cato Major, Lælius, Brutus, sive de Senectute, de Amicitia, de Claris Oratoribus Dialogi.** With Notes by W. BROWNRIGG SMITH M.A., F.R.G.S. 2s.
16. **Livy:** History of Rome. Notes by H. YOUNG and W. B. SMITH, M.A. Part 1. Books i., ii., 1s. 6d.
16*. ——— Part 2. Books iii., iv., v., 1s. 6d.
17. ——— Part 3. Books xxi., xxii., 1s. 6d.
19. **Latin Verse Selections,** from Catullus, Tibullus, Propertius, and Ovid. Notes by W. B. DONNE, M.A., Trinity College, Cambridge. 2s.
20. **Latin Prose Selections,** from Varro, Columella, Vitruvius, Seneca, Quintilian, Florus, Velleius Paterculus, Valerius Maximus Suetonius, Apuleius, &c. Notes by W. B. DONNE, M.A. 2s.
21. **Juvenalis Satiræ.** With Prolegomena and Notes by T. H. S. ESCOTT, B.A., Lecturer on Logic at King's College, London. 2s.

GREEK.

14. **Greek Grammar**, in accordance with the Principles and Philological Researches of the most eminent Scholars of our own day. By HANS CLAUDE HAMILTON. 1s. 6d.
15,17. **Greek Lexicon.** Containing all the Words in General Use, with their Significations, Inflections, and Doubtful Quantities. By HENRY R. HAMILTON. Vol. 1. Greek-English, 2s. 6d.; Vol. 2. English-Greek, 2s. Or the Two Vols. in One, 4s. 6d.: cloth boards, 5s.
14,15. **Greek Lexicon** (as above). Complete, with the GRAMMAR, in
17. One Vol., cloth boards, 6s.

GREEK CLASSICS. With Explanatory Notes in English.

1. **Greek Delectus.** Containing Extracts from Classical Authors, with Genealogical Vocabularies and Explanatory Notes, by H. YOUNG. New Edition, with an improved and enlarged Supplementary Vocabulary, by JOHN HUTCHISON, M.A., of the High School, Glasgow. 1s. 6d.
2, 3. **Xenophon's Anabasis**; or, The Retreat of the Ten Thousand. Notes and a Geographical Register, by H. YOUNG. Part 1. Books i. to iii., 1s. Part 2. Books iv. to vii., 1s.
4. **Lucian's Select Dialogues.** The Text carefully revised, with Grammatical and Explanatory Notes, by H. YOUNG. 1s. 6d.
5-12. **Homer, The Works of.** According to the Text of BAEUMLEIN. With Notes, Critical and Explanatory, drawn from the best and latest Authorities, with Preliminary Observations and Appendices, by T. H. L. LEARY, M.A., D.C.L.

THE ILIAD: Part 1. Books i. to vi., 1s.6d. | Part 3. Books xiii. to xviii., 1s. 6d
Part 2. Books vii. to xii., 1s.6d. | Part 4. Books xix. to xxiv., 1s. 6d.
THE ODYSSEY: Part 1. Books i. to vi., 1s. 6d | Part 3. Books xiii. to xviii., 1s. 6d.
Part 2. Books vii. to xii., 1s. 6d. | Part 4. Books xix. to xxiv., and Hymns, 1s.

13. **Plato's Dialogues:** The Apology of Socrates, the Crito, and the Phædo. From the Text of C. F. HERMANN. Edited with Notes, Critical and Explanatory, by the Rev. JAMES DAVIES, M.A. 2s.
14-17. **Herodotus, The History of,** chiefly after the Text of GAISFORD. With Preliminary Observations and Appendices, and Notes, Critical and Explanatory, by T. H. L. LEARY, M.A., D.C.L.
Part 1. Books i., ii. (The Clio and Euterpe), 2s.
Part 2. Books iii., iv. (The Thalia and Melpomene), 2s.
Part 3. Books v.-vii. (The Terpsichore, Erato, and Polymnia), 2s.
Part 4. Books viii., ix. (The Urania and Calliope) and Index, 1s. 6d.
18. **Sophocles:** Œdipus Tyrannus. Notes by H. YOUNG. 1s.
20. **Sophocles:** Antigone. From the Text of DINDORF. Notes, Critical and Explanatory, by the Rev. JOHN MILNER, B.A. 2s.
23. **Euripides:** Hecuba and Medea. Chiefly from the Text of DINDORF. With Notes, Critical and Explanatory, by W. BROWNRIGG SMITH, M.A., F.R.G.S. 1s. 6d.
26. **Euripides:** Alcestis. Chiefly from the Text of DINDORF. With Notes, Critical and Explanatory, by JOHN MILNER, B.A. 1s. 6d.
30. **Æschylus:** Prometheus Vinctus: The Prometheus Bound. From the Text of DINDORF. Edited, with English Notes, Critical and Explanatory, by the Rev. JAMES DAVIES, M.A. 1s.
32. **Æschylus:** Septem Contra Thebes: The Seven against Thebes. From the Text of DINDORF. Edited, with English Notes, Critical and Explanatory, by the Rev. JAMES DAVIES, M.A. 1s.
40. **Aristophanes:** Acharnians. Chiefly from the Text of C. H. WEISE. With Notes, by C. S. T. TOWNSHEND, M.A. 1s. 6d.
41. **Thucydides:** History of the Peloponnesian War. Notes by H. YOUNG. Book 1. 1s. 6d.
42. **Xenophon's Panegyric on Agesilaus.** Notes and Introduction by Lt. F. W. JEWITT. 1s. 6d.
43. **Demosthenes.** The Oration on the Crown and the Philippics. With English Notes. By Rev. T. H. L. LEARY, D.C.L., formerly Scholar of Brasenose College, Oxford. 1s. 6d.

www.ingramcontent.com/pod-product-compliance
Lightning Source LLC
Chambersburg PA
CBHW051724300426
44115CB00007B/460